TExES Mathematics 8-12
135 Teacher Certification Exam

By: Sharon Wynne, M.S.
Southern Connecticut State University

"And, while there's no reason yet to panic, I think it's only prudent that we make preparations to panic."

XAMonline, INC.
Boston

Copyright © 2009 XAMonline, Inc.

All rights reserved. No part of the material protected by this copyright notice may be reproduced or utilized in any form or by any means, electronic or mechanical, including photocopying, recording or by any information storage and retrievable system, without written permission from the copyright holder.

To obtain permission(s) to use the material from this work for any purpose including workshops or seminars, please submit a written request to:

> XAMonline, Inc.
> 25 First St. Suite 106
> Cambridge, MA 02141
> Toll Free 1-800-509-4128
> Email: info@xamonline.com
> Web www.xamonline.com
> Fax: 1-781-662-9268

Library of Congress Cataloging-in-Publication Data

Wynne, Sharon A.
    Mathematics 8-12 135: Teacher Certification / Sharon A. Wynne. -2nd ed.
    ISBN 978-1-58197-339-6
    1. Mathematics 8-12 135.   2. Study Guides.   3. TExES
    4. Teachers' Certification & Licensure.   5. Careers

Disclaimer:

The opinions expressed in this publication are the sole works of XAMonline and were created independently from the National Education Association, Educational Testing Service, or any State Department of Education, National Evaluation Systems or other testing affiliates.

Between the time of publication and printing, state specific standards as well as testing formats and website information may change that is not included in part or in whole within this product. Sample test questions are developed by XAMonline and reflect similar content as on real tests; however, they are not former tests. XAMonline assembles content that aligns with state standards but makes no claims nor guarantees teacher candidates a passing score. Numerical scores are determined by testing companies such as NES or ETS and then are compared with individual state standards. A passing score varies from state to state.

Printed in the United States of America   œ-1

TExES: Mathematics 8-12 135
ISBN: 978-1-58197-339-6

TEACHER CERTIFICATION STUDY GUIDE

About the Subject Assessments

TEXES™: Subject Assessment in the Mathematics 8-12 examination

Purpose: The assessments are designed to test the knowledge and competencies of prospective secondary level teachers. The question bank from which the assessment is drawn is undergoing constant revision. As a result, your test may include questions that will not count towards your score.

Test Version: There are two versions of subject assessment for Mathematics in Texas. The Mathematics 4-8 (115) exam emphasizes comprehension in Number Concepts; Patterns and Algebra; Geometry and Measurement; Probability and Statistics; Mathematical Processes and Statistics; Mathematical Learning, Instruction, and Assessment .The Mathematics 8-12 (135) exam emphasizes comprehension in Number Concepts; Patterns and Algebra; Geometry and Measurement; Probability and Statistics; Mathematical Processes and Statistics; Mathematical Learning, Instruction, and Assessment. The Mathematics 8-12 study guide is based on a typical knowledge level of persons who have completed a _bachelor's degree program_ in Mathematics.

Time Allowance: You will have 5 hours to finish the exam. There are approximately 90 multiple-choice questions in the exam.

Weighting: Approximately 14% of the tests material consists of Number Concepts; 33% consists of Patterns Algebra; 19% consists of Geometry and Measurement; 14% consists of Probability and Statistics; 10% consists of Mathematical Processes and Perspectives; 10% consists of Mathematical Learning, Instruction, and Assessment.

Additional Information about the TEXES Assessments: The TEXES series subject assessments are developed by _National Evaluation Systems._ They provide additional information on the TEXES series assessments, including registration, preparation and testing procedures and study materials such topical guides that have about 72 pages of information including approximately 34 additional sample questions.

TEACHER CERTIFICATION STUDY GUIDE

TABLE OF CONTENTS

COMPETENCY #   PG #

DOMAIN I.   **NUMBER CONCEPTS**

Competency 1. The teacher understands the real number system and its structure, operations, algorithms, and representations 1

Competency 2. The teacher understands the complex number system and its structure, operations, algorithms, and representations 15

Competency 3. The teacher understands number theory concepts and principles and uses numbers to model and solve problems in a variety of situations 26

DOMAIN II.   **PATTERNS AND ALGEBRA**

Competency 4. The teacher uses patterns to model and solve problems and formulate conjectures 40

Competency 5. The teacher understands attributes of functions, relations, and their graphs 54

Competency 6. The teacher understands linear and quadratic functions, analyzes their algebraic and graphical properties, and uses them to model and solve problems 72

Competency 7. The teacher understands polynomial, rational, radical, absolute value, and piecewise functions, analyzes their algebraic and graphical properties, and uses them to model and solve problems 98

Competency 8. The teacher understands exponential and logarithmic functions, analyzes their algebraic and graphical properties, and uses them to model and solve problems 126

Competency 9. The teacher understands trigonometric and circular functions, analyzes their algebraic and graphical properties, and uses them to model and solve problems 148

Competency 10. The teacher understands and solves problems using differential and integral calculus 166

MATHEMATICS 8-12

TEACHER CERTIFICATION STUDY GUIDE

DOMAIN III.   **GEOMETRY AND MEASUREMENT**

Competency 11. The teacher understands measurement as a process 205

Competency 12. The teacher understands geometries, in particular Euclidean geometry, as axiomatic systems 222

Competency 13. The teacher understands the results, uses, and applications of Euclidean geometry 245

Competency 14. The teacher understands coordinate, transformational, and vector geometry and their connections 261

DOMAIN IV.   **PROBABILITY AND STATISTICS**

Competency 15. The teacher understands how to use appropriate graphical and numerical techniques to explore data, characterize patterns, and describe departures from patterns 293

Competency 16. The teacher understands concepts and applications of probability 314

Competency 17. The teacher understands the relationships among probability theory, sampling, and statistical inference, and how statistical inference is used in making and evaluating predictions 336

DOMAIN V.   **MATHEMATICAL PROCESSES AND PERSPECTIVES**

Competency 18. The teacher understands mathematical reasoning and problem solving 354

Competency 19. The teacher understands mathematical connections both within and outside of mathematics and how to communicate mathematical ideas and concepts 363

DOMAIN VI.   **MATHEMATICAL LEARNING, INSTRUCTION, AND ASSESSMENT**

Competency 20. The teacher understands how children learn mathematics and plans, organizes, and implements instruction using knowledge of students, subject matter, and statewide curriculum (Texas Essential Knowledge and Skills [TEKS]) 371

Competency 21. The teacher understands assessment and uses a variety of formal and informal assessment techniques to monitor and guide mathematics instruction and to evaluate student progress 380

CURRICULUM AND INSTRUCTION 385

SAMPLE TEST 395

ANSWER KEY 410

RIGOR TABLE 411

RATIONALES FOR SAMPLE QUESTIONS 412

TEACHER CERTIFICATION STUDY GUIDE

Great Study and Testing Tips!

What to study in order to prepare for the subject assessments is the focus of this study guide, but equally important is *how* you study.

You can increase your chances of truly mastering the information by taking some simple but effective steps.

Study Tips:

1. Some foods aid the learning process. Foods such as milk, nuts, seeds, rice and oats help your study efforts by releasing natural memory enhancers called CCKs (*cholecystokinin*) composed of *tryptophan*, *choline* and *phenylalanine*. All of these chemicals enhance the neurotransmitters associated with memory. Before studying, try a light, protein-rich meal of eggs, turkey and fish. All of these foods release the memory-enhancing chemicals. The better the connections, the more you comprehend.

Likewise, before you take a test, stick to a light snack of energy-boosting and relaxing foods. A glass of milk, a piece of fruit or some peanuts all release various memory-boosting chemicals and help you to relax and focus on the subject at hand.

2. Learn to take great notes. A by-product of our modern culture is that we have grown accustomed to getting our information in short doses (e.g., TV news sound bites or USA Today–style newspaper articles).

Consequently, we've subconsciously trained ourselves to assimilate information better in neat little packages. If your notes are scrawled all over the paper, it fragments the flow of the information. Strive for clarity. Newspapers use a standard format to achieve clarity. Your notes can be much clearer through use of proper formatting. A very effective format is called the *"Cornell Method."*

> Take a sheet of loose-leaf lined notebook paper and draw a line all the way down the paper about 1–2" from the left-hand edge.
>
> Draw another line across the width of the paper about 1–2" up from the bottom. Repeat this process on the reverse side of the page.

Look at the helpful result. You have ample room for notes, a left-hand margin for special emphasis items or inserting supplementary data from the textbook, a large area at the bottom for a brief summary, and a little rectangular space for just about anything you want.

3. <u>Get the concept, then the details</u>. Too often we focus on the details and don't gather an understanding of the concept. However, if you simply memorize only dates, places or names, you may well miss the whole point of the subject.

A good way to understand concepts is to put them in your own words. If you are working from a textbook, automatically summarize each paragraph in your mind. If you are outlining text, don't simply copy the author's words.

Rephrase them in your own words. You remember your own thoughts and words much better than someone else's, and you subconsciously tend to associate the important details with the core concepts.

4. <u>Ask Why?</u> Pull apart written material paragraph by paragraph, and don't forget the captions under the illustrations.

Example: If the heading is "Stream Erosion," flip it around to read "Why do streams erode?" then answer the question.

If you train your mind to think in a series of questions and answers, not only will you learn more, but you will also help lessen the test anxiety because you are used to answering questions.

5. <u>Read for reinforcement and future needs</u>. Even if you only have 10 minutes, put your notes or a book in your hand. Your mind is similar to a computer; you must input data to have it processed. *By reading, you are creating the neural connections for future retrieval.* The more times you read something, the more you reinforce the learning of ideas.

Even if you don't fully understand something on the first pass, *your mind stores much of the material for later recall.*

6. <u>Relax to learn, and go into exile.</u> Our bodies respond to an inner clock called *biorhythms*. Burning the midnight oil works well for some people, but not for everyone.

If possible, set aside a particular place to study that is free of distractions. Turn off the television, cell phone and pager, and exile your friends and family during your study period.

If you really are bothered by silence, try background music. Light classical music at a low volume has been shown to aid in concentration over other types. Music that evokes pleasant emotions without lyrics are highly suggested. Try just about anything by Mozart. This relaxes you.

7. <u>Use arrows, not highlighters</u>. At best, it's difficult to read a page full of yellow, pink, blue and green streaks. Try staring at a neon sign for a while and you'll soon see that the horde of colors obscures the message.

A quick note, a brief dash of color, an underline or an arrow pointing to a particular passage is much clearer than a horde of highlighted words.

8. <u>Budget your study time</u>. Although you shouldn't ignore any of the material, *allocate your available study time in the same ratio that topics may appear on the test.*

TEACHER CERTIFICATION STUDY GUIDE

Testing Tips:

1. Get smart, play dumb. Don't read anything into the question. Don't make an assumption that the test writer is looking for something else other than what is asked. Stick to the question as written and don't read anything into it.

2. Read the question and all the choices *twice* before answering the question. You may miss something by not carefully reading and then re-reading both the question and the answers.

If you really don't have a clue as to the right answer, leave it blank on the first time through. Go on to the other questions, as they may provide a clue as to how to answer the skipped questions.

If, later on, you still can't answer the skipped questions . . . *Guess.* The only penalty for guessing is that you *might* get it wrong. Only one thing is certain: if you don't put anything down, you will get it wrong!

3. Turn the question into a statement. Look at the way the questions are worded. The syntax of the question usually provides a clue. Does it seem more familiar as a statement rather than as a question? Does it sound strange?

By turning a question into a statement, you may be able to spot whether an answer sounds right, and it may also trigger memories of material you have read.

4. Look for hidden clues. It's actually very difficult to compose multiple-foil (choice) questions without giving away part of the answer in the options presented.

In most multiple-choice questions you can often readily eliminate one or two of the potential answers. This leaves you with only two real possibilities, and automatically your odds increase to fifty-fifty for very little work.

5. Trust your instincts. For every fact that you have read, you subconsciously retain something of that knowledge. On questions that you aren't really certain about, go with your basic instincts. **Your first impression on how to answer a question is usually correct.**

6. Mark your answers directly in the test booklet. Don't bother trying to fill in the optical scan sheet on the first pass through the test. *Just be very careful not to miss-mark your answers when you eventually transcribe them to the scan sheet.*

7. Watch the clock! You have a set amount of time to answer the questions. Don't get bogged down trying to answer a single question at the expense of 10 questions you can more readily answer.

TEACHER CERTIFICATION STUDY GUIDE

DOMAIN I.	NUMBER CONCEPTS

Competency 001 The teacher understands the real number system and its structure, operations, algorithms, and representations.

This competency section reviews some of the fundamental concepts of the real number system, including subsets of the real numbers and their associated properties, various representations of real numbers and operations, and the use of irrational numbers in solving problems.

The beginning teacher understands the concepts of place value, number base, and decimal representations of real numbers.

Place Value

In a number, every digit has a face value and a place value. The face values of the digits in the number 3467 are 3, 4, 6 and 7. The place value of a digit depends on its position in the number.

Whole number place values are where the digits fall to the left of the decimal point. Consider the number 792; reading from left to right, the first digit (7) represents the hundreds place. Thus, there are 7 sets of one hundred in the number 792. The second digit (9) represents the tens place. The last digit (2) represents the ones place.

Decimal place value is where the digits fall to the right of the decimal point. Consider the number 4.873; reading from left to right, the first digit (4) is in the ones place. After the decimal, 8 is in the tenths place and indicates that the number contains 8 tenths. The digit 7 is in the hundredths' place and tells us the number contains 7 hundredths. The same pattern applies to the rest of the number, with each successive digit to the right of the decimal point decreasing progressively in powers of ten.

<u>Example:</u> 12345.6789 occupies the following powers of ten positions:

10^4	10^3	10^2	10^1	10^0		10^{-1}	10^{-2}	10^{-3}	10^{-4}
1	2	3	4	5	.	6	7	8	9

MATHEMATICS 8-12

Number Systems Bases

The standard method of writing numbers is the decimal (or base 10) system, where the digits represent powers of 10. Other bases can also be used to represent a number. The binary (or base 2) system, for instance, uses the powers of 2 (2^0, 2^1, 2^2, and so on) to represent a number. Base 2 only uses the digits 0 and 1.

Decimal Binary Conversion		
Decimal	Binary	Place Value
1	1	2^0
2	10	2^1
4	100	2^2
8	1000	2^3

Thus, the number 9 in base 10 is equal to 1001 in base 2. Fractions, ratios and other functions alter in the same way.

Base	Number System
2	Binary
3	Ternary
4	Quarternary
5	Quinary
6	Senary
8	Octal
16	Hexadecimal

Fundamentally, computers use a binary system, because the transistors and logic circuitry that compose these machines uses two basic states (which can be interpreted as 0 and 1, off and on, false and true, or a similar representation). Particular computers may be implemented such that they perform arithmetic on groups of bits (eight bits, for instance, which is called a byte), but the fundamental operation of the machine is still binary.

The beginning teacher understands the algebraic structure and properties of the real number system and its subsets (e.g., real numbers as a field, integers as an additive group).

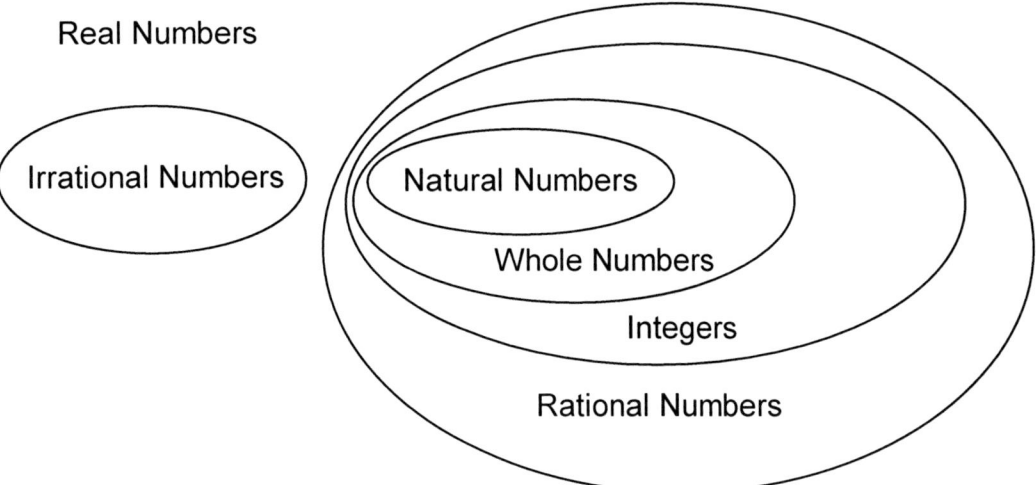

Real numbers are denoted by ℝ and are numbers that can be shown by an infinite decimal representation such as 3.286275347.... Real numbers include rational numbers, such as 242 and −23/129, and irrational numbers, such as $\sqrt{2}$ and π, and they can be represented as points along an infinite number line. Real numbers are also known as "the unique complete Archimedean *ordered field.*" Real numbers are to be distinguished from imaginary numbers, which involve a factor of $\sqrt{-1}$.

Real numbers are classified as follows:

A. Natural numbers, denoted by ℕ: the counting numbers, 1, 2, 3,…

B. Whole numbers: the counting numbers along with zero, 0, 1, 2, 3,…

C. Integers, denoted by ℤ: the counting numbers, their negatives, and zero, …,−2, −1, 0, 1, 2,…

D. Rationals, denoted by ℚ: all of the fractions that can be formed using whole numbers. Zero cannot be the denominator. In decimal form, these numbers will either be terminating or repeating decimals. Simplify square roots to determine if the number can be written as a fraction.

E. Irrationals: Real numbers that cannot be written as a fraction. The decimal forms of these numbers are neither terminating nor repeating. Examples include π, e and $\sqrt{2}$.

Fields, Rings, and Groups

A useful property that can describe arbitrary sets of numbers (including fields, rings, and groups) is **closure**. A set is closed under an operation if the operation performed on any given elements of the set always yields a result that is likewise an element of the set. For instance, the set of real numbers is closed under multiplication, because for any two real numbers a and b, the product ab is also a real number.

Any set that includes at least two nonzero elements that satisfies the field axioms for addition and multiplication is a **field**. The real numbers, $\mathbb{R}$, as well as the complex numbers, $\mathbb{C}$, are each a field, with the real numbers being a subset of the complex numbers. The field axioms are summarized below.

Addition:

Commutativity	$a+b = b+a$
Associativity	$a+(b+c) = (a+b)+c$
Identity	$a+0 = a$
Inverse	$a+(-a) = 0$

Multiplication:

Commutativity	$ab = ba$
Associativity	$a(bc) = (ab)c$
Identity	$a \cdot 1 = a$
Inverse	$a \cdot \frac{1}{a} = 1 \quad (a \neq 0)$

Addition and multiplication:

Distributivity	$a(b+c) = (b+c)a = ab+ac$

Note that both the real numbers and the complex numbers satisfy the axioms summarized above.

A **ring** is an integral domain with two binary operations (addition and multiplication) where, for every non-zero element a and b in the domain, the product ab is non-zero. A field is a ring where multiplication is commutative, or $a \cdot b = b \cdot a$, and all non-zero elements have a multiplicative inverse. The set $\mathbb{Z}$ (integers) is a ring that is not a field in that it does not have the multiplicative inverse; therefore, integers are not a field. A polynomial ring is also not a field, as it also has no multiplicative inverse. Furthermore, matrix rings do not constitute fields because matrix multiplication is not generally commutative.

Note: Multiplication is implied when there is no symbol between two variables. Thus, $a \times b$ can be written ab. Multiplication can also be indicated by a raised dot ($\cdot$).

A **group** is a set of numbers that obeys certain axioms with respect to a particular binary operation (such as addition). For a set G to be a group, G must be closed under the defined operation, the operation must obey associativity, G must contain an identity element, and each element in G must have an inverse element also in G. These rules are summarized below for elements a, b, and c in G for the binary operation $*$.

Closure	$a * b \in G$
Associativity	$a * (b * c) = (a * b) * c$
Identity	$I \in G$ such that $I * a = a * I = a$
Inverse	$a_{inv} \in G$ such that $a_{inv} * a = a * a_{inv} = I$

In the inverse rule, I is the same as the identity element. An example of a group is the set of integers under addition. For any two integers a and b, the sum $a + b$ is also an integer—thus, the set of integers is closed under addition. Also, associativity applies, since $a + (b + c) = (a + b) + c$ for any integers a, b, and c. The identity element 0 (zero) is also in the set of integers and $0 + a = a + 0 = a$ for all a. Furthermore, the inverse element $-a$ for element a leads to $a + (-a) = 0$. Thus, the set of integers also obeys the identity and inverse axioms, meaning that the integers are a group under the binary operation of addition.

Real numbers are an ordered field and can be ordered. As such, an ordered field F must contain a subset P (such as the positive numbers) such that if a and b are elements of P, then both $a + b$ and ab are also elements of P. (In other words, the set P is closed under addition and multiplication.) Furthermore, it must be the case that for any element c contained in F, exactly one of the following conditions is true: c is an element of P, $-c$ is an element of P or $c = 0$.

Likewise, **the rational numbers also constitute an ordered field**. The set P can be defined as the positive rational numbers. For each a and b that are elements of the set ☐ (the rational numbers), $a + b$ is also an element of P, as is ab. (The sum $a + b$ and the product ab are both rational if a and b are rational.) Since P is closed under addition and multiplication, ☐ constitutes an ordered field.

The beginning teacher describes and analyzes properties of subsets of the real numbers (e.g., closure, identities).

The preceding skill describes various properties of subsets of real numbers.

The beginning teacher selects and uses appropriate representations of real numbers (e.g., fractions, decimals, percents, roots, exponents, scientific notation) for particular situations.

Real numbers can be represented in a variety of formats. Some of these formats are more amenable to certain problems than others, and it is important to be able to select the proper representation of a real number for a given situation.

For instance, if exact calculations are required, **decimal representations** (or, similarly, **percent representations**—which are simply the decimal representation multiplied by 100) of irrational numbers are not appropriate. The use of a decimal necessarily requires use of a finite representation; thus, the decimal form of an irrational number must be rounded to some digit, leading to inaccuracies in calculations. Thus, irrational numbers such as the number π and square roots of certain integers should often be left in their symbolic or square root forms. If inexact calculations are acceptable, then a decimal or approximate fractional representation may be suitable.

If the decimal is repeating (such as 0.1111111...), a **fractional representation** may be the best approach. A fraction can be manipulated easily, and it is sometimes more conducive to exact calculations than are repeating decimals (or even long non-repeating decimals in some cases).

In other instances, an **exponential form** is useful. Exponentials (or their inverses, **logarithms**) may be a preferred representation of real numbers in various cases. In addition to considering whether exact or inexact calculations are needed for a particular problem, the simplicity of the calculation is also important when selecting an appropriate representation of a number. For hand/mental calculations, simplicity may be paramount, for instance.

Scientific notation is a convenient method for writing very large and very small numbers. It employs two factors: the first factor is a number between –10 and 10, and the second factor is a power of 10. This notation is a shorthand way to express large numbers (like the weight in kilograms of 100 freight cars) or small numbers (like the weight in grams of an atom).

For example, 356.73 can be written in various forms.

$$356.73 = 3567.3 \times 10^{-1} \quad (1)$$
$$= 356.73 \times 10^{-0} \quad (2)$$
$$= 35.673 \times 10^{1} \quad (3)$$
$$= 3.5673 \times 10^{2} \quad (4)$$
$$= 0.35673 \times 10^{3} \quad (5)$$

Only (4) is written in proper scientific notation format.

Example: Write 46,368,000 in scientific notation.

1) Introduce a decimal point. 46,368,000 = 46,368,000.0

2) Move the decimal place to **left** until only one nonzero digit is in front of it, in this case between the 4 and 6.

3) Count the number of digits the decimal point moved, in this case seven. This is the nth the power of 10 and is **positive** because the decimal point moved **left**.

Therefore, $46,368,000 = 4.6368 \times 10^{7}$.

Example: Write 0.00397 in scientific notation.

1) Decimal point is already in place.

2) Move the decimal point to the **right** until there is only one nonzero digit in front of it, in this case between the 3 and 9.

3) Count the number of digits the decimal point moved, in this case three. This is the nth the power of ten and is **negative** because the decimal point moved **right**.

Therefore, $0.00397 = 3.97 \times 10^{-3}$.

Thus, there are a number of possible representations for a given real number depending on the problem or situation under consideration. The following examples illustrate some problems where certain representations of given real numbers are better than others.

Example: A particular material has a mass of 0.01 grams in one liter. What is the material's density in grams per milliliter?

A cursory examination of this problem shows that it will be necessary to divide a small number (0.01 grams) by a large number (1000 milliliters = 1 liter) to get the density. Thus, scientific notation is a helpful representation of the numbers in the problem. The density d is then the following:

$$d = \frac{1 \times 10^{-2} \, g}{1 \times 10^{3} \, mL} = 1 \times 10^{-5} \, \frac{g}{mL}$$

The calculation and the result in this case are simplified considerably through the use of scientific notation. The solution is in a much neater form than 0.00001.

Example: Express the repeating decimal $0.\overline{254}$ as a number in closed form.

This problem calls for selecting an appropriate closed-form representation in the real number system for a repeating decimal. First, note that because the decimal repeats, the three repeating digits can be isolated as follows. Let d be equal to the repeating decimal $0.\overline{254}$.

$$1000d = 254.\overline{254} = 254 + d$$
$$999d = 254$$
$$d = \frac{254}{999}$$

Thus, this repeating decimal can be expressed in closed form using a fractional representation.

The beginning teacher uses a variety of models (e.g., geometric, symbolic) to represent operations, algorithms, and real numbers.

Number operations can be represented in various ways, for instance to aid understanding of the operation or to simplify the solution of a particular problem. Multiplication of two numbers, for example, can be represented geometrically as the area of a rectangle with width and length equal to the two factors. Consider the operation 2×3:

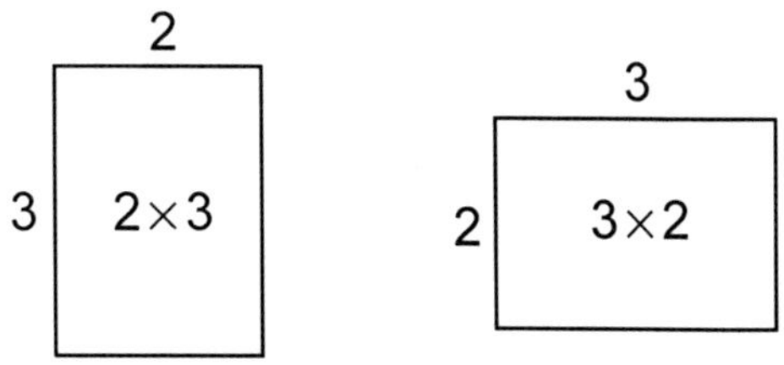

Notice that these two rectangles are identical, and therefore 2×3 is the same as 3×2. This fact is another way to understand the commutativity of multiplication for the set of real (and complex) numbers.

Addition and subtraction, as another example, can be expressed using arbitrary symbols to represent integral (or fractional) values. The operation $6 - 2$ can be represented as shown below using circles.

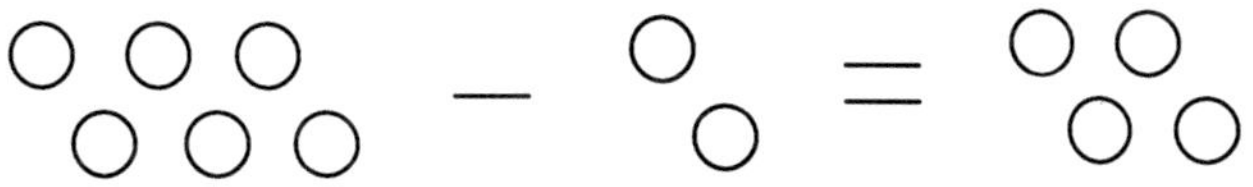

In such a case, negative numbers could be represented as filled circles (with a negative value conceptually corresponding, for example, to "debt"). The example symbolic representation below illustrates the case of 2 − 6.

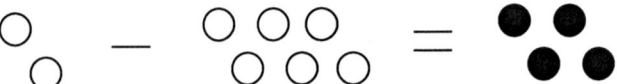

Likewise, algorithms can be represented using various representations. Algorithms intended to run on a computer can be written in a specific programming language (such as FORTRAN, C, or Java) or in "pseudocode," which is an unspecified approach to writing algorithms in a manner that is similar to but more general than specified programming languages. A simple example is an algorithm that generates the Fibonacci numbers. The following "pseudocode" illustrates such an algorithm for generating the first N Fibonacci numbers.

$F_i \leftarrow 0$
$F_{i+1} \leftarrow 1$
If $N \geq 0$, Output F_i
If $N \geq 1$, Output F_{i+1}
$i \leftarrow 2$
While $i \leq N$:
$\quad F_{i+2} \leftarrow F_i + F_{i+1}$
Output F_{i+2}
$F_i \leftarrow F_{i+1}$
$\quad F_{i+1} \leftarrow F_{i+2}$
$\quad i \leftarrow i+1$
End

This algorithm "outputs" the first N Fibonacci numbers regardless of the value chosen for N (assuming N is an integer greater than or equal to zero).

An algorithm can also be represented using a block diagram or flow chart. The diagram below illustrates the Fibonacci algorithm.

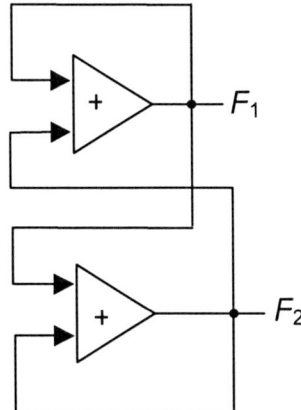

If F_1 and F_2 are initialized to 0 and 1, respectively, this diagram represents a recursive algorithm that produces the Fibonacci numbers (specifically, if F_1 is followed). The triangular blocks represent summation of the inputs. Each time the system is "updated," the next successive Fibonacci number is produced (the full sequence is only produced at the F_1 output).

Likewise, operations and algorithms can be represented in a variety of other ways.

The beginning teacher uses real numbers to model and solve a variety of problems.

The set of real numbers is a crucial component of innumerable mathematical problems. Selection of the appropriate subset and representation of the real numbers, as well as accurate calculation in accordance with the characteristics of that subset, are critical to correctly modeling and solving problems. The other skill sections in this competency, as well as numerous other sections throughout the guide, provide example problems that illustrate the proper use of real numbers in modeling and solving problems in a variety of mathematical contexts. The example problem below also illustrates the use of real numbers.

Example: If a, b, and c are positive real numbers, prove that $c(a+b) = (b+a)c$.

Use the properties of the set of real numbers.

$$c(a+b) = c(b+a) \quad \text{Additive commutativity}$$
$$= cb + ca \quad \text{Distributivity}$$
$$= bc + ac \quad \text{Multiplicative commutativity}$$
$$= (b+a)c \quad \text{Distributivity}$$

The beginning teacher uses deductive reasoning to simplify and justify algebraic processes.

The properties of real and complex numbers (see **Competency 002** for more on complex numbers) can be applied to the construction of various mathematical arguments. A **mathematical argument** proves that a proposition is true (or false). **Deductive reasoning** involves making particular inferences based on general premises or axioms. Application of these premises and the rules of deductive logic can be very helpful when solving problems. The example problems below illustrate the use of some of the fundamental principles of basic deductive logic (such as syllogisms).

Example: Prove that for every integer y, if y is an even number, then y^2 is even.

The definition of even implies that for each integer y there is at least one integer x such that $y = 2x$.

$$y = 2x$$
$$y^2 = 4x^2$$

Since $4x^2$ is always evenly divisible by two ($2x^2$ is an integer), y^2 is even for all values of y.

Example: Given real numbers a, b, c and d where ad = –bc, prove that (a + bi)(c + di) is real.

Expand the product of the complex numbers.

$$(a+bi)(c+di) = ac + bci + adi + bdi^2$$

Use the definition of i^2.

$$(a+bi)(c+di) = ac - bd + bci + adi$$

Apply the fact that ad = –bc.

$$(a+bi)(c+di) = ac - bd + bci - bci = ac - bd$$

Since a, b, c and d are all real, ac – bd must also be real.

Example: Determine if the set of irrational numbers is closed under addition.

One option for solving this problem is to attempt to prove conclusively the positive assertion that the set is closed under addition. An alternative, however, is to attempt to prove the opposite of an assertion through counterexample. In this case, attempt to find two irrational numbers whose sum is not irrational. Consider, for instance, $\sqrt{2}$ and $-\sqrt{2}$.

$$\sqrt{2} + \left(-\sqrt{2}\right) = 0$$

Since 0 is a rational number in this counterexample, the set of irrational numbers is not closed under addition.

The beginning teacher demonstrates how some problems that have no solution in the integer or rational number systems have solutions in the real number system.

The set of real numbers is composed entirely of the union of two mutually exclusive sets: the set of rational numbers and the set of irrational numbers. In many cases, the solutions to certain problems may not be found among the rational numbers, yet the solutions are indeed real numbers. Consider the following examples.

Example: Find the hypotenuse of a right triangle with legs of length 2 and 3.

The diagram below illustrates the problem.

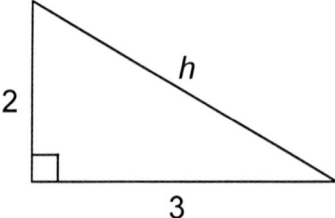

The length h of the hypotenuse can be found using the Pythagorean theorem as follows.

$$h^2 = 2^2 + 3^2 = 4 + 9 = 13$$
$$h = \sqrt{13}$$

The result, $\sqrt{13}$, is not a rational number. Thus, in this case, although the length h is obviously in the set of real numbers, it is not in the set of rational numbers. Logically, the result must be an irrational number then, as calculated above.

Example: Calculate the volume of a circular cylinder with a diameter of 4 centimeters and a height of 10 centimeters.

The volume of a circular cylinder is $\pi r^2 h$. In this case, the radius r is 2cm and the height h is 10cm. The volume is then the following.

$$V = \pi r^2 h = \pi (2)^2 (10) = 40\pi$$

Since π is an irrational number, the product 40π is likewise irrational. Here, again, although the volume of the cylinder is not an integer or otherwise a rational number, it is a real number (and hence irrational).

Competency 002   **The teacher understands the complex number system and its structure, operations, algorithms, and representations.**

The set of complex numbers includes the real numbers but is expanded through use of the factor $i = \sqrt{-1}$. This section reviews the fundamental concepts and applications of complex numbers, including operations, representations of numbers and operations, and problems involving the complex domain.

The beginning teacher understands the properties of complex numbers (e.g., complex conjugate, magnitude/modulus, multiplicative inverse).

The set of complex numbers is denoted by ☐. The set ☐ is defined as $\{a+bi : a, b \in ☐\}$ ($\in$ means "element of"). In other words, complex numbers are an extension of real numbers made by attaching an imaginary number i, which satisfies the equality $i^2 = -1$. Complex numbers are of the form $a + bi$, where a and b are *real* numbers and $i = \sqrt{-1}$. Thus, a is the real part of the number and b is the imaginary part of the number. When i appears in a fraction, the fraction is usually simplified so that i is not in the denominator. The set of complex numbers includes the set of real numbers, where any real number n can be written in its equivalent complex form as $n + 0i$. In other words, it can be said that ☐ $\subseteq$ ☐ (or ☐ is a subset of ☐).

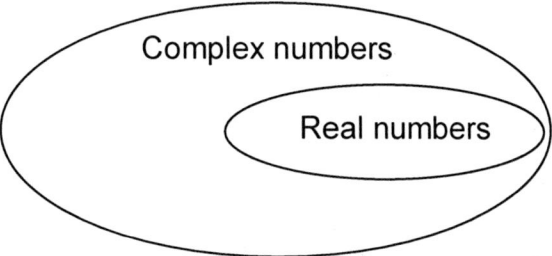

The number $3i$ has a real part 0 and imaginary part 3; the number 4 has a real part 4 and an imaginary part 0. As another way of writing complex numbers, we can express them as **ordered pairs**:

Complex number	Ordered pair
$3 + 2i$	$(3, 2)$
$\sqrt{3} + \sqrt{3}i$	$(\sqrt{3}, \sqrt{3})$
$7i$	$(0, 7)$
$\dfrac{6 + 2i}{7}$	$\left(\dfrac{6}{7}, \dfrac{2}{7}\right)$

The basic operations for complex numbers can be summarized as follows, where $z_1 = a_1 + b_1 i$ and $z_2 = a_2 + b_2 i$. Note that the operations are performed in the standard manner, where i is treated as a standard radical value. The result of each operation is written in the standard form for complex numbers. Also note that the **complex conjugate** of a complex number $z = a + bi$ is denoted as $z^* = a - bi$.

$$z_1 + z_2 = (a_1 + a_2) + (b_1 + b_2)i$$
$$z_1 - z_2 = (a_1 - a_2) + (b_1 - b_2)i$$
$$z_1 z_2 = (a_1 a_2 - b_1 b_2) + (a_1 b_2 + a_2 b_1)i$$
$$\frac{z_1}{z_2} = \frac{z_1}{z_2} \frac{z_2^*}{z_2^*} = \frac{a_1 a_2 + b_1 b_2}{a_2^2 + b_2^2} + \frac{a_2 b_1 - a_1 b_2}{a_2^2 + b_2^2} i$$

Note that because the division operation above is defined, the **multiplicative inverse** of any complex number $z \neq 0$ is also defined (where z_1 is 1 and z_2 is z) in the set of complex numbers.

In addition to these operations, the **absolute value of a complex number** $z = a + bi$ (written $|z|$ or $|a + bi|$) is also defined. (The absolute value may also be termed the "magnitude" or the "modulus" of the number.)

$$|z| = \sqrt{zz^*} = \sqrt{a^2 + b^2}$$

The beginning teacher understands the algebraic structure of the complex number system and its subsets (e.g., complex numbers as a field, complex addition as vector addition).

The set of complex numbers forms a field, and thus, it obeys all the field axioms discussed in **Competency 001**. Nevertheless, **the complex numbers are not an ordered field**. Consider the number $i = \sqrt{-1}$ contained in the set ▢ of complex numbers. Assume that ▢ has a subset P (positive numbers) that is closed under both addition and multiplication. Assume that $i > 0$. A difficulty arises in that $i^2 = -1 < 0$, so i cannot be included in the set P. Likewise, assume $i < 0$. The problem once again arises that $i^4 = 1 > 0$, so i cannot be included in P. It is clearly the case that $i \neq 0$, so there is no place for i in an ordered field. Thus, the complex numbers cannot be ordered.

Because the complex numbers include the real numbers, they also include all the subsets of the real numbers (such as integers, rational numbers, and irrational numbers). Likewise, complex numbers can be organized into sets that contain, for instance, numbers with irrational imaginary parts, integer real parts, or other specific properties (or combinations thereof).

Representation of complex numbers using vectors is discussed below.

The beginning teacher selects and uses appropriate representations of complex numbers (e.g., vector, ordered pair, polar, exponential) for particular situations.

As mentioned above, complex numbers can be written as ordered pairs. Thus, complex numbers can be plotted graphically in a Cartesian complex plane, where the x-axis represents real numbers and the y-axis represents imaginary numbers. An equivalent representation, following a graphical approach, is to use **polar coordinates** (a magnitude and an angle). These two representations are shown below for the example of 2 + 3i.

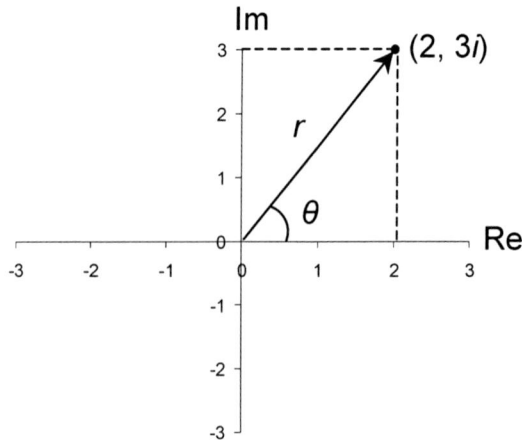

Thus, a complex number a + bi can be represented using a vector length r and an angle θ, measured in the counter-clockwise direction from the x-axis. Using trigonometry, the complex number a + bi can be written as follows (for more on trigonometry, see **Competency 009**).

$$a + bi = r(\cos\theta + i\sin\theta)$$

Note that

$$r = \sqrt{a^2 + b^2}$$
$$\theta = \arctan\frac{b}{a}$$

Because complex numbers can be written as ordered pairs, they can also be treated as **vectors** in the two-dimensional (complex) plane. (For more on vectors, see Competencies 003 and 014.) For instance, the sum (or difference) of two complex numbers is the same as the sum (difference) of two vectors corresponding to the same ordered pairs. Thus, (1 – 2i) + (2 + 2i) can be found graphically as shown on the next page.

TEACHER CERTIFICATION STUDY GUIDE

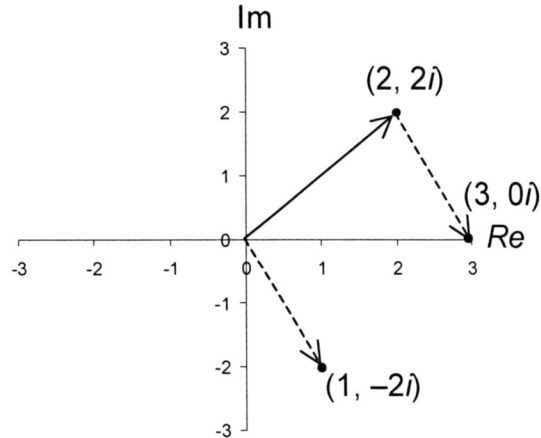

The result, found in the graph above using vector addition, is 3 + 0*i*.

In addition to the use of trigonometric functions for a polar representation, complex numbers can also be represented in polar form using exponentials. To derive Euler's formula, define *z* as follows.

$$z = \cos\theta + i\sin\theta$$

Differentiate *z* with respect to θ and form a differential equation (for more on differential and integral calculus, see **Competency 010**).

$$\frac{dz}{d\theta} = \frac{d}{d\theta}(\cos\theta + i\sin\theta)$$
$$\frac{dz}{d\theta} = -\sin\theta + i\cos\theta$$
$$dz = (-\sin\theta + i\cos\theta)d\theta$$

Note that *dz* can be rewritten as follows using the definition of *i*.

$$dz = (i^2\sin\theta + i\cos\theta)d\theta$$
$$dz = i(\cos\theta + i\sin\theta)d\theta = iz\,d\theta$$
$$\frac{dz}{z} = i\,d\theta$$

Integrate the result.

$$\int\frac{dz}{z} = \int i\,d\theta$$
$$\ln z = i\theta$$

MATHEMATICS 8-12

Use the rules of exponentials to find z.

$$e^{\ln z} = e^{i\theta}$$
$$z = \cos\theta + i\sin\theta = e^{i\theta}$$

This is **Euler's formula**. Thus, complex numbers can be written in either trigonometric or exponential polar form.

$$a + bi = r(\cos\theta + i\sin\theta) = re^{i\theta}$$

This **exponential polar form** is particularly beneficial for fast multiplication and division of complex numbers, because these operations simply involve multiplying or dividing the multiplicative factors (r) and adding or subtracting the exponents ($i\theta$). In addition, the **square root operation** becomes more lucid when this representation is used, since

$$\sqrt{a+bi} = \sqrt{re^{i\theta}} = \sqrt{r}\,e^{i\frac{\theta}{2}}$$

Example: Write $2 - 5i$ in polar form.

This complex number can either be written in terms of sine and cosine or in terms of an exponential. In either case, first calculate r and θ.

$$r = \sqrt{2^2 + (-5)^2} = \sqrt{4+25} = \sqrt{29} \approx 5.385$$
$$\theta = \arctan\left(-\frac{5}{2}\right) \approx -1.190$$

Thus,

$$2 - 5i \approx 5.385\left[\cos(-1.190) + i\sin(-1.190)\right]$$
$$= 5.385\left[\cos 1.190 - i\sin 1.190\right]$$

or

$$2 - 5i \approx 5.385 e^{-1.190i}$$

In the first case, the result can be checked by evaluating the trigonometric functions and multiplying.

$$5.385[\cos 1.190 - i\sin 1.190] \approx 5.385[0.372 - i0.928]$$
$$5.385[\cos 1.190 - i\sin 1.190] \approx 2.003 - i4.998$$

Although a lack of numerical accuracy yields a result that is not exact, it is close enough to assume that it is correct. To be more certain, simply use greater numerical accuracy throughout the problem.

Example: Express $x + ix \tan x$ in polar form.

Calculate r and θ, noting that the real and imaginary parts of this complex number are functions instead of numbers.

$$r = \sqrt{x^2 + (x\tan x)^2} = \sqrt{x^2 + x^2 \tan^2 x} = \sqrt{x^2(1 + \tan^2 x)}$$
$$r = \sqrt{x^2 \sec^2 x} = x \sec x$$
$$\theta = \arctan\left(\frac{x \tan x}{x}\right) = \arctan(\tan x) = x$$

Write the polar form.

$$x + i \tan x = x \sec x (\cos x + i \sin x)$$
$$x + i \tan x = (x \sec x) e^{ix}$$

Expand the trigonometric polar form to test the result.

$$x \sec x (\cos x + i \sin x) = x \sec x \cos x + ix \sec x \sin x$$
$$x \sec x (\cos x + i \sin x) = x \frac{\cos x}{\cos x} + ix \frac{\sin x}{\cos x} = x + ix \tan x$$

The result is confirmed.

Example: Find the imaginary part of the expression $\frac{\sqrt{3+2i}}{-1+5i}$.

The imaginary part of this expression can be found by reducing it to the form $a + bi$. To do so, first evaluate the square root using the polar form.

$$3+2i = |3+2i|e^{i\arctan\frac{2}{3}} \approx \sqrt{(3+2i)(3-2i)}e^{0.588i}$$
$$3+2i \approx \sqrt{(3+2i)(3-2i)}e^{0.588i} = \sqrt{9+4}e^{0.588i} = \sqrt{13}e^{0.588i}$$

Evaluate the square root of this value

$$\sqrt{3+2i} \approx \sqrt{3.606e^{0.588i}} \approx 1.899e^{0.294i}$$

Return this result to $a + bi$ form:

$$\sqrt{3+2i} \approx 1.899e^{0.294i} = 1.899\cos 0.294 + i1.899\sin 0.294$$
$$\sqrt{3+2i} \approx 1.818 + 0.550i$$

Next, use the complex conjugate to eliminate the denominator of the original fraction given in the question.

$$\frac{\sqrt{3+2i}}{-1+5i} \approx \frac{1.818+0.550i}{-1+5i}\left(\frac{-1-5i}{-1-5i}\right)$$
$$\frac{\sqrt{3+2i}}{-1+5i} \approx \frac{-1.818-9.09i-0.550i+2.75}{1+25} = \frac{0.932-9.64i}{26}$$

Thus, the imaginary part of the expression is approximately $-\frac{9.64}{26}$, or about -0.371.

The beginning teacher demonstrates how some problems that have no solution in the real number system have solutions in the complex number system.

Numerous problems do not have solutions that are contained in the set of real numbers. In such cases, complex solutions may be required. Consider the following canonical example.

$$x^2 + 1 = 0$$

This equation cannot be solved for x in the real domain.

$$x^2 = -1$$
$$x = \pm\sqrt{-1}$$

Since there is no real number whose square is −1, there is no real solution for x. The imaginary number i is a solution, however; thus, this equation can be solved in the complex domain.

$$x = \pm i$$

More generally, the Fundamental Theorem of Algebra states that any polynomial of degree n must have n solutions (or roots), which may include real, complex, or non-distinct roots (or some combination thereof).

Other operations can likewise lead to imaginary results. Consider, for instance, the natural logarithm of a negative numbers (for more on logarithms, see **Competency 008**).

$$\ln(-1) = ?$$

In the real domain, this expression is undefined, because there is no real exponent of e that results in a negative number. If Euler's formula is applied, however, a solution from the set of complex numbers becomes apparent.

$$\ln(-1) = \ln(\cos\pi + i\sin\pi) = \ln(e^{i\pi})$$
$$\ln(-1) = i\pi\ln(e) = i\pi$$

Thus, the expression has an equivalent numerical expression from the set of complex numbers. (For precision, it is noteworthy that the general result is $in\pi$, where n is an integer.)

Example: Find all the roots of the polynomial $x^3 - x^2 + 3x - 3$.

This problem can be tackled in any of several ways. First, note that the Fundamental Theorem of Algebra requires that the polynomial must have three solutions. In this case, the polynomial can be factored. By inspection, it appears that $x = 1$ might be a solution:

$$(1)^3 - (1)^2 + 3(1) - 3 = 1 - 1 + 3 - 3 = 0$$

Thus, (x − 1) must be a factor in the expression. The remaining factor must be a polynomial of degree two.

$$x^3 - x^2 + 3x - 3 = (x-1)(ax^2 + bx + c)$$

Obviously, *a* must be equal to 1. Likewise, *c* must be equal to 3. Then

$$(x-1)(x^2 + bx + 3) = x^3 - x^2 + bx^2 - bx + 3x - 3$$
$$x^3 - x^2 + 3x - 3 = x^3 - (1-b)x^2 - (b-3)x - 3$$

Thus, *b* = 0.

$$x^3 - x^2 + 3x - 3 = (x-1)(x^2 + 3)$$

Factoring once more yields

$$0 = (x-1)(x + i\sqrt{3})(x - i\sqrt{3})$$

The roots are then 1, $i\sqrt{3}$, and $-i\sqrt{3}$.

The beginning teacher describes complex number operations (e.g., addition, multiplication, roots) using symbolic and geometric representations.

As with the example above for addition of complex numbers being represented as vector addition, other operations involving complex numbers can likewise be represented in various ways. Multiplication, for instance, can be represented geometrically as the product of certain radii (or lengths) and the sum of certain angles. As noted above, a complex number can be written in the polar form $re^{i\theta}$, where *r* is a length (or radius) and θ is an angle.

Thus, the product of two complex numbers $r_1 e^{i\theta_1}$ and $r_2 e^{i\theta_2}$ is just $r_1 r_2 e^{i(\theta_1+\theta_2)}$. This operation can be illustrated geometrically as shown below.

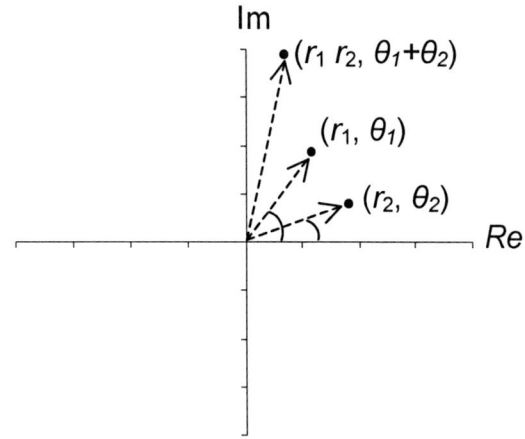

Notice that since the sum of the angles and the product of the radii are both commutative operations, and therefore, the product of two complex numbers is shown to be commutative as well. Based on the above representation, it is clear that a square root simply involves halving of the angle θ of a complex number and taking the square root of the real radius r. Other operations on complex numbers can be given similar geometric or graphical representations, similar to those provided above and in **Competency 001** for the case of real numbers.

TEACHER CERTIFICATION STUDY GUIDE

Competency 003 The teacher understands number theory concepts and principles and uses numbers to model and solve problems in a variety of situations.

This section reviews the application of number theory to various topics. Concepts such as the rules of divisibility for natural numbers and the Fundamental Theorem of Arithmetic are discussed, followed by a number of example problems involving proofs and real-world situations. Finally, estimation and reasonableness are reviewed.

The beginning teacher applies ideas from number theory (e.g., prime numbers and factorization, the Euclidean algorithm, divisibility, congruence classes, modular arithmetic, the fundamental theorem of arithmetic) to solve problems.

Divisibility Rules

a. A number is **divisible by 2** if that number is an even number (which means the last digit is 0, 2, 4, 6 or 8).

Consider a number $abcd$ defined by the digits a, b, c and d (for instance, 1,234). Rewrite the number as follows.

$$10abc + d = abcd$$

Note that $10abc$ is divisible by 2. Thus, the number $abcd$ is only divisible by 2 if d is divisible by two; in other words, $abcd$ is divisible by two only if it is an even number. For example, the last digit of 1,354 is 4, so it is divisible by 2. On the other hand, the last digit of 240,685 is 5, so it is not divisible by 2.

b. A number is **divisible by 3** if the sum of its digits is evenly divisible by 3.

Consider a number $abcd$ defined by the digits a, b, c and d. The number can be written as

$$abcd = 1000a + 100b + 10c + d$$

The number can also be rewritten as

$$abcd = (999+1)a + (99+1)b + (9+1)c + d$$
$$abcd = 999a + 99b + 9c + (a+b+c+d)$$

Note that the first three terms in the above expression are all divisible by 3. Thus, the number is evenly divisible by 3 only if $a + b + c + d$ is divisible by 3. The same logic applies regardless of the size of the number. This proves the rules for divisibility by 3.

The sum of the digits of 964 is 9+6+4 = 19. Since 19 is not divisible by 3, neither is 964. The digits of 86,514 are 8+6+5+1+4 = 24. Since 24 is divisible by 3, 86,514 is also divisible by 3.

c. A number is **divisible by 4** if the number in its last 2 digits is evenly divisible by 4.

Let a number $abcd$ be defined by the digits a, b, c and d.

$$ab(100) + cd = abcd$$

Since 100 is divisible by 4, $100ab$ is also divisible by 4. Thus, $abcd$ is divisible by 4 only if cd is divisible by 4.

$$25ab + \frac{cd}{4} = \frac{abcd}{4}$$

The number 113,336 ends with the number 36 for the last 2 digits. Since 36 is divisible by 4, 113,336 is also divisible by 4. The number 135,627 ends with the number 27 for the last 2 digits. Since 27 is not evenly divisible by 4, 135,627 is also not divisible by 4.

d. A number is **divisible by 5** if the number ends in either a 5 or a 0.

Use the same number $abcd$.

$$100ab + cd = abcd$$

The first term is evenly divisible by 5, but the second term is only evenly divisible by 5 if it is 0, 5, 10, 15,...,95. In other words, $abcd$ is divisibly by 5 only if it ends in a 0 or 5. For instance, 225 ends with a 5, so it is divisible by 5. The number 470 is also divisible by 5 because its last digit is a 0. The number 2,358 is not divisible by 5 because its last digit is an 8.

e. A number is **divisible by 6** if the number is even and the sum of its digits is evenly divisible by 3 or 6.

Let a number *efgh* be defined by the digits *e*, *f*, *g* and *h*. If *efgh* is even, then it is divisible by 2. Write *abcd* as follows.

$$\frac{efgh}{2} = abcd = 999a + 99b + 9c + (a+b+c+d)$$

$$efgh = 2abcd = 2(999)a + 2(99)b + 2(9)c + 2(a+b+c+d)$$

Then divide *efgh* by 6.

$$\frac{efgh}{6} = \frac{2(999)}{6}a + \frac{2(99)}{6}b + \frac{2(9)}{6}c + \frac{2}{6}(a+b+c+d)$$

$$\frac{efgh}{6} = 333a + 33b + 3c + \frac{2}{6}(a+b+c+d)$$

Notice that *efgh* is divisible by 6 only if the sum of the digits is divisible by 3 or by 6. The number *efgh* must also be even, since 2 is a factor of 6. For instance, 4,950 is an even number and its digits add to 18 (4 + 9 + 5 + 0 = 18). Since the number is even and the sum of its digits is 18 (which is divisible by 3 and 6), then 4,950 is divisible by 6. On the other hand, 326 is an even number, but its digits add up to 11. Since 11 is not divisible by 3 or 6, then 326 is not divisible by 6.

f. A number is **divisible by 8** if the number in its last 3 digits is evenly divisible by 8.

The logic for the proof of this case follows that of numbers divisible by 2 and 4. The number 113,336 ends with the 3-digit number 336 in the last 3 columns. Since 336 is divisible by 8, then 113,336 is also divisible by 8. The number 465,627 ends with the number 627 in the last 3 columns. Since 627 is not evenly divisible by 8, then 465,627 is also not divisible by 8.

g. A number is **divisible by 9** if the sum of its digits is evenly divisible by 9.

The logic for the proof of this case follows that for the case of numbers that are divisible by 3 and 6. The sum of the digits of 874, for example, is 8 + 7 + 4 = 19. Since 19 is not divisible by 9, neither is 874. The sum of the digits of 116,514 is 1 + 1 + 6 + 5 + 1 + 4 = 18. Since 18 is divisible by 9, 116,514 is also divisible by 9.

The Fundamental Theorem of Arithmetic

Every integer greater than 1 can be written uniquely in the form

$$p_1^{e_1} p_2^{e_2} \cdots p_k^{e_k},$$

The p_i are distinct prime numbers and the e_i are positive integers. Any integer $n > 1$ that is divisible by at least one positive integer that is not equal to one or n is called a **composite number**. A natural number n that is only divisible by one and n is called a **prime number**.

GCF is the abbreviation for the **greatest common factor**. The GCF is the largest number that is a factor of all the numbers given in a problem. The GCF can be no larger than the smallest number given in the problem. If no other number is a common factor, then the GCF will be the number 1. To find the GCF, list all possible factors of the smallest number given (include the number itself). Starting with the largest factor (which is the number itself), determine if it is also a factor of all the other given numbers. If so, that is the GCF. If that factor does not work, try the same method on the next smaller factor. Continue until a common factor is found. This is the GCF. Note: There can be other common factors besides the GCF.

Example: Find the GCF of 12, 20, and 36.

The smallest number in the problem is 12. The factors of 12 are 1, 2, 3, 4, 6 and 12. 12 is the largest factor, but it does not divide evenly into 20. Neither does 6, but 4 will divide into both 20 and 36 evenly. Therefore, 4 is the GCF.

Example: Find the GCF of 14 and 15.

Factors of 14 are 1, 2, 7 and 14. 14 is the largest factor, but it does not divide evenly into 15. Neither does 7 or 2. Therefore, the only factor common to both 14 and 15 is the number 1, which is the GCF.

The **Euclidean Algorithm** is a formal method for determining the **greatest common devisor** (GCD) (another name for GCF) of two positive integers. The algorithm can be formulated in a recursive manner that simply involves repetition of a few steps until a terminating point is reached. The algorithm can be summarized as follows, where a and b are the two integers for which determination of the GCD is to be undertaken. (Assign a and b such that $a > b$.)

1. If $b = 0$, a is the GCD.
2. Calculate $c = a$ mod b.
3. If $c = 0$, b is the GCD.
4. Go back to step 2, replacing a with b and b with c.

Note that the "**mod**" operator in this case is simply a remainder operator. Thus, a mod b is the remainder of division of a by b.

Example: Find the GCD of 299 and 351.

To find the GCD, first let $a = 351$ and $b = 299$. Begin the algorithm as follows.

1. $b \neq 0$.
2. $c = 351$ mod $299 = 52$
3. $c \neq 0$

Perform the next iteration, starting with step 2.

2. $c = 299$ mod $52 = 39$
3. $c \neq 0$

Continue to iterate recursively until a solution is found.

2. $c = 52$ mod $39 = 13$
3. $c \neq 0$

2. $c = 39$ mod $13 = 0$
3. $c = 0$: GCD = 13

Thus, the GCD of 299 and 351 is thus 13.

LCM is the abbreviation for **least common multiple**. The least common multiple of a group of numbers is the smallest number that all of the given numbers will divide into. The least common multiple will always be the largest of the given numbers or a multiple of the largest number.

Example: Find the LCM of 20, 30 and 40.

The largest number given is 40, but 30 will not divide evenly into 40. The next multiple of 40 is 80 (2 × 40), but 30 will not divide evenly into 80 either. The next multiple of 40 is 120. 120 is divisible by both 20 and 30, so 120 is the LCM (least common multiple).

Example: Find the LCM of 96, 16 and 24.

The largest number is 96. 96 is divisible by both 16 and 24, so 96 is the LCM.

The fundamental theorem of arithmetic can be used to show that **every fraction is equivalent to a unique fraction where the numerator and denominator are relatively prime.**

Given a fraction $\frac{a}{b}$, the integers a and b can both be written uniquely as a product of prime factors.

$$\frac{a}{b} = \frac{p_1^{x_1} p_2^{x_2} p_3^{x_3} \ldots p_n^{x_n}}{q_1^{y_1} q_2^{y_2} q_3^{y_3} \ldots q_m^{y_m}}$$

When all the common factors are cancelled, the resulting numerator a_1 (the product of remaining factors $p_n^{x_n}$) and the resulting denominator b_1 (the product of remaining factors $q_m^{y_m}$) have no common divisor other than 1; i.e., they are **relatively prime**.

Since, according to the Fundamental Theorem of Arithmetic, the initial prime decomposition of the integers a and b is unique, the new reduced fraction $\frac{a_1}{b_1}$ is also **unique**. Hence, any fraction is equivalent to a unique fraction where the numerator and denominator are relatively prime.

The proof that the square root of any integer, not a perfect square number, is irrational may also be demonstrated using prime decomposition.

Let n be an integer. Assuming that the square root of n is rational, we can write

$$\sqrt{n} = \frac{a}{b}$$

Since every fraction is equivalent to a unique fraction where the numerator and denominator are relatively prime (shown earlier), we can reduce the fraction $\frac{a}{b}$ to the fraction $\frac{a_1}{b_1}$ and write

$$\sqrt{n} = \frac{a_1}{b_1}; n = \frac{a_1^2}{b_1^2}$$

where a_1 and b_1 are relatively prime.

Since a_1 and b_1 are relatively prime, a_1^2 and b_1^2 must also be relatively prime. Also, since n is an integer, $\frac{a_1^2}{b_1^2}$ must be an integer. The only way the above two conditions can be satisfied is if the denominator $b_1^2 = 1$. Thus, $n = a_1^2$

As a result, the square root of an integer can be rational only if the integer is a perfect square. Stated in an alternative manner, the square root of an integer, not a perfect square, is irrational.

Modular Arithmetic

Modular arithmetic involves numbers and operations that have a circular or wrap-around characteristic, such as that shown by a clock or even the odometer of an automobile. Given a set of integers from 0 to *n* (or, identically, 1 to *n* + 1), any result of an operation is actually the remainder of the quotient of the result divided by *n* + 1. Thus, in terms of a clock, which involves "modulo 12" arithmetic,

$$7 + 10 = 5$$

Alternatively, this operation can be expressed as follows:

$$7 + 10 \equiv 5 \,(\text{mod}\,12)$$

In other words, 10 hours after 7 o'clock is 5 o'clock (in the context of time on a standard 12-hour clock). The number 12 in this case is called the modulus. The symbol $\equiv$ indicates "**congruence**." Generally, for any value *c* and arbitrary integer *i* in modulo *n* arithmetic, the following statement is always true:

$$c + ni \equiv c \,(\text{mod}\,n)$$

This rule is useful when a negative number is encountered. Simply add some multiple (one is generally sufficient) of the modulus to find the correct result of the operation.

Likewise,

$$ni \equiv 0 \,(\text{mod}\, n)$$

The case of modulo 2 arithmetic (which involves the integers 0 and 1) has particular application to binary and Boolean systems.

Example: Calculate the following in modulo 5 arithmetic: 3 + 8, (4)(7), 7 − 8.

In each case, use the remainder of the result divided by the modulus to find the correct answer.

$$\text{remainder}\left(\frac{3+8}{5}\right) = 1 \qquad 3+8 \equiv 1\,(\text{mod}\, 5)$$

$$\text{remainder}\left(\frac{4 \cdot 7}{5}\right) = 3 \qquad 4 \cdot 7 \equiv 3\,(\text{mod}\, 5)$$

For the last operation, use the rule $c + ni \equiv c\,(\text{mod}\, n)$ to find the answer:

$$\text{remainder}\left(\frac{7-8}{5}\right) = -1 \qquad (7-8)+5 \equiv 4\,(\text{mod}\, 5)$$

Example: A bicycle has a simple odometer that goes up to 999 miles. If the current odometer reading is 431 miles, what will it read 1500 miles later?

The odometer has readings that range from 0 to 999. Upon reaching 999, the odometer returns to 0 after the next mile. Thus, use modulo 1000 (or 999 + 1) arithmetic to solve this problem.

$$431 + 1500 \equiv 931\,(\text{mod}\, 1000)$$

Thus, the odometer will read 931.

The beginning teacher applies number theory concepts and principles to justify and prove number relationships.

The concepts and principles of number theory can be used to justify and prove various relationships among numbers. The best way to prepare for particular problems is to thoroughly learn and understand number theory. The following examples, however, provide some illustrations of problem-solving approaches.

Example: Prove that every fraction is equivalent to a unique fraction where the numerator and denominator are relatively prime.

The Fundamental Theorem of Arithmetic can be used to prove this statement. Given a fraction $\frac{a}{b}$, the integers a and b can both be written uniquely as a product of prime factors.

$$\frac{a}{b} = \frac{p_1^{x_1} p_2^{x_2} p_3^{x_3} \ldots p_n^{x_n}}{q_1^{y_1} q_2^{y_2} q_3^{y_3} \ldots q_m^{y_m}}$$

When all the common factors are cancelled, the resulting numerator a_1 (the product of remaining factors $p_i^{x_i}$) and the resulting denominator b_1 (the product of remaining factors $q_j^{y_j}$) have no common divisor other than 1; i.e., they are **relatively prime**.

Since, according to the Fundamental Theorem of Arithmetic, the initial prime decomposition of the integers a and b is unique, **the new reduced fraction $\frac{a_1}{b_1}$ is also unique**. Hence, any fraction is equivalent to a unique fraction where the numerator and denominator are relatively prime.

Example: Prove that the square root of any integer that is not a perfect square is irrational.

The proof of this proposition can also be demonstrated using prime decomposition. Let n be an integer. Assuming that the square root of n is rational, we can write

$$\sqrt{n} = \frac{a}{b}$$

Since every fraction is equivalent to a unique fraction where the numerator and denominator are relatively prime (shown earlier), we can reduce the fraction $\frac{a}{b}$ to the fraction $\frac{a_1}{b_1}$ and write

$$\sqrt{n} = \frac{a_1}{b_1}; \quad n = \frac{a_1^2}{b_1^2}$$

where a_1 and b_1 are relatively prime. Since a_1 and b_1 are relatively prime, a_1^2 and b_1^2 must also be relatively prime. Also, since n is an integer, $\frac{a_1^2}{b_1^2}$ must be an integer. The only way the above two conditions can be satisfied is if the denominator $b_1^2 = 1$. Thus, $n = a_1^2$

As a result, the square root of an integer can be rational only if the integer is a perfect square. Stated in an alternative manner, the square root of an integer, not a perfect square, is irrational.

Example: Determine if the set of integers is closed under division.

For the set of integers to be closed under addition, it must be the case that $\frac{a}{b}$ is an integer for any integers a and b. Consider $a = 2$ and $b = 3$.

$$\frac{a}{b} = \frac{2}{3}$$

This result is not an integer. Therefore, the set of integers is not closed under division.

The beginning teacher uses properties of numbers (e.g., fractions, decimals, percents, ratios, proportions) to model and solve real-world problems.

The following examples illustrate the use of the properties of numbers (such as integers, fractions, decimals, percents, and ratios) in problem solving.

Example: The sum of two consecutive integers is 51, find the integers.

Let the consecutive integers be x and $(x + 1)$. Then,

$x + (x + 1) = 51$
$2x + 1 = 51$
$2x = 50$
$x = 25$

Thus, the numbers are 25 and 26.

Example: Find the item with the best unit price: $1.79 for 10 ounces, $1.89 for 12 ounces, or, $5.49 for 32 ounces.

For each price and weight, find the ratio of these values. The item with the lowest price-to-unit ratio is the best value (i.e., it has the best unit price).

$$\frac{\$1.79}{10 oz} = \$0.179 / oz$$
$$\frac{\$1.89}{12 oz} = \$0.158 / oz$$
$$\frac{\$5.49}{32 oz} = \$0.172 / oz$$

Thus, $1.89 for 12 ounces is the best price.

Example: An item that is on sale costs $18. The sale is 10% off the original price. What was the original price?

Let the original price of the item be x. Write the following equation to solve for x:

$x - (10\%)x = x - 0.10x = \$18$
$0.90x = \$18$
$x = \$20$

Thus, the original price of the item was $20.

Example: Two numbers have a ratio of 3:5. Find the numbers if the difference between their squares is 144.

Let the numbers be x and y. Then

$$\frac{y}{x} = \frac{3}{5} \text{ and } x^2 - y^2 = 144$$

Substituting $y = \frac{3}{5}x$ into the second equation yields

$$x^2 - \frac{9}{25}x^2 = 144$$

$$\frac{16}{25}x^2 = 144$$

$$x^2 = \frac{(144)(25)}{16} = 225$$

$$x = 15$$

Based on this result, $y = 9$. Thus, the numbers are 9 and 15.

The beginning teacher compares and contrasts properties of vectors and matrices with properties of number systems (e.g., existence of inverses, non-commutative operations).

Sets of vectors and matrices behave in many ways like sets of scalars (numbers). The characteristics of vectors and matrices are discussed in Competencies 006 and 014.

The beginning teacher applies counting techniques such as permutations and combinations to quantify situations and solve problems.

Application of permutations, combinations, and various counting techniques to problems is discussed in **Competency 016**.

The beginning teacher uses estimation techniques to solve problems and judges the reasonableness of solutions.

Estimation and approximation can be used to check the reasonableness of answers or to speed up a calculation where exact answers are not required. Estimation can be particularly important when calculators are used. Estimation requires good mental math skills.

There are several different ways of estimating. A common estimation strategy involves replacing numbers with simpler numbers that make for simpler computations. These methods include rounding, front-end digit estimation, and compensation. Although rounding is done to a specific place value (e.g., the nearest ten or hundred), front-end estimation involves rounding or truncating to the place value of the first digit in the number. The following example uses front-end estimation.

Example: Estimate the result of the calculation $\frac{58 \times 810}{1989}$.

To simplify the calculation, round each number to the highest place value. Thus, the calculation becomes

$$\frac{58 \times 810}{1989} \approx \frac{60 \times 800}{2000} = \frac{48000}{2000} = 24$$

A more precise result is 23.62. In this case, the estimated value is close to the exact result.

Another estimation technique, compensation, involves replacing different numbers in different ways so that one change can more or less compensate for the other.

Example: Calculate 32 + 53.

Although this example is simple, it is noteworthy that compensation can make the numbers easier to handle mentally.

$$32 + 53 = 30 + 55 = 85$$

Here both numbers are replaced in a way that minimizes the change; one number is increased and the other is decreased. This technique can also be applied to algebraic expressions in addition to numbers. In such cases, compensation or other similar forms of reorganization of numbers can drastically simplify certain operations.

A third estimation strategy is calculating a range for the result.

Example: Estimate 458 + 873.

Again, this is a very simple example, but the principle can be applied to a range of different calculations. A simple range containing the answer can be found as follows.

$$458 + 873 > 400 + 800 = 1200$$
$$458 + 873 < 500 + 900 = 1400$$

Thus, the correct answer is in the following range:

$$1200 < 458 + 873 < 1400$$

Various other estimation techniques can be applied as well, depending on the problem being solved.

The reasonableness of an estimate can be judged in various ways depending on the context of the problem. For instance, if the exact result of a calculation or the exact solution to a problem is known, then the estimated value can be considered reasonable if it varies from the exact value by a sufficiently small percentage. Often times, a variation of a few percent is acceptable; in some contexts, up to 10% variation is likewise acceptable. Large variations, however, may indicate that the estimate is not reasonable.

If the exact result or solution is not known, the reasonableness of an estimate can be judged to some extent by the expected effect of rounding, regrouping, or whatever the procedure used in the estimate. For example, if a number is rounded in a calculation and the process of rounding has a small effect on the value of the number (only a percent or two, for instance), then it may be likely that the estimate is reasonable. The particular operations involved in a calculation are important to making a good judgment, however. Consider the calculation $10.4929 - 10.5103$. If these numbers are rounded to the nearest one, or even to the nearest tenth, then the estimated result is zero instead of -0.0174. In some contexts, this difference may be negligible, but in other contexts, it may be very important. Thus, estimating the reasonableness of an estimate requires consideration of the numbers and the operations involved in the problem, as well as consideration of the context of the problem.

DOMAIN II.   PATTERNS AND ALGEBRA

Competency 004   **The teacher uses patterns to model and solve problems and formulate conjectures.**

This competency reviews the use of patterns in the form of tables, graphs, sequences, and series to solve problems using techniques such as iteration and recursion, mathematical induction, and other forms of pattern analysis.

The beginning teacher ecognizes and extends patterns and relationships in data presented in tables, sequences, or graphs.

Tabular, numeric, graphical, symbolic or even pictorial data sometimes contain inherent patterns and relationships. Identifying these patterns allows one to extend them and make predictions about data outside the given range.

A numerical sequence is a pattern of numbers arranged in a particular order. Inspection of a sequence sometimes reveals a particular rule that is followed in creating it. For instance, 1, 4, 9, 16... is a series that consists of the squares of the natural numbers. Using this rule, the next term in the series, 25, can be found by squaring the next natural number: 5.

Example: Find the next term in the series 1, 1, 2, 3, 5, 8,…

Inspecting the terms in the series, one finds that this pattern is neither arithmetic nor geometric. Every term in the series is the sum of the previous two terms. Thus, the next term is 5 + 8 = 13.

This particular sequence is a well-known series named the Fibonacci sequence.

Example: Kepler discovered a relationship between the average distance of a planet from the sun and the time it takes the planet to orbit the sun.

TEACHER CERTIFICATION STUDY GUIDE

The following table shows the data for the six planets closest to the sun:

	Mercury	Venus	Earth	Mars	Jupiter	Saturn
Average distance, x	0.387	0.723	1	1.523	5.203	9.541
x^3	0.058	0.378	1	3.533	140.852	868.524
Time, y	0.241	0.615	1	1.881	11.861	29.457
y^2	0.058	0.378	1	3.538	140.683	867.715

Looking at the data in the table, we can assume that $x^3 = y^2$. We can conjecture the following function for Kepler's relationship:

$$y = \sqrt{x^3}$$

For graphical representation of patterns, see the discussion on functions and relations in **Competency 005**.

The beginning teacher uses methods of recursion and iteration to model and solve problems.

A **recurrence relation** is an equation that defines a sequence recursively; in other words, each term of the sequence is defined as a function of the preceding terms. For instance, the formula for the balance of an interest-bearing savings account after t years, which is given above in closed form (that is, explicit form), can be expressed recursively as follows.

$$A_t = A_{t-1}\left(1 + \frac{r}{n}\right)^n \quad \text{where} \quad A_0 = P$$

Here, r is the annual interest rate and n is the number of times the interest is compounded per year. Mortgage and annuity parameters can also be expressed in recursive form. Calculation of a past or future term by applying the recursive formula multiple times is called **iteration**.

Sequences of numbers can be defined by iteratively applying a recursive pattern. For instance, the Fibonacci sequence is defined as follows.

$$F_i = F_{i-1} + F_{i-2} \quad \text{where} \quad F_0 = 0 \text{ and } F_1 = 1$$

Applying this recursive formula gives the sequence {0, 1, 1, 2, 3, 5, 8, 13, 21,...}.

It is sometimes difficult or impossible to write recursive relations in explicit or closed form. In such cases, especially where computer programming is involved, the recursive form can still be helpful. When the elements of a sequence of numbers or values depend on one or more previous values, then it is possible that a recursive formula could be used to summarize the sequence.

If a value or number from a later point in the sequence (that is, other than the beginning) is known and it is necessary to find previous terms, then the indices of the recursive relation can be adjusted to find previous values instead of later ones. Consider, for instance, the Fibonacci sequence.

$$F_i = F_{i-1} + F_{i-2}$$
$$F_{i+2} = F_{i+1} + F_i$$
$$F_i = F_{i+2} - F_{i+1}$$

Thus, if any two consecutive numbers in the Fibonacci sequence are known, then the previous numbers of the sequence can be found (in addition to the later numbers).

Example: Write a recursive formula for the following sequence: {2, 3, 5, 9, 17, 33, 65...}.

By inspection, it can be seen that each number in the sequence is equal to twice the previous number, less one. If the numbers in the sequence are indexed such that for the first number $i = 1$ and so on, then the recursion relation is the following.

$$N_i = 2N_{i-1} - 1$$

Example: If a recursive relation is defined by $N_i = N_{i-1}^2$ and the fourth term is 65,536, what is the first term?

Adjust the indices of the recursion and then solve for N_i.

$$N_{i+1} = N_i^2$$
$$N_i = \sqrt{N_{i+1}}$$

Use this relationship to backtrack to the first term.

$$N_3 = \sqrt{N_4} = \sqrt{65,536} = 256$$
$$N_2 = \sqrt{N_3} = \sqrt{256} = 16$$
$$N_1 = \sqrt{N_2} = \sqrt{16} = 4$$

The first term of the sequence is thus 4.

It is helpful in some situations to convert between the recursive form and closed form of a function. Given a closed-form representation of a function, the recursive form can be found by writing out the corresponding series or sequence and then determining a pattern or formula that accurately represents that series or sequence. Consider, for instance, the mortgage principal formula in recursive form:

$$A_i = A_{i-1}\left(1 + \frac{r}{n}\right) - M \quad \text{where} \quad A_0 = P$$

Here, A_i is the remaining principal on the mortgage after the ith payment, r is the annual interest rate, which is compounded n times annually, and M is the monthly payment. The initial value P is the original loan amount for the mortgage. To obtain a closed-form expression for this recursive formula, first write out the terms of the corresponding sequence.

$$A_0 = P$$

$$A_1 = P\left(1 + \frac{r}{n}\right) - M$$

$$A_2 = A_1\left(1 + \frac{r}{n}\right) - M = P\left(1 + \frac{r}{n}\right)^2 - M\left(1 + \frac{r}{n}\right) - M$$

$$A_3 = A_2\left(1 + \frac{r}{n}\right) - M = P\left(1 + \frac{r}{n}\right)^3 - M\left(1 + \frac{r}{n}\right)^2 - M\left(1 + \frac{r}{n}\right) - M$$

This pattern continues until $i = k$, where k is the total number of payments in the mortgage term. Note that the term A_k can be written as follows, where $z = 1 + \frac{r}{n}$:

$$A_k = Pz^k - M\{z^{k-1} + z^{k-2} + \ldots z^2 + z + 1\}$$

But the expression in the curly brackets is simply a geometric series, which can be written in closed form as

$$z^{k-1} + z^{k-2} + \ldots z^2 + z + 1 = \frac{1-z^k}{1-z}$$

Thus, the closed form expression for the principle remaining after k payments on the mortgage is

$$A_k = Pz^k - M\frac{1-z^k}{1-z} \quad \text{where} \quad z = 1 + \frac{r}{n}$$

The process for converting from closed form to recursive form is similar (it is essentially the reverse of the process described above). Simply write out the terms, determine the pattern and then write the ith value in terms of the $(i-1)^{st}$ or $(i+1)^{st}$ value.

The beginning teacher uses the principle of mathematical induction.

Mathematical induction is a method for proving certain results (such as formulas or relations) in number theory. This method essentially involves showing first that a formula or relation works for an initial case ($i = 1$) and then showing that it works for the $(i + 1)$st case, assuming it works for the ith case.

Let $p(n)$ denote the relation involving the integer variable n. If $p(1)$ is true and $p(k + 1)$ is true for each integer $k \geq 1$ whenever $p(k)$ is true, then $p(n)$ is true for all $n \geq 1$. In other words, the relation is true for all numbers if the following two statements are true:

1. The statement is true for $n = 1$.
2. If the statement is true for $n = k$, then it is also true for $n = k + 1$.

The four basic components of mathematical induction proofs are:

1. Identify the statement to be proved
2. Prove the initial case ("let $n = 1$")
3. Make the assumption ("let $n = k$ and assume the statement is true for k")
4. Prove the induction step ("let $n = k+1$")

Example: Prove that the sum of all numbers from 1 to n is equal to $\dfrac{n(n+1)}{2}$.

First, identify the relation to be proved.

$$\sum_{i=1}^{n} i = \frac{n(n+1)}{2}$$

Next, show that this relation is valid for $n = 1$.

$$\sum_{i=1}^{1} i = 1$$

$$\frac{n(n+1)}{2} = \frac{1(1+1)}{2} = \frac{2}{2} = 1$$

Assume that this relation works for $n = k$. For the induction step, prove that the relation works for $n = k + 1$. Since the relation is assumed to work for $n = k$,

$$\sum_{i=1}^{k+1} i = \sum_{i=1}^{k} i + (k+1) = \frac{k(k+1)}{2} + (k+1)$$

Manipulate this expression to find the equivalent form of $\dfrac{n(n+1)}{2}$ for $k + 1$.

$$\sum_{i=1}^{k+1} i = \frac{k(k+1)}{2} + \frac{2(k+1)}{2} = \frac{k^2 + k + 2k + 2}{2}$$

$$\sum_{i=1}^{k+1} i = \frac{k^2 + 3k + 2}{2} = \frac{(k+1)(k+2)}{2}$$

$$\sum_{i=1}^{k+1} i = \frac{(k+1)([k+1]+1)}{2}$$

Thus, it has been proven that the formula works for $n = k + 1$. The induction is complete, and $\dfrac{n(n+1)}{2}$ is the correct formula for the sum $1 + 2 + 3 + \ldots + n$.

The beginning teacher analyzes the properties of sequences and series (e.g., Fibonacci, arithmetic, geometric) and uses them to solve problems involving finite and infinite processes.

Sequences and series can take on a vast range of different forms and patterns. Sequences and series are essentially two different representations of a set of numbers: a sequence is the set of numbers, and a series is the sum of the terms. That is, a sequence such as

$$a_1, a_2, a_3, \ldots$$

has a corresponding series S such that

$$S = a_1 + a_2 + a_3 + \ldots$$

Two of the most common forms of series are the arithmetic and geometric series, both of which are discussed below.

A finite series of numbers where the difference between successive terms is constant is called an **arithmetic series**. An arithmetic series with n terms can be expressed as follows, where a and d are constants. (The constant a is the first term, and d is the difference between successive terms.)

$$a + (a+d) + (a+2d) + (a+3d) + \ldots (a+[n-1]d)$$

To derive the general formula, examine the series sum for several small values of n.

n	Sum
1	a
2	$2a + d$
3	$3a + 3d$
4	$4a + 6d$
5	$5a + 10d$
6	$6a + 15d$
$\vdots$	$\vdots$
n	$na + d\sum_{i=1}^{n-1} i$

The result in the table for n terms is found by examining the pattern of the previous series. All that is necessary, then, is to determine a closed expression for the summation.

By inspection, it can be seen that the product of n and (n + 1), divided by 2, is the expression for the sum of 1 + 2 + 3 + 4 + 5 + ... + n. Then:

$$\sum_{i=1}^{n} i = \frac{1}{2}n(n+1)$$

A simple derivation of this relationship may be made as follows:

$$S_n = 1 + 2 + 3 + \ldots + n$$

Writing the terms in reverse order:

$$S_n = n + (n-1) + (n-2) + \ldots + 1$$

Adding the two expressions for S_n term by term we get

$$2S_n = (1+n) + (2+n-1) + (3+n-2) + \ldots + (n+1)$$
$$= (1+n) + (1+n) + (1+n) + \ldots + (n+1)$$
$$= n(n+1)$$

Therefore, $S_n = \frac{n(n+1)}{2}$.

For the general case (with the first term a and common difference d) therefore, the sum for a series with n terms is given by

$$na + d\sum_{i=1}^{n-1} i = na + d\frac{(n-1)(n)}{2} = \frac{1}{2}n(2a + d(n-1))$$

Often times, closed formulas for series such as the arithmetic series must be found by inspection, as a more rigorous derivation is difficult. The result can be proven using mathematical induction, however.

Example: Find the eighth term of the arithmetic sequence 5, 8, 11,...

$$a_n = a_1 + (n-1)d$$
$$a_1 = 5 \quad \text{Identify first term.}$$
$$d = 3 \quad \text{Find d.}$$
$$a_8 = 5 + (8-1)3 \quad \text{Substitute.}$$
$$a_8 = 26$$

Example: Calculate the sum of the series 1 + 5 + 9 +...+ 57.

This is an arithmetic series, as the difference between successive terms, d, is constant ($d = 4$). Determine the total number of terms by subtracting the first term from the last term, dividing by d and adding 1.

$$n = \frac{57-1}{4} + 1 = \frac{56}{4} + 1 = 14 + 1 = 15$$

That this approach works can be seen by testing simple examples. For instance, if the series is 1 + 5 + 9, then

$$n = \frac{9-1}{4} + 1 = \frac{8}{4} + 1 = 2 + 1 = 3$$

There are indeed 3 terms in this simple series. Next, apply the formula, noting that $a = 1$.

$$\frac{1}{2}n[2a + d(n-1)] = \frac{1}{2}(15)[2(1) + (4)(15-1)]$$
$$= \frac{15}{2}[2 + 4(14)] = \frac{15}{2}(58) = 435$$

Thus, the answer is 435.

A **geometric series** is a series whose successive terms are related by a common factor (rather than the common difference of the arithmetic series). Assuming a is the first term of the series and r is the common factor, the general n-term geometric series can be written as follows.

$$a + ar + ar^2 + ar^3 + \ldots + ar^{n-1}$$

The geometric series can also be written using sum notation.

$$a + ar + ar^2 + \ldots + ar^{n-1} = \sum_{i=0}^{n-1} ar^i$$

To derive the closed-form expression for this finite series, let the sum for *n* terms be defined as S_n. Multiply S_n by *r*.

$$S_n = a + ar + ar^2 + \ldots + ar^{n-1}$$
$$rS_n = ar + ar^2 + ar^3 + \ldots + ar^n$$

Note that if *a* is added to this new series, the result is the sum S_{n+1}, which has *n* + 1 terms.

$$a + rS_n = a + ar + ar^2 + ar^3 + \ldots + ar^n = S_{n+1}$$

But S_{n+1} is simply $S_n + ar^n$, so the above expression can be written solely in terms of S_n.

$$a + rS_n = S_{n+1} = S_n + ar^n$$

Rearrange the result to obtain a simple formula for the geometric series.

$$a + rS_n = S_n + ar^n$$
$$a - ar^n = S_n - rS_n$$
$$a(1 - r^n) = S_n(1 - r)$$
$$S_n = a\frac{1 - r^n}{1 - r}$$

The infinite geometric series is the limit of S_n as *n* approaches infinity.

$$a + ar + ar^2 + \ldots = \lim_{n \to \infty} a\frac{1 - r^n}{1 - r}$$

Three cases are of interest: $r \geq 1$, $r \leq -1$ and $-1 < r < 1$. To determine the limit in each case, first apply L'Hopital's rule.

$$\lim_{n \to \infty} a \frac{1-r^n}{1-r} = a \lim_{n \to \infty} \frac{\frac{d}{dr}(1-r^n)}{\frac{d}{dr}(1-r)} = a \lim_{n \to \infty} \frac{-nr^{n-1}}{-1}$$

$$\lim_{n \to \infty} a \frac{1-r^n}{1-r} = a \lim_{n \to \infty} nr^{n-1}$$

Thus, it can be seen that if r is either 1 or −1, the limit goes to infinity due to the factor n. The same reasoning applies if r is greater than 1 or less than −1. For $-1 < r < 1$, rearrange the original form of the limit.

$$\lim_{n \to \infty} a \frac{1-r^n}{1-r} = a \frac{1-r^\infty}{1-r}$$

Since the magnitude of r is less than 1, r^∞ must be zero. This yields a closed form for the infinite geometric series, which converges only if $-1 < r < 1$.

$$a + ar + ar^2 + \ldots = \frac{a}{1-r}$$

Example: Find the 8th term of the geometric sequence 2, 8, 32,...

$r = \dfrac{a_{n+1}}{a_n}$ Use the common ratio formula to find ratio r.

$r = \dfrac{8}{2} = 4$ Substitute $a_n = 2$ $a_{n+1} = 8$

$r = 4$

$a_n = a_1 \times r^{n-1}$ Use $r = 4$ to solve for the 8th term.
$a_8 = 2 \times 4^{8-1}$
$a_8 = 32,768$

Example: Evaluate the following series: $1+\frac{1}{2}+\frac{1}{4}+\frac{1}{8}+\ldots$.

Note that this series is an infinite geometric series with $a = 1$ and $r = \frac{1}{2}$ (or 0.5). Use the formula to evaluate the series.

$$1+\frac{1}{2}+\frac{1}{4}+\frac{1}{8}+\ldots = \frac{a}{1-r} = \frac{1}{1-0.5} = \frac{1}{0.5} = 2$$

The answer is thus 2.

The beginning teacher understands how sequences and series are applied to solve problems in the mathematics of finance (e.g., simple, compound, and continuous interest rates; annuities).

Arithmetic and geometric sequences and series can be used to model various phenomena and are especially useful for financial mathematics. Compound interest, annuities and mortgages can all be modeled using sequences and series (also see discussion above concerning recursion and iteration). Growth and decay problems, such as those that arise in physics and other sciences, can also be modeled using sequences and series.

Compound interest can be treated by deriving a general formula for the total amount of money available after a principal deposit P has accrued interest at rate of r (a decimal) compounded n times annually for t years. The same sequence-based approach to modeling compound interest is illustrative and can be applied in a similar manner to the various aspects of annuities and mortgages as well. Interest is compounded n times per year, so, after each interval, the total balance accrues interest at a rate $\frac{r}{n}$. For the first instance,

$$A_1 = P + \frac{r}{n}P = P\left(1+\frac{r}{n}\right)$$

In the second instance,

$$A_2 = A_1 + \frac{r}{n}A_1 = A_1\left(1+\frac{r}{n}\right)$$

The pattern continues. Note that if the expression is written out fully, then:

$$A_2 = A_1\left(1 + \frac{r}{n}\right) = P\left(1 + \frac{r}{n}\right)^2$$

The pattern that emerges from this approach can be expressed generally for a sequence with n terms:

$$A_n = P\left(1 + \frac{r}{n}\right)^n$$

A_n is the total amount (principal plus accrued interest) after one year. Thus, for t years, the total balance is the following.

$$A = P\left(1 + \frac{r}{n}\right)^{nt}$$

If the interest is compounded continuously, then the formula becomes the limit of the above expression as n approaches infinity.

$$A = \lim_{n \to \infty}\left[P\left(1 + \frac{r}{n}\right)^{nt}\right]$$

By rearranging this expression, a simpler form of the limit can be found.

$$A = P\lim_{n \to \infty}\left[\left(1 + \frac{1}{n/r}\right)^{n/r}\right]^{rt} = P\lim_{(n/r) \to \infty}\left[\left(1 + \frac{1}{n/r}\right)^{n/r}\right]^{rt}$$

Evaluating the limit reveals the balance after t years with continuously compounded interest.

$$A = Pe^{rt}$$

Example: Calculate the interest accrued after 3 years for $1,000 in a savings account that compounds the interest monthly at an annual rate of 3%.

Use the formula derived above, where $t = 3$, $n = 12$, $P = 1{,}000$ and $r = 0.03$.

$$A = P\left(1+\frac{r}{n}\right)^{nt} = \$1000\left(1+\frac{.03}{12}\right)^{(12)(3)} = \$1000(1.0025)^{36}$$
$$A = \$1094.05$$

Thus, the total interest accrued is $94.05.

Example: What is the total balance in a savings account after 5 years with an annual continuously compounded interest rate of 1% if the initial deposit is $500?

Use the expression for continuously compounded interest, substituting all the relevant information described in the problem.

$$A = Pe^{rt} = \$500e^{(0.01)(5)} = \$500e^{0.05}$$
$$A = \$525.64$$

Competency 005   **The teacher understands attributes of functions, relations, and their graphs.**

This section reviews the general concepts of relations and functions, the domain and range of functions and relations, and the graphs (and associated symmetry in terms of evenness or oddness) of functions. The discussion then covers transformations of and operations on functions as well as the use of graphs in making conjectures about identities.

The beginning teacher understands when a relation is a function.

A **relation** is any set of ordered pairs. The **domain** of a relation is the set containing all the first coordinates of the ordered pairs, and the **range** of a relation is the set containing all the second coordinates of the ordered pairs.

A **function** is a relation in which each value in the domain corresponds to only one value in the range. It is notable, however, that a value in the range may correspond to any number of values in the domain. Thus, although a function is necessarily a relation, not all relations are functions, since a relation is not bound by this rule.

Example: Which set illustrates a function?

{ (0,1) (0,2) (0,3) (0,4) }
{ (3,9) (−3,9) (4,16) (− 4,16) }
{ (1,2) (2,3) (3,4) (1,4) }
{ (2,4) (3,6) (4,8) (4,16) }

Each number in the domain can only be matched with one number in the range. A is not a function because 0 is mapped to 4 different numbers in the range. In C, 1 is mapped to two different numbers. In D, 4 is also mapped to two different numbers. So the answer is B.

A relation can also be described algebraically. An equation such as $y = 3x + 5$ describes a relation between the independent variable x and the dependent variable y. Thus, y is written as $f(x)$ or a "function of x."

On a graph, use the **vertical line test** to check whether a relation is a function. If any vertical line intersects the graph of a relation in more than one point, then the relation is not a function.

Example: Determine whether the following graph depicts a function.

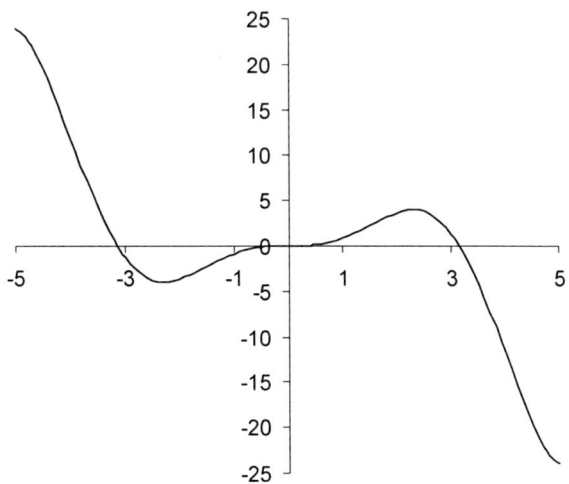

Use the vertical line test on the graph, as shown below. For every location of the vertical line, the plotted curve crosses the line only once. Therefore, the graph depicts a function.

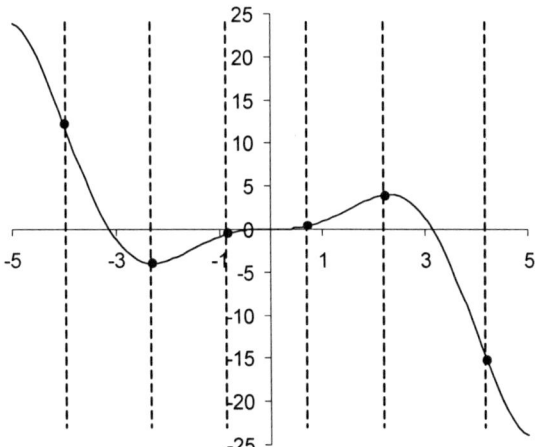

A **mapping** is essentially the same as a function. Mappings (or maps) can be depicted using diagrams with arrows drawn from each element of the domain to the corresponding element (or elements) of the range. If two arrows originate from any single element in the domain, then the mapping is not a function. Likewise, for a function, if each arrow is drawn to a unique value in the range (that is, there are no cases where more than one arrow is drawn to a given value in the range), then the relation is one-to-one.

Example: Are the mappings shown below true functions?

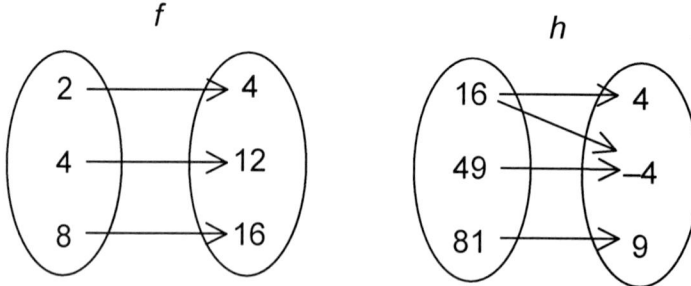

The mapping f is a function, but h is not.

The beginning teacher identifies the mathematical domain and range of functions and relations and determines reasonable domains for given situations.

The domain and range of a function may be determined by inspecting the graph of the function or by analyzing the algebraic formula for the function.

Example: Determine the domain and range of the following graph:

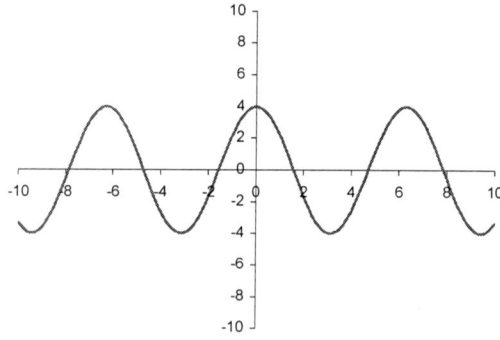

The domain of the function shown in the graph $= -\infty, \infty$.

Since the function is periodic and the y values vary between +4 and −4, the range of the function is −4 to +4.

Example: Determine the domain of the function depicted in the following graph.

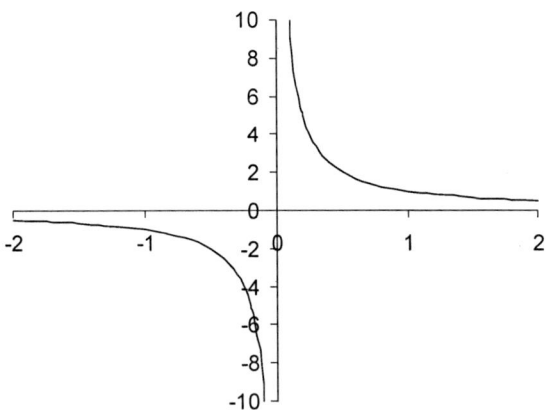

Note that this function is not **continuous**. It has two asymptotes: one for $y = 0$ and one for $x = 0$. It is apparent that the function is not defined for $x = 0$, but that it has finite values everywhere else. Thus, the domain of the function is all real numbers except 0.

The function plotted here is $y = \dfrac{1}{x}$; thus, by way of the function, it is clear that the range includes all real values except 0, for which the function goes to either positive or negative infinity in the limit (depending on the direction).

Example: Give the domain for the function over the set of real numbers: $y = \dfrac{3x + 2}{2x^2 - 3}$

Find the values of x for which the denominator is 0. These values are excluded from the domain.

$$2x^2 - 3 = 0$$
$$2x^2 = 3$$
$$x^2 = 3/2$$
$$x = \pm\sqrt{\dfrac{3}{2}} = \pm\sqrt{\dfrac{3}{2}} \cdot \sqrt{\dfrac{2}{2}} = \pm\dfrac{\sqrt{6}}{2}$$

Example: Determine the domain and range of this mapping.

domain: {4, −5}

range: {8}

The beginning teacher understands that a function represents a dependence of one quantity on another and can be represented in a variety of ways (e.g., concrete models, tables, graphs, diagrams, verbal descriptions, symbols).

A relationship between two quantities can be represented in a variety of ways, including as a symbolic expression (for instance, $f(x) = 3x^2 + \sin x$), a graph, a table of values and a common-language expression (for example, "the speed of the car increases linearly from zero to 100 miles per hour in 12 seconds"). In the following example, the rule $y = 9x$ describes the relationship between the total amount earned, y, and the total amount of $9 sunglasses sold, x. In a relationship of this type, one of the quantities (e.g., total amount earned) is dependent on the other (e.g., number of glasses sold). They are known as the **dependent** and **independent** variables, respectively.

A table using this data would appear as follows:

number of sunglasses sold	1	5	10	15
total dollars earned	9	45	90	135

Each (x,y) relationship between a pair of values is called the coordinate pair and can be plotted on a graph. The coordinate pairs (1,9), (5,45), (10,90), and (15,135) are plotted on the graph below. The graph shows a linear relationship. A linear relationship is one in which two quantities are proportional to each other. Doubling x also doubles y. On a graph, a straight line depicts a linear relationship.

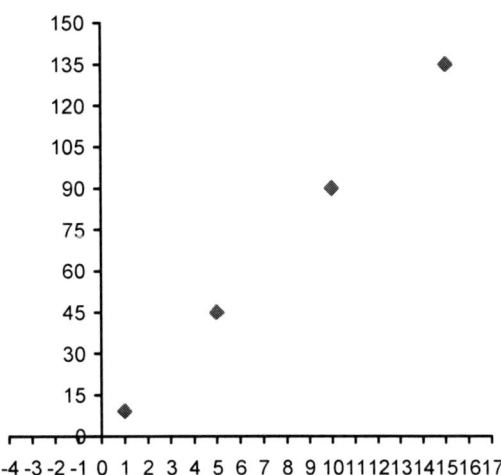

The function or relationship between two quantities can be analyzed to determine how one quantity depends on the other. For example, the function $y = 2x + 1$ shows a linear relationship between y and x. The function is written as a symbolic rule. The same relationship is also shown in the table below:

x	0	2	3	6	9
y	1	5	7	13	19

A relationship could be written in words by saying "The value of y is equal to two times the value of x, plus one." This relationship could be shown on a graph by plotting given points such as the ones shown in the table above.

The ability to convert among various representations of a function depends on how much information is provided. For instance, although a graph of a function can provide some clues as to its symbolic representation, it is often difficult or impossible to obtain an exact symbolic form based only on a graph. The same difficulty applies to tables.

Converting from a symbolic form to a graph or table, however, is relatively simple, especially if a computer is available. The symbolic expression need simply be evaluated for a representative set of points that can be used to produce a sufficiently detailed graph or table.

It is important to notice that the particular symbols used in function notation are not important, as long as they are used consistently. Thus, the following are all the same functions, but simply use different symbols to represent the erstwhile *x* and *y* notation.

$$f(\alpha) = \alpha^2$$
$$\beta(r) = r^2$$
$$\Pi(w) = w^2$$

The fundamental principle for this function notation is to represent that for each *x* (or, generally, variable) value, the function (be it *f*, *β*, or any other symbol) has only one value.

The beginning teacher identifies and analyzes even and odd functions, one-to-one functions, inverse functions, and their graphs.

Symmetries in a function can also be described in terms of reflections or "mirror images." A function can be symmetric about the the *y*-axis (but not about the *x*-axis, except for the function *f(x)* = 0, since every function must pass the vertical line test). A function is symmetric about the *y*-axis if for every point (*x*, *y*) that is included in the graph of the function, the point (–*x*, *y*) is also included in the graph. Consider the function $f(x) = x^2$. Note that for each point (x, x^2) in the graph, the point $(-x, x^2)$ is also in the graph. The symmetry of the function about the *y*-axis can also be seen in the graph below.

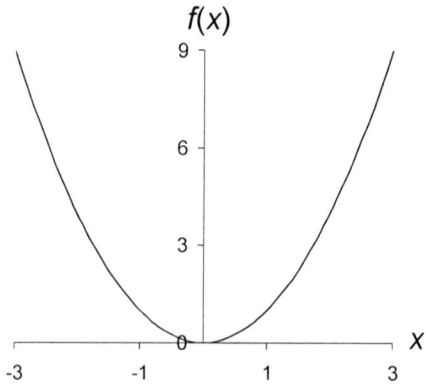

A function that is symmetric about the *y*-axis is also called an **even function**.

Although functions cannot be symmetric about the x-axis, relations that do not obey the vertical line test can be symmetric in this way. A relation is symmetric about the x-axis if for every point (x, y) in the graph of the relation, the point (x, −y) is also in the graph. Consider, for instance, the relation $g(x) = \pm\sqrt{x}$. For every value of x in the domain, the points $(x, \sqrt{x})$ and $(x, -\sqrt{x})$ are both in the graph, as shown below.

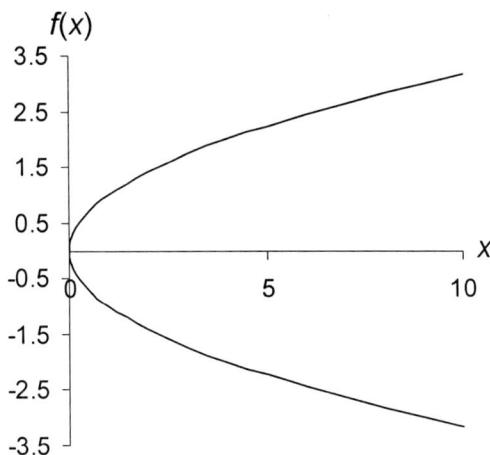

Functions may also be symmetric with respect to the origin. Such functions are also called **odd** or **antisymmetric functions**, and are defined by the property that for any point (x, y) in the graph of the function, the point (−x, −y) is also in the graph of the function. The function $f(x) = x^3$, for instance, is symmetric with respect to the origin, as shown in the graph below.

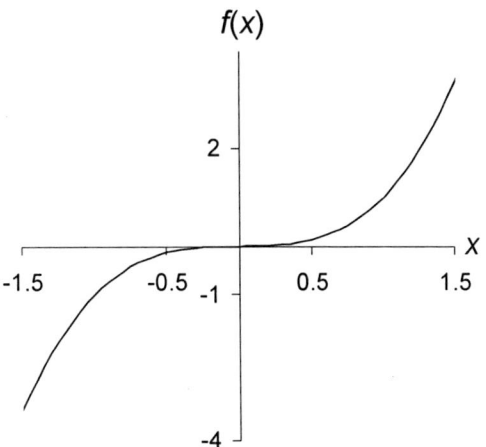

Definition: A function f is even if $f(-x) = f(x)$ and odd if $f(-x) = -f(x)$ for all x in the domain of f.

Example: Determine if the given function is even, odd, or neither even nor odd.

1. $f(x) = x^4 - 2x^2 + 7$
   $f(-x) = (-x)^4 - 2(-x)^2 + 7$
   $f(-x) = x^4 - 2x^2 + 7$

   1. Find $f(-x)$.
   2. Replace x with $-x$.
   3. Since $f(-x) = f(x)$, $f(x)$ is an even function.

2. $f(x) = 3x^3 + 2x$
   $f(-x) = 3(-x)^3 + 2(-x)$
   $f(-x) = -3x^3 - 2x$

   $-f(x) = -3x^3 - 2x$

   1. Find $f(-x)$.
   2. Replace x with $-x$.
   3. Since $f(x)$ is not equal to $f(-x)$, $-f(x) = -(3x^3 + 2x)$
   4. Try $-f(x)$.
   5. Since $f(-x) = -f(x)$, $f(x)$ is an odd function.

3. $g(x) = 2x^2 - x + 4$
   $g(-x) = 2(-x)^2 - (-x) + 4$
   $g(-x) = 2x^2 + x + 4$

   $g(x) = -2x^2 + x - 4$

   1. First find $g(-x)$.
   2. Replace x with $-x$.
   3. Since $g(x)$ does not equal $g(-x)$, $-g(x) = -(2x^2 - x + 4)$, $g(x)$ is not an even function
   4. Try $-g(x)$.
   5. Since $-g(x)$ does not equal $g(-x)$, $g(x)$ is not an odd function.

Thus, $g(x)$ is neither even nor odd.

A relation is considered **one-to-one** if each value in the domain corresponds to only one value in the range and if each value in the range corresponds to only one value in the domain. Thus, a one-to-one relation is also a function, but it adds an additional condition.

In the same way that the graph of a relation can be examined using the vertical line test to determine whether it is a function, the **horizontal line test** can be used to determine if a function is a one-to-one relation. If no horizontal lines superimposed on the plot intersect the graph of the relation in more than one place, then the relation is one-to-one (assuming it also passes the vertical line test and, therefore, is a function). Inverse functions are discussed later in this competency.

The beginning teacher applies basic transformations [e.g., $kf(x)$, $f(x) + k$, $f(x - k)$, $f(kx)$, $|f(x)|$] to a parent function, f, and describes the effects on the graph of $y = f(x)$.

Different types of **function transformations** affect the graph and characteristics of a function in predictable ways. The basic types of transformation are horizontal and vertical shift, horizontal and vertical scaling, and reflection. As an example of the types of transformations, we will consider transformations of the functions $f(x) = x^2$.

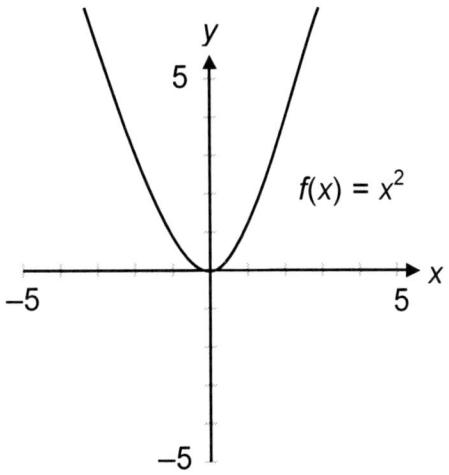

Horizontal shifts take the form $g(x) = f(x \pm c)$. For example, we obtain the graph of the function $g(x) = (x + 2)^2$ by shifting the graph of $f(x) = x^2$ two units to the left. The graph of the function $h(x) = (x - 2)^2$ is the graph of $f(x) = x^2$ shifted two units to the right.

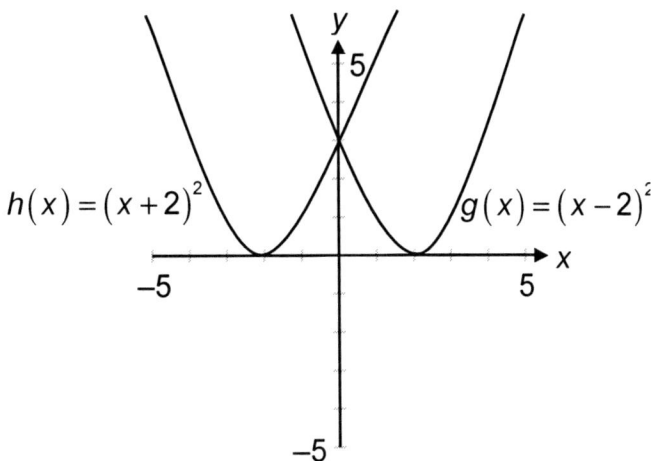

Vertical shifts take the form $g(x) = f(x) \pm c$. For example, we obtain the graph of the function $g(x) = (x^2) - 2$ by shifting the graph of $f(x) = x^2$ two units down. The graph of the function $h(x) = x^2 + 2$ is the graph of $f(x) = x^2$ shifted two units up.

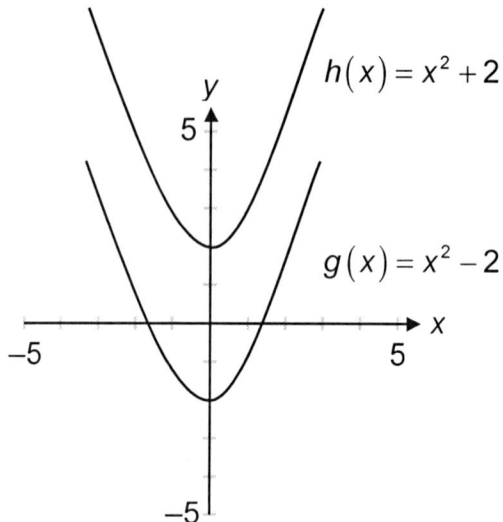

Horizontal scaling takes the form $g(x) = f(cx)$. For example, we obtain the graph of the function $g(x) = (2x)^2$ by compressing the graph of $f(x) = x^2$ in the x-direction by a factor of two. If $c > 1$ the graph is compressed in the x-direction, while if $1 > c > 0$ the graph is stretched in the x-direction.

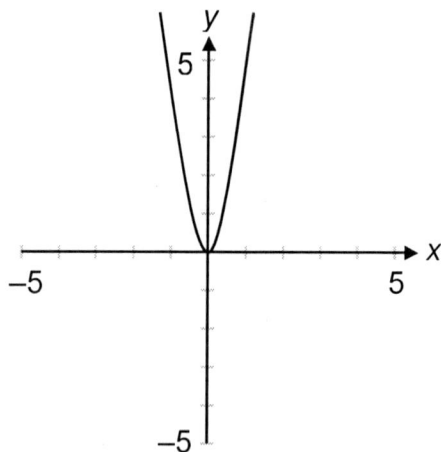

The beginning teacher performs operations (e.g., sum, difference, composition) on functions, finds inverse relations, and describes results symbolically and graphically.

Addition of Functions

$(f + g)(x) = f(x) + g(x)$

To add two functions, add together the solution for each.
For example, if $f(x) = x^2 + 1$ and $g(x) = 6x - 1$, then

$(f + g)(1) = f(1) + g(1) = 2 + 5 = 7$
$(f + g)(x) = f(x) + g(x) = (x^2 + 1) + (6x - 1) = x^2 + 6x$.

Subtraction of Functions

$(f - g)(x) = f(x) - g(x)$

To subtract two functions, subtract the solution of each.
For example, if $f(x) = 3x - 4$ and $g(x) = 5x + 2$, then

$(f - g)(2) = f(2) - g(2) = 2 - 12 = -10$.
$(f - g)(x) = f(x) - g(x) = (3x - 4) - (5x + 2) = -2x - 6$.

Example: Given $f(x) = 3x + 1$ and $g(x) = 4 + 3x$, find $(f + g)(x)$, $(f - g)(x)$, $(f \times g)(x)$, and $(f / g)(x)$.

To find the answers, just apply the operations (plus, minus, times, and divide)

$(f + g)(x) = f(x) + g(x) = [3x + 1] + [4 + 3x]$
$\qquad = 3x + 3x + 1 + 4$
$\qquad = 6x + 5$

$(f - g)(x) = f(x) - g(x) = [3x + 1] - [4 + 3x]$
$\qquad = 3x - 3x + 1 - 4$
$\qquad = -3$

$(f \times g)(x) = [f(x)][g(x)] = (3x + 1)(4 + 3x)$
$\qquad = 12x + 4 + 9x^2 + 3x$
$\qquad = 9x^2 + 15x + 4$

$\left(\dfrac{f}{g}\right)x = \dfrac{f(x)}{g(x)} = \dfrac{3x + 1}{4 + 3x}$

Example: Given $f(x) = 2x$, $g(x) = x + 4$, and $h(x) = 5 - x^3$, find $(f + g)(2)$, $(h - g)(2)$, $(f \times h)(2)$, and $(h / g)(2)$.

$$f(2) = 2(2) = 4$$
$$g(2) = (2) + 4 = 6$$
$$h(2) = 5 - (2)^3 = 5 - 8 = -3$$

Evaluate the following:

$$(f + g)(2) = f(2) + g(2) = 4 + 6 = 10$$
$$(h - g)(2) = h(2) - g(2) = -3 - 6 = -9$$
$$(f \times h)(2) = f(2) \times h(2) = (4)(-3) = -12$$
$$(h / g)(2) = h(2) \div g(2) = -3 \div 6 = -0.5$$

The addition of functions is commutative and associative:

$$f + g = g + f \text{ and } (f + g) + h = f + (g + h).$$

Composition of functions is a way of combining functions such that the range of one function is the domain of another. For instance, the composition of functions f and g can be either $f \circ g$ (the composite of f with g) or $g \circ f$ (the composite of g with f). Another way of writing these compositions is $f(g(x))$ and $g(f(x))$. The domain of $f(g(x))$ includes all values x such that $g(x)$ is in the domain of $f(x)$.

Example: What is the composition $f \circ g$ for functions $f(x) = ax$ and $g(x) = bx^2$?

The correct answer can be found by substituting the function $g(x)$ into $f(x)$.

$$f(g(x)) = ag(x) = abx^2$$

On the other hand, the composition $g \circ f$ would yield a different answer.

$$g(f(x)) = b(f(x))^2 = b(ax)^2 = a^2bx^2$$

Example: If f(x) = sqrt(x) and g(x) = x + 2, find the composition functions f ∘ g and g ∘ f and state their domains.

$$(f \circ g)(x) = f(g(x)) = f(x + 2) = \text{sqrt}(x + 2)$$
$$(g \circ f)(x) = g(f(x)) = g(\text{sqrt}(x)) = \text{sqrt}(x) + 2$$

The domain of f(g(x)) is x ≥ -2 because x + 2 must be non-negative in order to take the square root.

The domain of g(f(x)) is x ≥ 0 because x must be non-negative in order to take the square root.

Note that defining the domain of composite functions is important when square roots are involved.

The **inverse of a function** f(x) is typically labeled $f^{-1}(x)$ and satisfies the following two relations:

$$f(f^{-1}(x)) = x$$
$$f^{-1}(f(x)) = x$$

For a function f(x) to have an inverse, it must be one-to-one. This fact is easily seen since both f(x) and $f^{-1}(x)$ must satisfy the vertical line test (that is, both must be functions). A function takes each value in a domain and relates it to only one value in the range. Logically, then, the inverse must do the same, only backwards: relate each value in the range to a single value in the domain.

Finding the inverse of a function can be a difficult or impossible task, but there are some simple approaches that can be followed in many cases. The simplest method for finding the inverse of a function is to interchange the variable and the function symbols and then solve to find the inverse. The approach is summarized in the outline below, given a one-to-one function f(x).

1. Replace the symbol f(x) with x.
2. Replace all instances of x in the function definition with $f^{-1}(x)$ (or y or some other symbol).
3. Solve for $f^{-1}(x)$.
4. Check the result using $f(f^{-1}(x)) = x$ or $f^{-1}(f(x)) = x$.

Example: Determine if the function $f(x) = x^2$ has an inverse. If so, find the inverse.

First, determine if f(x) is one-to-one. Note that f(1) = f(–1) = 1, so f(x) is not one-to-one, and it therefore has no inverse function.

Example: Determine if the function $f(x) = x^3 + 1$ has an inverse. If so, find the inverse.

The function $f(x) = x^3 + 1$ has an inverse because it increases monotonically for x > 0 and decreases monotonically for x < 0. As a result, it is one-to-one, and the inverse exists. To calculate the inverse, let y be $f^{-1}(x)$. Replace f(x) with x and replace x with y.

$$f(x) = x^3 + 1 \Rightarrow x = y^3 + 1$$

Solve for y.

$$x - 1 = y^3$$
$$y = \sqrt[3]{x - 1}$$
$$f^{-1}(x) = \sqrt[3]{x - 1}$$

Test the result.

$$f^{-1}(f(x)) = \sqrt[3]{(x^3 + 1) - 1}$$
$$f^{-1}(f(x)) = \sqrt[3]{x^3 + 1 - 1} = \sqrt[3]{x^3} = x$$

The result is thus correct.

The beginning teacher uses graphs of functions to formulate conjectures of identities [e.g., y = x^2 – 1 and y = (x – 1)(x + 1), y = log x^3 and y = 3 log x, y = sin(x + 2) and y = cos x].

In mathematics, identities are necessary truths. In other words, **an identity relates two functions that are equal for all variable values.** We can often identify identities by constructing a table of values for each function under consideration and plotting the graph from the table of values. Functions that are identities have identical graphs. The following are three examples of identity functions with corresponding tables of values and graphs.

Example: Consider the functions $y = x^2 - 1$ and $y = (x - 1)(x + 1)$. Determine by graphing if the functions represent an identity.

Using the definition of identity, if the functions represent an identity, $x^2 - 1 = (x - 1)(x + 1)$ for all values of x.

x	$x^2 - 1$	$(x - 1)(x + 1)$
0	−1	−1
1	0	0
−1	0	0
2	3	3
−2	3	3

Note that for all values x, $x^2 - 1 = (x - 1)(x + 1)$. The graph of $y = x^2 - 1$ and $y = (x - 1)(x + 1)$ are identical.

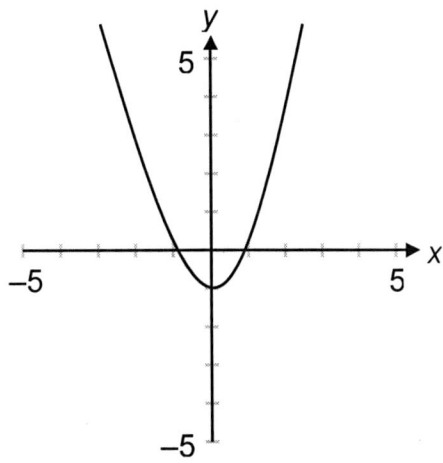

Example: Consider the functions $y = \log x^3$ and $y = 3 \log x$. Determine if the functions represent an identity by graphing.

Determine if $\log x^3 = 3 \log x$ for all values of x.

x	$\log x^3$	$3 \log x$
1	0	0
10	3	3
100	6	6

It is the case that $\log x^3 = 3 \log x$ for all values of x. The functions form an identity, and the graphs of $y = \log x^3$ and $y = 3 \log x$ are identical.

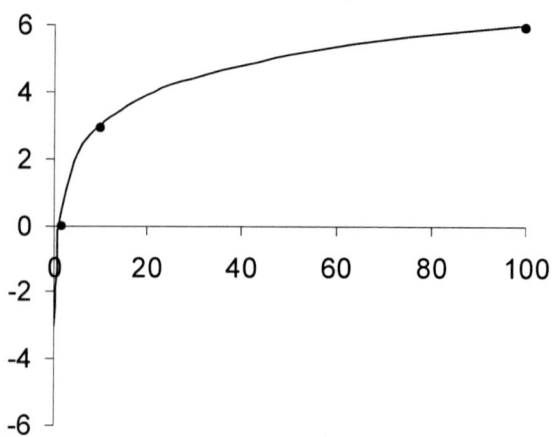

Example: Consider the functions $y = \sin(x + \frac{\pi}{2})$ and $y = \cos x$. Determine if the functions represent an identity by graphing.

x	$\sin(x + \frac{\pi}{2})$	$\cos x$
0	1	1
$\frac{\pi}{4}$	0.707	0.707
$\frac{\pi}{2}$	0	0
$\frac{3\pi}{4}$	−0.707	−0.707
π	−1	−1

The functions $y = \sin(x + \frac{\pi}{2})$ and $y = \cos x$ form an identity because they are equal for all values of x. Shifting the graph of $y = \sin x$ to the right by $\frac{\pi}{2}$ units yields the graph of $y = \cos x$.

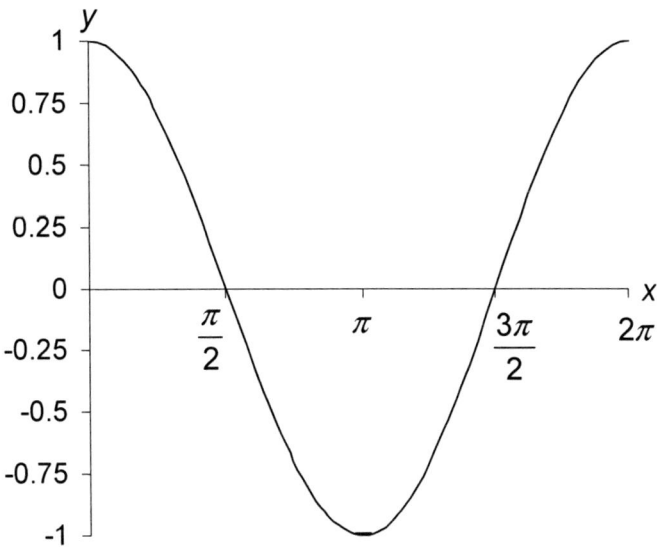

Competency 006   The teacher understands linear and quadratic functions, analyzes their algebraic and graphical properties, and uses them to model and solve problems.

This competency reviews the concept of slope and goes on to discuss the equation of a line and solution of systems of linear equations, as well as the equations, graphs, and solutions of complex functions (or equations).

The beginning teacher understands the concept of slope as a rate of change and interprets the meaning of slope and intercept in a variety of situations.

A **linear function** is a function defined by the equation $f(x) = mx + b$ where "m" is the **slope** of the line representing the function and "b" is the **y-intercept** or the y-coordinate at which the line crosses the y-axis. Many real world situations involve linear relationships. One example is the relationship between distance and time traveled when a car is moving at a constant speed. The relationship between the price and quantity of a bulk item bought at a store is also linear assuming that the unit price remains constant. These relationships can be expressed using the equation of a straight line and the slope is often used to describe a constant or average rate of change expressed in miles per hour or dollars per year for instance. Where the line intercepts the x- and y- axis indicates a starting point or a point at which values change from positive to negative or negative to positive.

Example: A man drives a car at a speed of 30 mph along a straight road. Express the distance d traveled by the man as a function of the time t assuming the man's initial position is d_0. The equation relating d and t is given by

$d = 30t + d_0$

Notice that this equation is in the familiar slope-intercept form $y = mx + b$. In this case, time t (in hours) is the independent variable, the distance d (in miles) is the dependent variable. The **slope** is the **rate of change** of distance with time: i.e., the speed (in mph). The **y-intercept** or intercept on the distance axis d_0 represents the **initial position** of the car at the start time $t = 0$.

The above equation is plotted below with $d_0 = 15$ miles (the point on the graph where the line crosses the y-axis).

$$d = 30t + 15$$

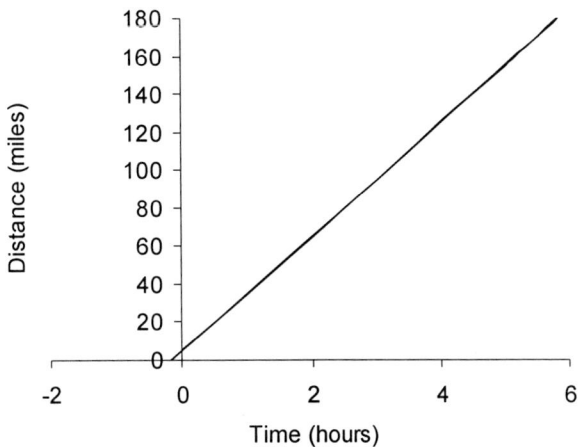

The **x-intercept** or intercept on the time axis represents the time at which the car would have been at $d = 0$ assuming it was traveling with the same speed before $t = 0$. This value can be found by setting $d = 0$ in the equation:

$$0 = 30t + 15$$
$$30t = -15$$
$$t = \frac{-15}{30} = -\frac{1}{2} \text{hr}$$

This simply means that if the car was at $d = 15$ miles when we started measuring the time ($t = 0$), it was at $d = 0$ miles half an hour before that.

Example: A model for the distance traveled by a migrating monarch butterfly looks like $f(t) = 80t$, where t represents time in days.

We interpret this to mean that the average speed of the butterfly is 80 miles per day and distance traveled may be computed by substituting the number of days traveled for t. In a linear function, there is a **constant** rate of change.

Example: The town of Verdant Slopes has been experiencing a boom in population growth. By the year 2000, the population had grown to 45,000, and by 2005, the population had reached 60,000. Using the formula for slope as a model, find the average rate of change in population growth, expressing your answer in people per year. Then using the average rate of change determined, predict the population of Verdant Slopes in the year 2010.

Let t represent the time and p represent population growth. The two observances are represented by (t_1, p_1) and (t_2, p_2)

1st observance = (t_1, p_1) = (2000, 45000)
2nd observance = (t_2, p_2) = (2005, 60000)

Use the formula for slope to find the average rate of change.

$$\text{Rate of change} = \frac{p_2 - p_1}{t_2 - t_1}$$

$$= \frac{60000 - 45000}{2005 - 2000}$$

$$= \frac{15000}{5} = 3000 \text{ people / year}$$

The average rate of change in population growth for Verdant Slopes between the years 2000 and 2005 was 3,000 people/year. The population of Verdant Slopes can be predicted using the following:

3,000 people/year x 5 years = 15,000 people
60,000 people + 15,000 people = 75,000 people

At a continuing average rate of growth of 3000 people/year, the population of Verdant Slopes could be expected to reach 75,000 by the year 2010.

The beginning teacher writes equations of lines given various characteristics (e.g., two points, a point and slope, slope and y-intercept)

The equation of a line can be found from its graph by finding its slope and y-intercept and substituting them in the slope-intercept form $y = mx + b$ (m is the slope, b is the y-intercept).

An alternate form of a linear equation is the point-slope form given below. Given the slope of a line and any one point (x_a, y_a) the line passes through, its equation may be written as

$$y - y_a = m(x - x_a)$$

Example: Find the equation of a line that has a slope of −1.5 and passes through the point (3,2).

Substituting the values of the slope m and the point (x_a, y_a) in the point-slope form of the equation we get

$$y - 2 = -1.5(x - 3) \quad \text{(point-slope form)}$$

Rearranging the terms,

$$y = -1.5x + 6.5 \quad \text{(slope-intercept form)}$$

or, multiplying by 2 and moving the x-term to the left hand side,

$$3x + 2y = 13 \quad \text{(standard form)}$$

The equation of a line may be expressed in any one of the above forms. All of them are equally valid.

Given two points on a line, the first thing to do is to find the slope of the line. If 2 points on the graph are (x_1, y_1) and (x_2, y_2), then the slope is found using the formula:

$$\text{slope} = \frac{y_2 - y_1}{x_2 - x_1}$$

The slope will now be denoted by the letter m. To write the equation of a line, choose either point. Substitute them into the formula:

$$Y - y_a = m(X - x_a)$$

Remember, (x_a, y_a) can be (x_1, y_1) or (x_2, y_2) If m, the value of the slope, is distributed through the parentheses, the equation can be rewritten into other forms of the equation of a line.

If the graph is a **vertical line**, then the equation solves to **x = the x coordinate of any point on the line**. If the graph is a **horizontal line**, then the equation solves to **y = the y coordinate of any point on the line**.

Example: Find the equation of a line that passes through the points (9, –6) and (–1, 2).

$$\text{slope} = \frac{y_2 - y_1}{x_2 - x_1} = \frac{2 - (-6)}{-1 - 9} = \frac{8}{-10} = -\frac{4}{5}$$

The y-intercept may be found by substituting the slope (m) and the coordinates (x, y) for one of the data points in the slope-intercept form of the equation y = mx+b giving

$$-6 = -\frac{4}{5} \times 9 + b \quad \text{where } b \text{ is the } y\text{-intercept}$$

$$b = \frac{6}{5}$$

Thus the slope-intercept form of the equation is:

$$y = -\frac{4}{5}x + \frac{6}{5}$$

Multiplying by 5 to eliminate fractions, it is:

$$5y = -4x + 6 \rightarrow 4x + 5y = 6 \quad \text{(standard form)}$$

The beginning teacher applies techniques of linear and matrix algebra to represent and solve problems involving linear systems

Matrices are often used to solve systems of equations. They are also used by physicists, mathematicians and biologists to organize and study data such as population growth, and they are used in finance for such purposes as investment growth and portfolio analysis. Matrices are easily translated into computer code in high-level programming languages and can be easily expressed in electronic spreadsheets.

The following is a simple financial example of using a matrix to solve a problem. A company has two stores. The income and expenses (in dollars) for the two stores, for three months, are shown in the matrices.

$$\begin{array}{c}\text{April}\\ \\ \text{Store 1}\\ \\ \text{Store 2}\end{array}\begin{array}{cc}\text{Income} & \text{Expenses}\\ \left[\begin{array}{cc}190{,}000 & 170{,}000\\ \\ 100{,}000 & 110{,}000\end{array}\right]\end{array} \qquad \begin{array}{c}\text{May}\\ \\ \text{Store 1}\\ \\ \text{Store 2}\end{array}\begin{array}{cc}\text{Income} & \text{Expenses}\\ \left[\begin{array}{cc}210{,}000 & 200{,}000\\ \\ 125{,}000 & 120{,}000\end{array}\right]\end{array}$$

$$\begin{array}{c}\text{June}\\ \\ \text{Store 1}\\ \\ \text{Store 2}\end{array}\begin{array}{cc}\text{Income} & \text{Expenses}\\ \left[\begin{array}{cc}220{,}000 & 215{,}000\\ \\ 130{,}000 & 115{,}000\end{array}\right]\end{array}$$

The owner wants to know what his first-quarter income and expenses were, so he adds the three matrices.

$$\begin{array}{c}\text{1st Quarter}\\ \\ \text{Store 1}\\ \\ \text{Store 2}\end{array}\begin{array}{cc}\text{Income} & \text{Expenses}\\ \left[\begin{array}{cc}620{,}000 & 585{,}000\\ \\ 355{,}000 & 345{,}000\end{array}\right]\end{array}$$

Then, to find the profit for each store:

Profit for Store 1 = $620,000 - $585,000 = $35,000
Profit for Store 2 = $355,000 - $345,000 = $10,000

When given the following system of equations:

$ax + by = e$
$cx + dy = f$

The matrix equation is written in the form:

$$\begin{pmatrix} a & b \\ c & d \end{pmatrix}\begin{pmatrix} x \\ y \end{pmatrix} = \begin{pmatrix} e \\ f \end{pmatrix}$$

The solution is found using the inverse of the matrix of coefficients. The inverse of a 2×2 matrix can be written as follows:

$$A^{-1} = \frac{1}{|A|}\begin{pmatrix} d & -b \\ -c & a \end{pmatrix}$$

Example: Write the matrix equation of the following system of equations and solve for x and y.

$$3x - 4y = 2$$
$$2x + y = 5$$

$$\begin{pmatrix} 3 & -4 \\ 2 & 1 \end{pmatrix} \begin{pmatrix} x \\ y \end{pmatrix} = \begin{pmatrix} 2 \\ 5 \end{pmatrix}$$ Definition of matrix equation

$$\begin{pmatrix} x \\ y \end{pmatrix} = \frac{1}{11} \begin{pmatrix} 1 & 4 \\ -2 & 1 \end{pmatrix} \begin{pmatrix} 2 \\ 5 \end{pmatrix}$$ Multiply by the inverse of the coefficient matrix.

$$\begin{pmatrix} x \\ y \end{pmatrix} = \frac{1}{11} \begin{pmatrix} 22 \\ 11 \end{pmatrix}$$ Matrix multiplication.

$$\begin{pmatrix} x \\ y \end{pmatrix} = \begin{pmatrix} 2 \\ 1 \end{pmatrix}$$ Scalar multiplication.

The solution is then x = 2 and y = 1.

The beginning teacher analyzes the zeros (real and complex) of quadratic functions

For a quadratic function of the form $ax^2 + bx + c$, the **discriminant** is the portion of the quadratic formula (see the discussion on graphing quadratics later in this competency) which is found under the square root sign; that is $b^2 - 4ac$.

According to the quadratic formula, the zeros of the function are given by

$$x = \frac{-b \pm \sqrt{b^2 - 4ac}}{2a}$$

Note that the radical sign is **not** part of the discriminant. Determine the value of the discriminant by substituting the values of a, b, and c from $ax^2 + bx + c = 0$.

The discriminant can be used to determine the nature of the solution of a quadratic equation.

1) If $b^2 - 4ac < 0$, there are **no real roots** and **two complex roots** that include the imaginary number I (square root of -1).

2) If $b^2 - 4ac = 0$, there is only **one real rational root**.

3) If $b^2 - 4ac > 0$ and also a perfect square, there are **two real rational roots**. (There are no longer any radical signs.)

4) If $b^2 - 4ac > 0$ and not a perfect square, then there are **two real irrational roots**. (There are still unsimplified radical signs.)

Example: Find the value of the discriminant for the equation $2x^2 - 5x + 6 = 0$. Then determine the number and nature of the solutions of that quadratic equation.

The discriminant is the following:

$$b^2 - 4ac = (-5)^2 - 4(2)(6) = 25 - 48 = -23$$

Since −23 is a negative number, there are **no real roots** and **two complex roots**, which are given below.

$$x = \frac{5}{4} \pm \frac{i\sqrt{23}}{4}$$

Example: Find the value of the discriminant for the equation $3x^2 - 12x + 12 = 0$. Then determine the number and nature of the solutions of the quadratic equation.

The discriminant is

$$b^2 - 4ac = (-12)^2 - 4(3)(12) = 144 - 144 = 0$$

Since 0 is the value of the discriminant, there is only **1 real rational root**: $x = 2$.

The beginning teacher makes connections between the $y = ax^2 + bx + c$ and the $y = a(x - h)^2 + k$ representations of a quadratic function and its graph

See below for a discussion on graphing quadratics.

The beginning teacher solves problems involving quadratic functions using a variety of methods (e.g., factoring, completing the square, using the quadratic formula, using a graphing calculator)

A quadratic equation is expressed in the form $ax^2 + bx + c = 0$, where a, b, and c are real numbers and $a \neq 0$. The degree of a quadratic equation (i.e., the highest exponent of the unknown variable x) is 2. Examples of quadratic equations are $5x^2 + 6x + 7 = 0$, $9x^2 - 4 = 0$, $2x^7 - 3x = 0$.

If $p(x) = 0$ is a quadratic equation, then the zeros of the polynomial $p(x)$ are called the roots or solutions of equation $p(x) = 0$. Finding the roots of a quadratic equation is known as solving for "x". There are several different methods for solving quadratic equations:

1) Factoring
2) Completing the Square
3) Quadratic Formula
4) Graphing

Factoring

This method is only applicable for quadratic equations where the polynomial can be expressed as a product of linear factors. If a quadratic polynomial $ax^2 + bx + c = 0$ can be expressed as a product of two linear factors, say $(px + q)$ and $(rx + s)$, where p, q, r, s are real numbers, then $ax^2 + bx + c = 0$ can be rewritten as

$$(px + q)(rx + s) = 0$$

This implies that either of the two factors must be equal to zero:

$$(px + q) = 0 \text{ or } (rx + s) = 0$$

Solving these linear equations, we get the possible roots of the given quadratic equation as:

$$x = -\frac{q}{p} \text{ and } x = -\frac{s}{r}$$

Example: Solve the following equation.

$$x^2 + 10x - 24 = 0$$
$$(x+12)(x-2) = 0 \quad \text{Factor.}$$
$$x+12 = 0 \text{ or } x-2 = 0 \quad \text{Set each factor equal to 0.}$$
$$x = -12 \quad x = 2 \quad \text{Solve.}$$

Check:

$$x^2 + 10x - 24 = 0$$
$$(-12)^2 + 10(-12) - 24 = 0 \qquad (2)^2 + 10(2) - 24 = 0$$
$$144 - 120 - 24 = 0 \qquad 4 + 20 - 24 = 0$$
$$0 = 0 \qquad 0 = 0$$

Completing the Square

A quadratic equation can be solved by completing the square. To complete the square, the coefficient of the x^2 term must be 1.

To solve a quadratic equation using this method:

1. Isolate the x^2 and x terms.
2. Add half of the coefficient of the x term squared to both sides of the equation.
3. Finally, take the square root of both sides and solve for x.

Example: Solve the following equation: $x^2 - 6x + 8 = 0$.

$$x^2 - 6x = -8 \quad \text{Move the constant to the right side.}$$
$$x^2 - 6x + 9 = -8 + 9 \quad \text{Add the square of half the coefficient}$$
$$\qquad \qquad \qquad \text{of } x \text{ to both sides.}$$
$$(x-3)^2 = 1 \quad \text{Write the left side as a perfect square.}$$
$$x - 3 = \pm\sqrt{1} \quad \text{Take the square root of both sides.}$$
$$x - 3 = 1 \quad x - 3 = -1 \quad \text{Solve.}$$
$$x = 4 \quad \quad x = 2$$

Check:

$$x^2 - 6x + 8 = 0$$

$$4^2 - 6(4) + 8 = 0 \qquad\qquad 2^2 - 6(2) + 8 = 0$$
$$16 - 24 + 8 = 0 \qquad\qquad 4 - 12 + 8 = 0$$
$$0 = 0 \qquad\qquad\qquad\qquad 0 = 0$$

Quadratic Formula

To solve a quadratic equation using the quadratic formula, be sure that your equation is in the form $ax^2 + bx + c = 0$. Substitute the values of *a*, *b*, and *c* into the formula:

$$x = \frac{-b \pm \sqrt{b^2 - 4ac}}{2a}$$

Simplify the result to find the answers. (Remember, there could be two real answers, one real answer, or two complex answers that include *i*. See the preceding skill sections for details.)

<u>Example:</u> Solve the following equation using the quadratic formula: $3x^2 = 7 + 2x$.

Rearrange the equation and use the quadratic formula as follows.

$$3x^2 = 7 + 2x \rightarrow 3x^2 - 2x - 7 = 0$$
$$a = 3 \quad b = -2 \quad c = -7$$
$$x = \frac{-(-2) \pm \sqrt{(-2)^2 - 4(3)(-7)}}{2(3)}$$
$$x = \frac{2 \pm \sqrt{4 + 84}}{6}$$
$$x = \frac{2 \pm \sqrt{88}}{6}$$
$$x = \frac{2 \pm 2\sqrt{22}}{6}$$
$$x = \frac{1 \pm \sqrt{22}}{3}$$

Graphing

The general technique for graphing quadratics is the same as that for graphing linear equations. Graphing a quadratic equation, however, results in a parabola instead of a straight line.

The general form of a quadratic function is $y = ax^2 + bx + c$. Once a function is identified as quadratic, it is helpful to recognize several features that can indicate the form of the graph. The parabola has an axis of symmetry along $x = -\dfrac{b}{2a}$ which is the x-coordinate of the vertex (turning point) of the graph.

This can be understood more clearly if we consider an alternate form of a quadratic equation, the standard form for a parabola

$$y = a(x - h)^2 + k$$

where (h, k) denote the coordinates of the vertex of the parabola. Transforming the general form $y = ax^2 + bx + c$ into the above form,

$$y = a\left(x^2 + \dfrac{b}{a}x\right) + c$$

$$y = a\left(x^2 + 2\dfrac{b}{2a}x + \left(\dfrac{b}{2a}\right)^2\right) - \dfrac{b^2}{4a} + c$$

$$y = a\left(x + \dfrac{b}{2a}\right)^2 - \dfrac{b^2}{4a} + c$$

Thus the coordinates of the vertex are given by $\left(-\dfrac{b}{2a}, -\dfrac{b^2}{4a} + c\right)$.

Example: Graph $y = 3x^2 + x - 2$.

Expressing this function in standard form we get

$$y = 3\left(x + \dfrac{1}{6}\right)^2 - \dfrac{25}{12}$$

Thus, the graph is a parabola with an axis of symmetry $x = -\dfrac{1}{6}$ and the vertex is located at the point $\left(-\dfrac{1}{6}, -\dfrac{25}{12}\right)$.

x	$y = 3x^2 + x - 2$
-2	8
-1	0
0	-2
1	2
2	12

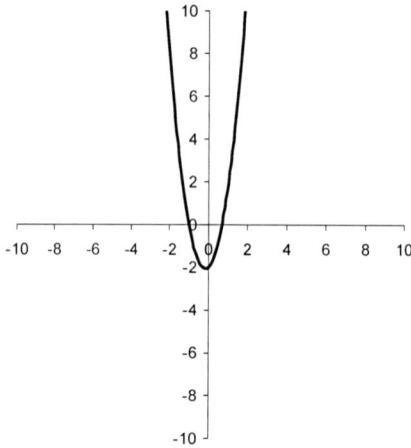

If the quadratic term is positive, then the parabola is concave up; if the quadratic term is negative, then the parabola is concave down. The function $-x^2 - 2x - 3$ is one such example and is shown below.

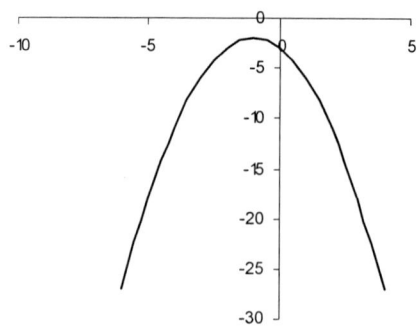

A quadratic function with two real roots (see example problems) will have two crossings of the x-axis. A quadratic function with one real root will graph as a parabola that is tangent to the x-axis. An example of such a quadratic function is shown in the example below for the function $x^2 + 2x + 1$. The function has a single real root at $x = -1$.

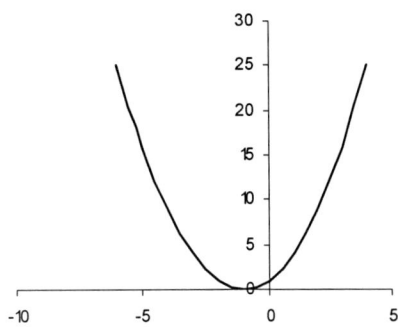

A quadratic function with no real roots will not cross the axis at any point. An example is the function $x^2 + 2x + 2$, which is plotted below.

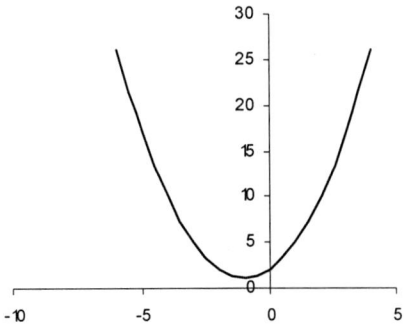

Example: Solve by graphing: $x^2 - 8x + 15 = 0$.

The roots of the polynomial $x^2 - 8x + 15$ are the x values for which the graph intersects the x-axis.

x	$y = x^2 - 8x + 15$
−2	35
−1	24
0	15
1	8
2	3

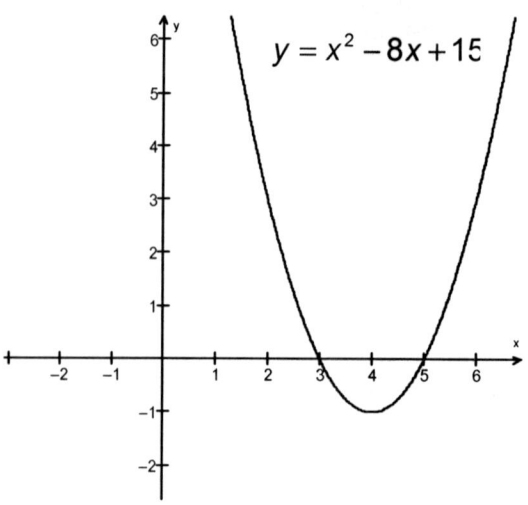

From the above graph, the x-intercepts or zeros are 3 and 5. So the solutions of the given quadratic equation are 3 and 5.

To **graph a quadratic inequality**, graph the quadratic as if it were an equation; however, if the inequality has just a > or < sign, then make the curve dotted. Shade above the curve for > or ≥. Shade below the curve for < or ≤.

Example: Graph the inequality $y < -x^2 + x - 2$.

The quadratic function $-x^2 + x - 2$ is plotted with a dotted line since the inequality sign is < and not ≤. Since y is "less than" this function, the shading is done below the curve.

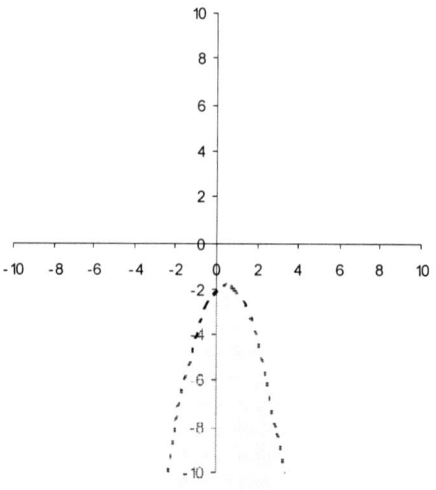

Example: Graph the inequality $y \geq x^2 - 2x - 9$.

The quadratic function $x^2 - 2x - 9$ is plotted with a solid line since the inequality sign is $\geq$. Since y is "greater than or equal to" this function, the shading is done above the curve.

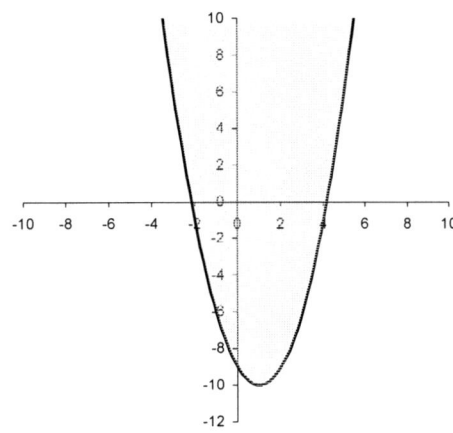

The beginning teacher models and solves problems involving linear and quadratic equations and inequalities using a variety of methods, including technology

For some examples of situations modeled using **linear equations** see the beginning of this competency.

Many word problems may be modeled and solved using **linear systems of equations and inequalities**. Some examples are given below.

Example: Farmer Greenjeans bought 4 cows and 6 sheep for $1,700. Mr. Ziffel bought 3 cows and 12 sheep for $2,400. If all the cows were the same price and all the sheep were another price, find the price charged for a cow or for a sheep.

Let x = price of a cow
Let y = price of a sheep

Then Farmer Greenjeans' equation would be $4x + 6y = 1700$
Mr. Ziffel's equation would be $3x + 12y = 2400$

To solve by **addition-subtraction**:

Multiply the first equation by $^-2$: $^-2(4x + 6y = 1700)$
Keep the other equation the same : $(3x + 12y = 2400)$

By doing this, the equations can be added to each other to eliminate one variable and solve for the other variable.

$$-8x - 12y = -3400$$
$$\underline{3x + 12y = 2400} \qquad \text{Add these equations.}$$
$$-5x = -1000$$

$$x = 200 \leftarrow \text{the price of a cow was \$200.}$$

Solving for y, $y = 150 \leftarrow$ the price of a sheep—$150. (This problem can also be solved by substitution or determinants.)

Example: Mrs. Allison bought 1 pound of potato chips, a 2-pound beef roast, and 3 pounds of apples for a total of $8.19. Mr. Bromberg bought a 3-pound beef roast and 2 pounds of apples for $9.05. Kathleen Kaufman bought 2 pounds of potato chips, a 3-pound beef roast, and 5 pounds of apples for $13.25. Find the per pound price of each item.

To solve by **substitution**:

Let x = price of a pound of potato chips
Let y = price of a pound of roast beef
Let z = price of a pound of apples

Mrs. Allison's equation is	$1x + 2y + 3z = 8.19$
Mr. Bromberg's equation is	$3y + 2z = 9.05$
K. Kaufman's equation is	$2x + 3y + 5z = 13.25$

Take the first equation and solve for x. (This equation was chosen because x is the easiest variable to get alone in this set of equations.) This equation becomes

$$x = 8.19 - 2y - 3z$$

Substitute this expression into the other equations in place of x:

$$3y + 2z = 9.05 \leftarrow \text{equation 2}$$
$$2(8.19 - 2y - 3z) + 3y + 5z = 13.25 \leftarrow \text{equation 3}$$

Simplify the equation by combining like terms:

$$3y + 2z = 9.05 \leftarrow \text{equation 2}$$
$$-1y - 1z = -3.13 \leftarrow \text{equation 3}$$

Solve equation 3 for either y or z:

$$y = 3.13 - z \quad (*)$$

Substitute this into equation 2 for y:

$$3(3.13 - z) + 2z = 9.05 \leftarrow \text{equation 2}$$
$$-1y - 1z = -3.13 \leftarrow \text{equation 3}$$

Combine like terms in equation 2:

$$9.39 - 3z + 2z = 9.05$$
$$z = \$0.34 \text{ per pound (price of apples)}$$

Substitute .34 for z in the equation marked with an asterisk (*) above to solve for y:

$$y = 3.13 - z$$
$$y = 3.13 - .34$$
$$y = \$2.79 = \text{per pound price of roast beef}$$

Substituting .34 for z and 2.79 for y in one of the original equations, solve for x:

$$1x + 2y + 3z = 8.19$$
$$1x + 2(2.79) + 3(.34) = 8.19$$
$$x + 5.58 + 1.02 = 8.19$$
$$x + 6.60 = 8.19$$
$$x = \$1.59 \text{ per pound of potato chips}$$
$$(x, y, z) = (\$1.59, \$2.79, \$0.34)$$

Example: Aardvark Taxi charges $4 initially plus $1 for every mile traveled. Baboon Taxi charges $6 initially plus $.75 for every mile traveled. Determine the mileage at which it becomes cheaper to ride with Baboon Taxi than it is to ride Aardvark Taxi.

Aardvark Taxi's equation: $y = 1x + 4$
Baboon Taxi's equation: $y = .75x + 6$
Use substitution: $.75x + 6 = x + 4$
Multiply both sides by 4: $3x + 24 = 4x + 16$
Solve for x: $8 = x$

This tells you that, at 8 miles, the total charge for the two companies is the same. If you compare the charge for 1 mile, Aardvark charges $5 and Baboon charges $6.75. Therefore, Aardvark Taxi is cheaper for distances up to 8 miles, but Baboon is cheaper for distances greater than 8 miles. This problem can also be solved by graphing the 2 equations.

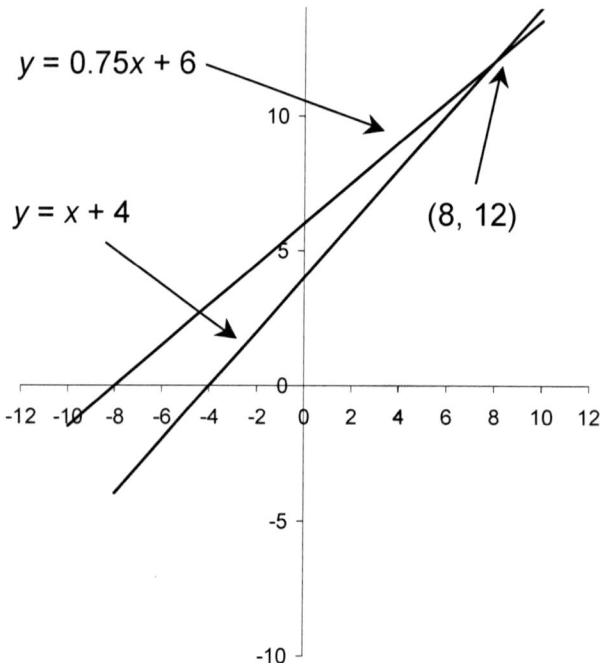

The lines intersect at (8, 12), therefore at 8 miles, both companies charge $12. For distances less than 8 miles, Aardvark Taxi charges less (the graph is below Baboon). For distances greater than 8 miles, Aardvark charges more (the graph is above Baboon).

TEACHER CERTIFICATION STUDY GUIDE

Linear programming is the optimization of a linear quantity that is subject to constraints expressed as linear equations or inequalities. It is often used in various industries, ecological sciences and governmental organizations to determine or project production costs, the amount of pollutants dispersed into the air, etc. The key to most linear programming problems is to organize the information in the word problem into a chart or graph of some type.

Example: The YMCA wants to sell raffle tickets to raise at least $32,000. If they must pay $7,250 in expenses and prizes out of the money collected from the tickets, how many tickets worth $25 each must they sell?

Since they want to raise at least $32,000, that means they would be happy to get $32,000 or more. This requires an inequality. Let x be number of tickets sold; then $25x$ is total money collected for x tickets. Total money minus expenses is greater than $32,000.

$$25x - 7250 \geq 32000$$
$$25x \geq 39250$$
$$x \geq 1570$$

If they sell 1,570 tickets or more, they will raise *at least* $32,000.

Example: A printing manufacturer makes two types of printers: a Printmaster and a Speedmaster printer. The Printmaster requires 10 cubic feet of space, weighs 5,000 pounds and the Speedmaster takes up 5 cubic feet of space and weighs 600 pounds. The total available space for storage before shipping is 2,000 cubic feet and the weight limit for the space is 300,000 pounds. The profit on the Printmaster is $125,000 and the profit on the Speedmaster is $30,000. How many of each machine should be stored to maximize profitability and what is the maximum possible profit?

First, let x represent the number of Printmaster units sold and let y represent the number of Speedmaster units sold. Then, the equation for the space required to store the units is the following.

$$10x + 5y \leq 2000$$
$$2x + y \leq 400$$

Since the number of units for both models must be no less than zero, also impose the restrictions that $x \geq 0$ and $y \geq 0$. The restriction on the total weight can be expressed as follows.

$$5000x + 600y \leq 300000$$
$$25x + 3y \leq 1500$$

The expression for the profit P from sales of the printer units is the following.

$$P = \$125,000x + \$30,000y$$

The solution to this problem, then, is found by maximizing P subject to the constraints given in the preceding inequalities, along with the constraints that $x \geq 0$ and $y \geq 0$. The equations are grouped below for clarity.

$$x \geq 0$$
$$y \geq 0$$
$$2x + y \leq 400$$
$$25x + 3y \leq 1500$$
$$P = \$125,000x + \$30,000y$$

The two inequalities in two variables are plotted in the graph below. The shaded region represents the set of solutions that obey both inequalities. (Note that the shaded region in fact only includes points where both x and y are whole numbers.)

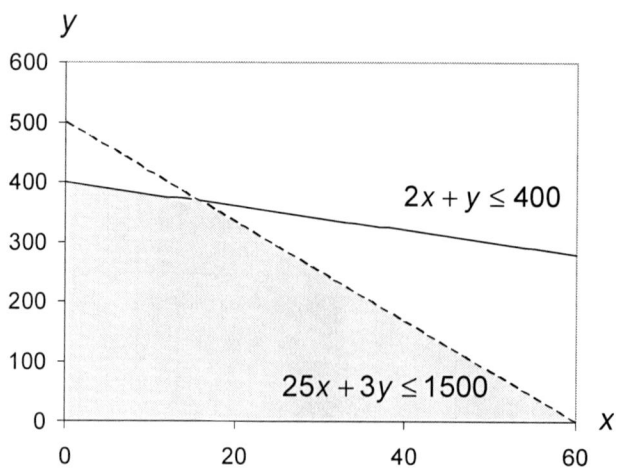

Note that the border of the shaded region that is formed by the two inequalities includes the solutions that constitute the maximum value of y for a given value of x. Note also that x cannot exceed 60 (since it would violate the second inequality). The solution to the problem, then, must lie on the border of the shaded region, since the border spans all the possible solutions that maximize the use of space and weight for a given number x.

To visualize the solution, plot the profit as a function of the solutions to the inequalities that lie along the border of the shaded area.

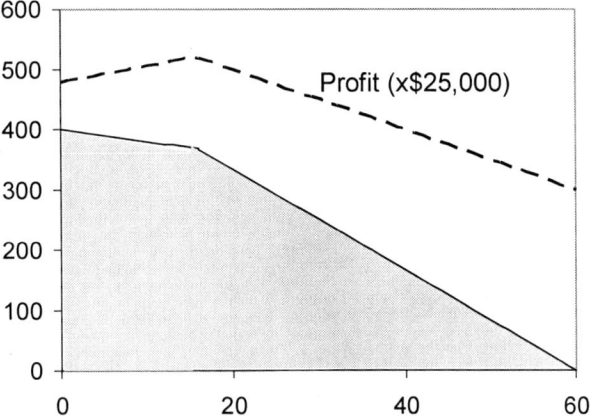

The profit curve shows a maximum at about x = 16. Test several values using a table to verify this result.

x	y	P (x$25,000)
15	370	519
16	366	519.2
17	358	514.6

Also double check to be sure that the result obeys the two inequalities.

$$2(16)+(366) = 398 \leq 400$$
$$25(16)+3(366) = 1498 \leq 1500$$

Thus, the optimum result is storage of 16 Printmaster and 366 Speedmaster printer units.

Example: Sharon's Bike Shoppe can assemble a 3-speed bike in 30 minutes and a 10-speed bike in 60 minutes. The profit on each bike sold is $60 for a 3 speed or $75 for a 10-speed bike. How many of each type of bike should it assemble during an 8-hour day (480 minutes) to maximize the possible profit? Total daily profit must be at least $300.

Let x be the number of 3-speed bikes and y be the number of 10-speed bikes. Since there are only 480 minutes to use each day, the first inequality is the following.

$$30x + 60y \leq 480$$
$$x + 2y \leq 16$$

Since the total daily profit must be at least $300, then the second inequality can be written as follows, where P is the profit for the day.

$$P = \$60x + \$75y \geq \$300$$
$$4x + 5y \geq 20$$

To visualize the problem, plot the two inequalities and show the potential solutions as a shaded region.

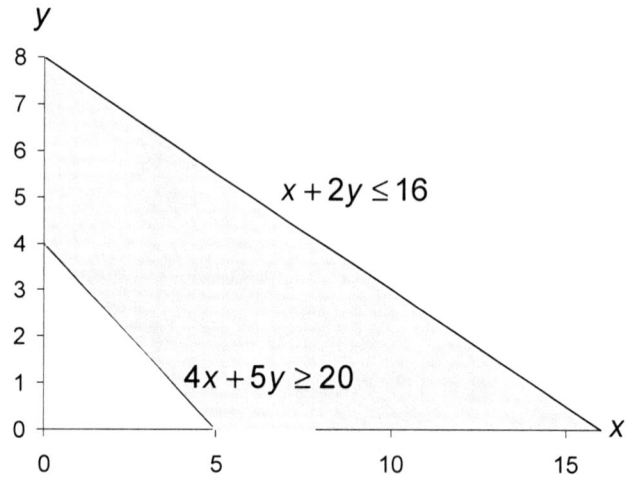

The solution to the problem is the ordered pair of whole numbers in the shaded area that maximizes the daily profit. The profit curve is added as shown below.

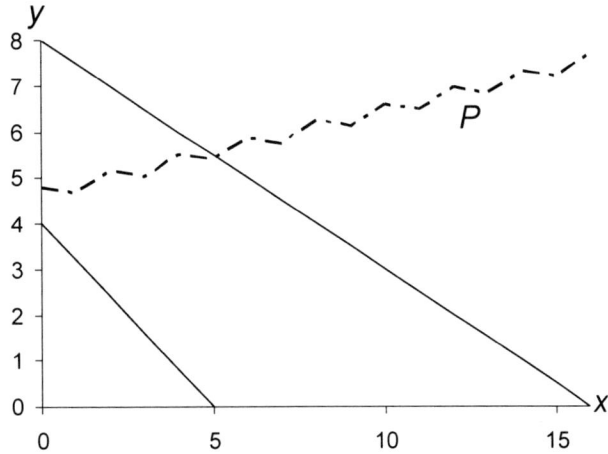

Based on the above plot, it is clear that the profit is maximized for the case where only 3-speed bikes (corresponding to x) are manufactured. Thus, the correct solution can be found by solving the first inequality for $y = 0$.

$$x + 2(0) \leq 16$$
$$x \leq 16$$

The manufacture of 16 3-speed bikes (and no 10-speed bikes) maximizes profit to $960 per day.

Other word problems may be modeled using **quadratic equations or inequalities**. Examples of this type of problem follow.

Example: A family is planning to add a new room to their house. They would like the room to have a length that is 10 ft more than the width and a total area of 375 sq. feet. Find the length and width of the room.

Let x be the width of the room. The length of the room is then $x + 10$. Thus,

$$x(x+10) = 375$$
$$x^2 + 10x - 375 = 0$$

Factor the quadratic expression to solve the equation:

$x^2 + 25x - 15x - 375 = 0$ Break up the middle
$x(x + 25) - 15(x + 25) = 0$ term using factors of 375
$(x + 25)(x - 15) = 0$
$x = -25$ or $x = 15$

Since the dimension of a room cannot be negative, we choose the positive solution x=15. Thus, the width of the room is 15 ft and the length of the room is 25ft.

Example: The height of a projectile fired upward at a velocity of *v* meters per second from an original height of *h* meters is $y = h + vx - 4.9x^2$. If a rocket is fired from an original height of 250 meters with an original velocity of 4800 meters per second, find the approximate time the rocket would drop to sea level (a height of 0).

Substituting the height and velocity into the equation yields:
$y = 250 + 4800x - 4.9x^2$. If the height at sea level is zero, then $y = 0$ so $0 = 250 + 4800x - 4.9x^2$. Solving for *x* could be done by using the quadratic formula.

$$x = \frac{-4800 \pm \sqrt{4800^2 - 4(-4.9)(250)}}{2(-4.9)}$$

$x \approx 979.53$ or $x \approx -0.05$ seconds

Since the time has to be positive, it will be approximately 980 seconds until the rocket reaches sea level.

Example: A family wants to enclose 3 sides of a rectangular garden with 200 feet of fence. A wall borders the fourth side of the garden. In order to have a garden with an area of **at least** 4800 square feet, find the dimensions the garden should be.

Existing Wall

Solution:
Let x = distance from the wall

Then 2x feet of fence is used for these 2 sides. The side opposite the existing wall would use the remainder of the 200 feet of fence, that is, 200 − 2x feet of fence. Therefore the width (w) of the garden is x feet and the length (l) is 200 − 2x feet.

The area is calculated using the formula a = lw = x(200 − 2x) = $200x - 2x^2$, and the area needs to be greater than or equal to 4800 sq. ft. This yields the inequality $4800 \le 200x - 2x^2$. Subtract 4800 from each side and the inequality becomes

$$2(-x^2 + 100x - 2400) \ge 0$$
$$-x^2 + 100x - 2400 \ge 0$$
$$(-x + 60)(x - 40) \ge 0$$
$$-x + 60 \ge 0$$
$$-x \ge -60$$
$$x \le 60$$
$$x - 40 \ge 0$$
$$x \ge 40$$

The area will be at least 4800 square feet if the width of the garden is from 40 up to 60 feet. (The length of the rectangle would vary from 120 feet to 80 feet depending on the width of the garden.)

Competency 007 The teacher understands polynomial, rational, radical, absolute value, and piecewise functions, analyzes their algebraic and graphical properties, and uses them to model and solve problems.

This competency reviews various polynomial (and similar) functions in terms of their respective representations, domains and ranges, and asymptotes. In addition, the solution of problems and the modeling of situations involving these functions are also discussed.

The beginning teacher recognizes and translates among various representations (e.g., written, tabular, graphical, algebraic) of polynomial, rational, radical, absolute value, and piecewise functions.

Polynomial Functions
A polynomial is a sum of terms where each term is a constant multiplied by a variable raised to a positive integer power. The general form of a polynomial $P(x)$ is

$$P(x) = a_n x^n + a_{n-1} x^{n-1} + \ldots + a_2 x^2 + a_1 x + a_0$$

Polynomials written in **standard form** have the terms written in decreasing exponent value, as shown above. The **degree of a polynomial function** in one variable is the value of the largest exponent to which the variable is raised. The above expression is a polynomial of degree n (assuming that $a_n \neq 0$). Any function that represents a line is a polynomial function of degree one. Quadratic functions are polynomials of degree two. $5x^2 - 4x - 6$, for instance, is a second degree polynomial whereas $2x^3 - 5x^2 + x$ is a polynomial of degree three. If a term has more than one variable (e.g., $2xy$) it is necessary to add the exponents of the variables within the term to get the degree of the polynomial. Since $1 + 1 = 2$, $2xy$ is a polynomial of the second degree.

A polynomial may also be represented in tabular or graphical form as shown in the example below.

Example: Express the polynomial $x^3 - 6x + 4$ in tabular and graphical form.

x	y
-3	-5
-2	8
-1	9
0	4
1	-1
2	0
3	13

Note the change in sign of the y value between $x = -3$ and $x = -2$. This indicates there is a zero between $x = -3$ and $x = -2$. Since there is another change in sign of the y value between $x = 0$ and $x = -1$, there is a second root there. When $x = 2$, $y = 0$ so $x = 2$ is an exact root of this polynomial.

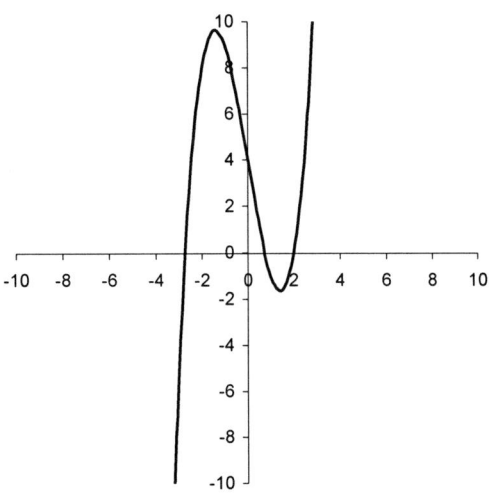

Rational functions

A rational function can be written as the ratio of two polynomial expressions. A rational function is given in the form $f(x) = p(x)/q(x)$. In the equation, $p(x)$ and $q(x)$ both represent polynomial functions ($q(x)$ does not equal zero).

Examples of rational functions are $r(x) = \dfrac{x^2 + 2x + 4}{x - 3}$ and $r(x) = \dfrac{x}{x - 3}$, which both are ratios of two polynomials.

The branches of rational functions approach asymptotes. Setting the denominator equal to zero and solving will give the value(s) of the **vertical asymptotes** since the function will be undefined at this point. If the value of f(x) approaches b as $|x|$ increases, the equation $y = b$ is a **horizontal asymptote**. To find the horizontal asymptote it is necessary to make a table of values for x that are to the right and left of the vertical asymptotes. The pattern for the horizontal asymptotes will become apparent as the $|x|$ increases.

If there is more than one vertical asymptote, remember to choose numbers to the right and left of each one in order to find the horizontal asymptotes and have sufficient points to graph the function.

Example: Graph $f(x) = \dfrac{3x+1}{x-2}$.

$x - 2 = 0$
$x = 2$

1. Set denominator $= 0$ to find the vertical asymptote.

x	f(x)
3	10
10	3.875
100	3.07
1000	3.007
1	⁻4
⁻10	2.417
⁻100	2.93
⁻1000	2.99

2. Make a table choosing numbers to the right and left of the vertical asymptote.

3. The pattern shows that as the $|x|$ increases f(x) approaches the value 3, therefore a horizontal asymptote exists at $y = 3$

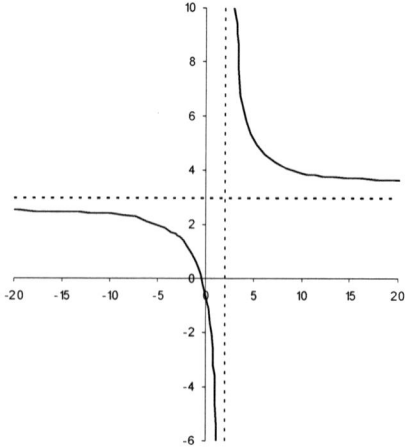

Note that $x = 2$ are excluded from the domain of the function and $y = 3$ is excluded from the range.

In some cases, the restriction on a rational function owing to the denominator being zero will not be a vertical asymptote but simply a **hole in the graph**. This happens when the value of x that reduces the denominator to zero is also a zero of the numerator.

Example: Plot the function $\frac{x-2}{x^2-4}$.

Factoring the denominator, we see that the denominator goes to zero at x = –2 and x = 2.

$$\frac{x-2}{x^2-4} = \frac{x-2}{(x+2)(x-2)}$$

There is a **vertical asymptote at x = h – 2**. Since the function can be simplified to the form 1/(x + 2) by canceling (x – 2) from the numerator and denominator, there is no asymptote at x = 2. The point x = 2, however, must be excluded from the function. Hence, there is a **hole in the graph at x = 2**.

Studying the function, we see that for large values of x the function goes to zero. Thus there is a **horizontal asymptote at y = 0**.

x	y	
–6	–1/4	
–4	–1/2	
–3	–1	
–2.5	–2	
–1.5	2	
–1	1	
0	1/2	
1	1/3	
2	1/4	(location of hole)
3	1/5	

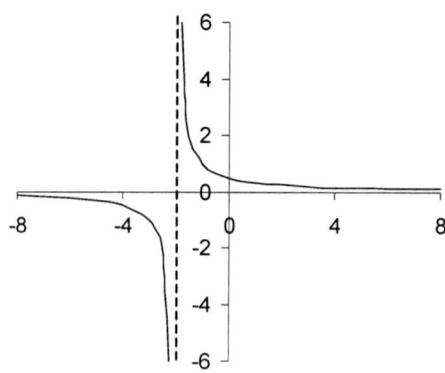

Absolute Value Functions

The absolute value function for a first-degree equation is of the form $y = m|x - h| + k$. Its graph is in the shape of a ∨. The point (h, k) is the location of the maximum/minimum point on the graph. "±m" are the slopes of the two sides of the ∨. The graph opens up if m is positive and down if m is negative.

The following are examples of graphs of absolute value functions.

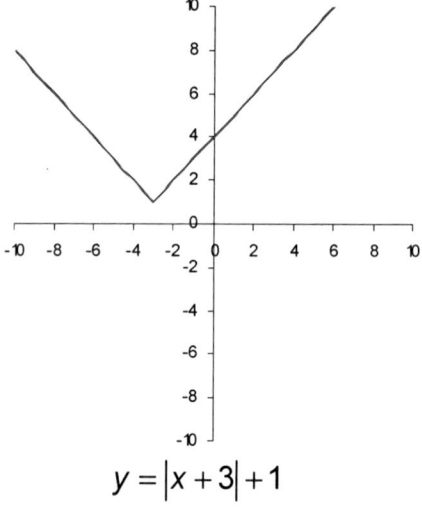

$$y = |x + 3| + 1$$

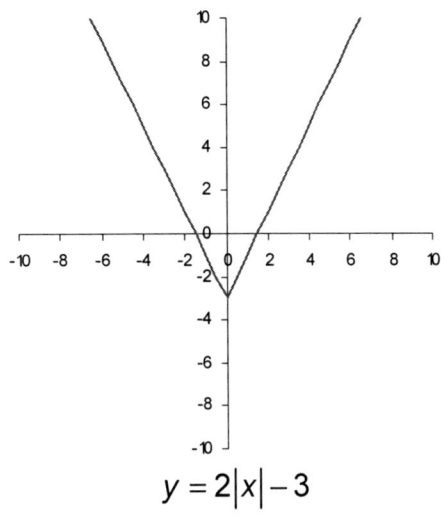

$$y = 2|x| - 3$$

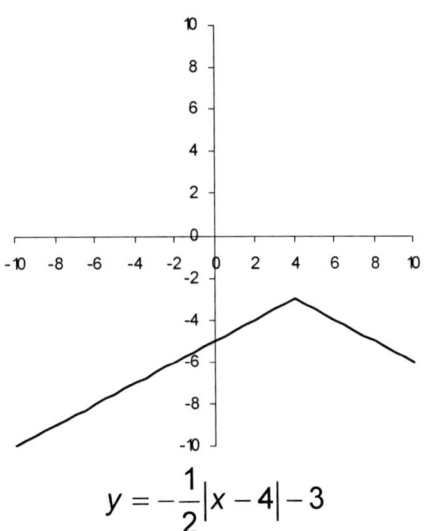

$$y = -\frac{1}{2}|x - 4| - 3$$

Note that on the first graph, the graph opens up since m is +1. It has (–3, 1) as its minimum point. The slopes of the two upward rays are ±1. The second graph also opens up since m is positive. The point (0, –3) is its minimum point. The slopes of the two upward rays are ±2. The third graph is a downward ∧ because m is $-\frac{1}{2}$.

The maximum point on the graph is at (4, –3). The slopes of the two downward rays are ±1/2.

Radical Functions

Radical functions are those that depend on a root of the independent variable x, typically the square root. Some examples are $3\sqrt{x} + 5$, $\sqrt{7 - x}$, and $2\sqrt{x + 3} - 5$.

Plotting a radical function follows a process similar to that of plotting virtually any other function. A set of representative points is needed, and prior knowledge of the domain of the function is helpful (for instance, if only real numbers are considered, the expression under the square root sign must always be positive). Typically, a calculator is needed to find the values of the function for specific variable values.

Example: Tabulate and plot the function $3\sqrt{x} + 5$.

x	y
0	5
1	8
4	11
9	14
16	17
25	20

Note that x must always be positive. Hence the domain of the function is $x > 0$.

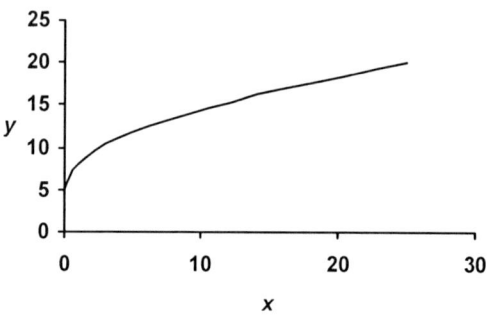

Piecewise Functions

Functions defined by two or more formulas are **piecewise functions**. The formula used to evaluate piecewise functions varies depending on the value of x. The graphs of piecewise functions consist of two or more pieces, or intervals, and are often discontinuous.

Example 1

$f(x) = \begin{cases} x+1 & \text{if } x > 2 \\ x-2 & \text{if } x \leq 2 \end{cases}$

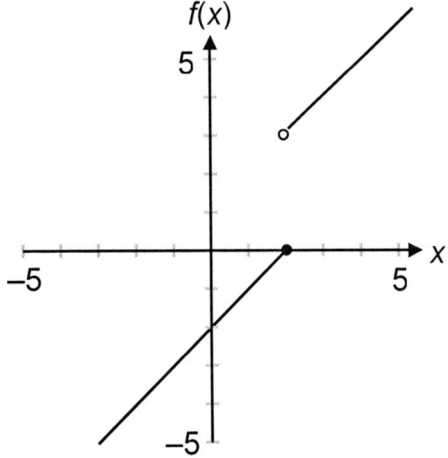

Example 2

$f(x) = \begin{cases} x & \text{if } x \geq 1 \\ x^2 & \text{if } x < 1 \end{cases}$

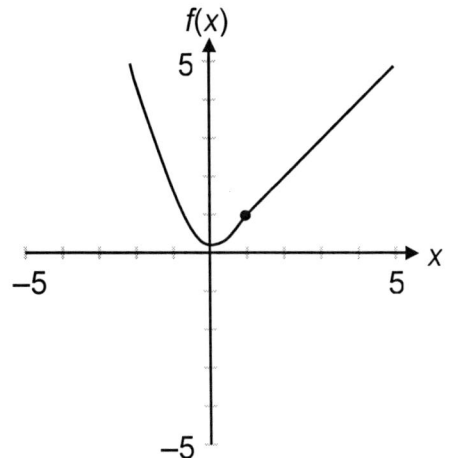

When graphing or interpreting the graph of piecewise functions it is important to note the points at the beginning and end of each interval because the graph must clearly indicate what happens at the end of each interval. Note that in the graph of Example 1, point (2, 3) is not part of the graph and is represented by an empty circle. On the other hand, point (2, 0) is part of the graph and is represented as a solid circle. Note also that the graph of Example 2 is continuous despite representing a piecewise function.

The beginning teacher describes restrictions on the domains and ranges of polynomial, rational, radical, absolute value, and piecewise functions.

The restrictions on the domain and range of a function are most easily understood when the function is viewed in graphical form.

The domain of a **polynomial function** includes all real values of x. The range depends on the form of the particular polynomial. For example, the polynomial $f(x) = -x^2 + 5$ has the range $f(x) \leq 5$, whereas $f(x) = -x^3 + 5$ has a range that includes all real numbers.

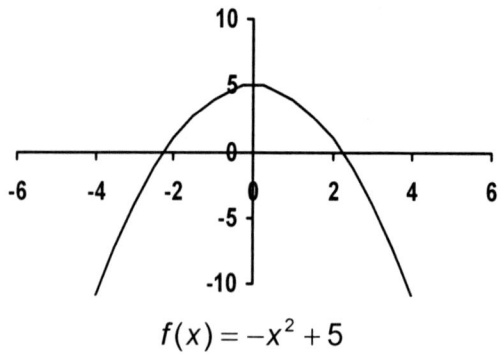

$f(x) = -x^2 + 5$

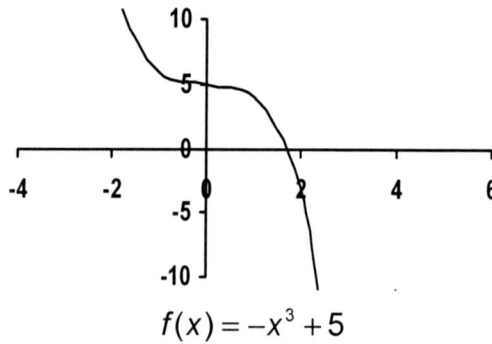

$f(x) = -x^3 + 5$

In general, the range of an even polynomial is restricted, and the range of an odd polynomial includes all real numbers.

The domain of a **rational function** $p(x)/q(x)$ includes all real values of x except those for which the denominator $q(x)$ is zero. The point $x = \pm 2$, for instance, is excluded from the domain of the function $\frac{x}{x^2 - 4}$. For more examples, see **Competency 005**.

Looking at a graph of a rational function, the points excluded from the domain are the x values for which vertical asymptotes or holes exist. The range of a rational function is determined by its horizontal asymptotes. In cases where the graph of the rational function does not cross the horizontal asymptote, the y value for the horizontal asymptote must be excluded from the range. For a detailed discussion of the asymptotes and holes of a rational function, see the discussion at the beginning of this competency.

The domain of a **radical function** is restricted to the x values that will result in a positive number under the square root sign. Thus, radical functions have an upper or lower limit placed on the domain. The restriction on the domain places a corresponding restriction on the range in the form of an upper or lower limit.

Example: Find the domain and range of the function
$y = 2\sqrt{x-3} - 5$

In order to limit the values within the radical sign to positive numbers, the domain of the function must be limited to $x \geq 3$.

The corresponding range is given by $y \geq -5$.

The graph plotted below displays the domain and range:

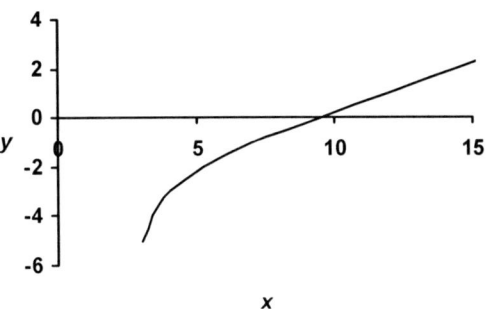

The domain of an **absolute value function** includes all real numbers. The vertex of the V-shaped function provides the upper or lower limit to the range of the function. For an upright V-shaped function, the vertex is the lower limit of the range. For an upside down V, the vertex is the upper limit of the range. See the beginning of this competency for examples.

The domain of a **piecewise function** is simply the union of all points that make up the domain of each piece. Likewise, the range of a piecewise function is the union of all values that make up the range of each piece.

The beginning teacher takes and uses connections among the significant points (e.g., zeros, local extrema, points where a function is not continuous or not differentiable) of a function, the graph of the function, and the function's symbolic representation.

> **The zeros or roots of a function f(x) are x values for which f(x) is 0.** Zeros may be real or complex. In graphical terms, the real zeros are the x values at which the graph intersects the x-axis.
>
> **The Fundamental Theorem of Algebra** states that a polynomial expression of degree n must have n roots (which may be real or complex and which may not be distinct).

Polynomial equations are in the form of P(x) given below, where n is the degree of the polynomial and the constant a_n is non-zero.

$$P(x) = a_n x^n + a_{n-1} x^{n-1} + \ldots + a_2 x^2 + a_1 x + a_0$$

If P(c) = 0 for some number c, then c is said to be a **zero** (or **root**) of the function. A zero is also called a **solution** to the equation.

The existence of n solutions can be seen by looking at a factorization of P(x). For instance, consider $P(x) = x^2 - x - 6$. This second-degree polynomial can be factored into

$$P(x) = (x+2)(x-3)$$

Note that P(x) has two roots in this case: x = –2 and x = 3. This corresponds to the degree of the polynomial, n = 2. The graph of this function crosses the x axis at -2 and 3.

In some cases, however, there may be non-distinct roots. Consider $P(x) = x^2$.

$$P(x) = (x)(x)$$

Note that the polynomial is factored in the same way as the previous example, but, in this case, the roots are identical: x = 0. Thus, although there are two roots for this second-degree polynomial, the roots are not distinct. The graph of this function touches the x-axis at just one point x = 0.

Likewise, roots of a polynomial may be complex. Consider $P(x) = x^2 + 1$. The range of this function is $P(x) \geq 1$, so there is no root in the sense that the function crosses the real x-axis. Nevertheless, if complex values of x are permitted, there are cases where P(x) is zero. Factor P(x) as before, but this time use complex numbers.

$$P(x) = (x+i)(x-i)$$

The solutions are x = i and x = –i. Thus, this second-degree polynomial still has two roots. The graph of this function does not intersect the x-axis.

See the discussion on graphing in **Competency 006** for examples of graphs corresponding to different types of zeros of quadratic functions.

For a general n^{th} degree polynomial, the function $P(x)$ can be factored in a similar manner.

$$P(x) = (x - c_n)(x - c_{n-1})\ldots(x - c_2)(x - c_1)$$

If a factor $(x-c)$ occurs k times in the factorization of a polynomial, the root c is said to have a **multiplicity** of k. For the polynomial $P(x) = x^2$, for instance, the root x=0 has a multiplicity of 2.

The zeros of a rational function $p(x)/q(x)$ are the zeros of the numerator $p(x)$ provided they are not also zeros of $q(x)$.

Consider functions $a(x) = \dfrac{x-2}{x+2}$ and $b(x) = \dfrac{x-2}{x^2-4} = \dfrac{x-2}{(x+2)(x-2)}$.
$x = 2$ is a zero of the function $a(x)$ but not of $b(x)$. This is because $x = 2$ makes the denominator of $b(x)$ zero and, therefore, lies outside the domain of $b(x)$.

For a detailed discussion of continuity and the extrema of a function see **Competency 010**.

The beginning teacher analyzes functions in terms of vertical, horizontal, and slant asymptotes.

A function may have one or more **asymptotes** that can be identified. An asymptote is a line for which the distance between it and a function or curve is arbitrarily small, especially as the function tends toward infinity in some direction. Asymptotes can be either vertical, horizontal or slant.

Consider, for instance, the plot of the hyperbola defined as follows.

$$g(x) = \pm\sqrt{x^2 + 1}$$

Note, for instance, that as x tends toward infinity, $g(x)$ gets arbitrarily close to x. The graph of $g(x)$ is shown below.

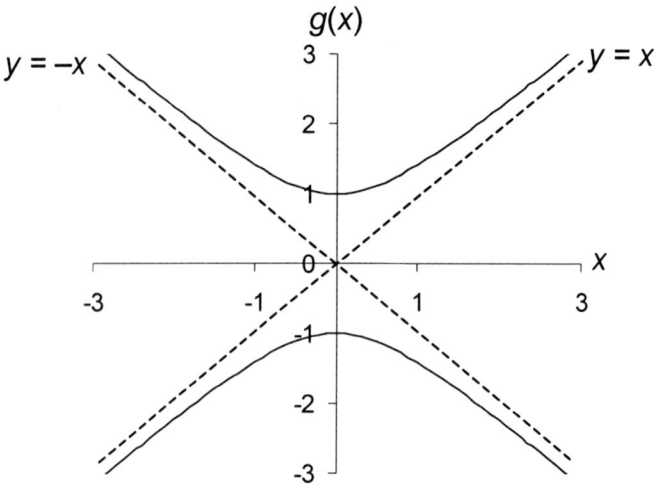

The (slant) asymptotes and their associated functions for this relation are displayed in the graph above as dashed lines.

Vertical and horizontal asymptotes for rational functions are discussed earlier in this competency. A rational function has slant asymptotes if the polynomial in the numerator is of a higher degree than the polynomial in the denominator.

Example: Find the equation of the slant asymptote for the function $\dfrac{x^2 + 3x + 4}{x - 2}$.

Using long division, we can rewrite the function as

$$\dfrac{x^2 + 3x + 4}{x - 2} = x + 5 + \dfrac{14}{x - 2}.$$

For large values of x, it is clear that the term $14/(x - 2)$ will become very small and the function will tend to follow the line $y = x + 5$. Thus, the function has two asymptotes, a vertical one at $x = 2$ and a slant asymptote along $y = x + 5$.

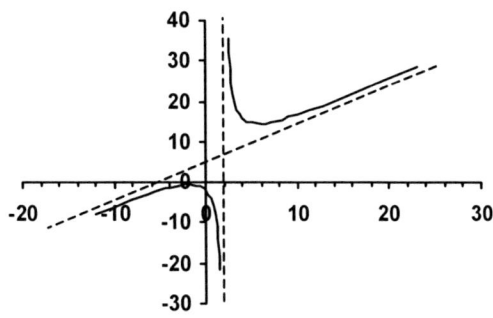

The beginning teacher analyzes and applies the relationship between inverse variation and rational functions.

An **inverse variation** can be expressed by the formula

$xy = k$, where k is a constant, $k \neq 0$.

It is clear that $y = \dfrac{k}{x}$ is a rational function with a vertical asymptote at $x = 0$ and horizontal asymptote at $y = 0$ as shown in the example below. An inverse variation is typically represented by a rational function with a constant in the numerator and a power of the independent variable in the denominator.

Example: On a 546 mile trip from Miami to Charlotte, one car drove 65 mph while another car drove 70 mph. How does this affect the driving time for the trip?

This is an inverse variation, since increasing your speed should decrease your driving time. Use the equation $t = \dfrac{d}{r}$ where t = driving time, r = speed and d = distance traveled.

(65 mph) t = 546 miles	and	(70 mph) t = 546 miles
t = 8.4 hours	and	t = 7.8 hours
slower speed, more time		faster speed, less time

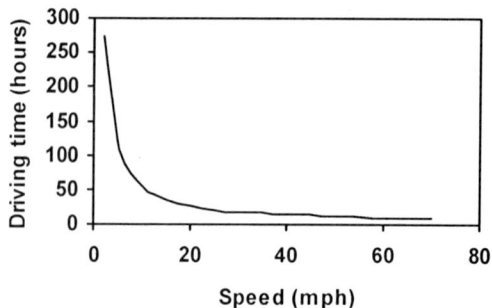

A quantity may also vary with different exponents of another quantity as shown in the examples below.

Example: A varies inversely as the square of R. When A = 2, R = 4. Find A if R = 10.

Since A varies inversely as the square of R,

$$A = \frac{k}{R^2} \text{ (equation 1)}, k \text{ is a constant.}$$

Use equation 1 to find k when A = 2 and R = 4.

$$2 = \frac{k}{4^2} \rightarrow 2 = \frac{k}{16} \rightarrow k = 32.$$

Substituting k = 32 into equation 1 with R = 10, we get:

$$A = \frac{32}{10^2} \rightarrow A = \frac{32}{100} \rightarrow A = 0.32$$

The beginning teacher solves equations and inequalities involving polynomial, rational, radical, absolute value, and piecewise functions using a variety of methods (e.g., tables, algebraic methods, graphs, use of a graphing calculator), and evaluates the reasonableness of solutions.

A range of methods can be used to solve polynomial equations. Several theorems, including the Fundamental Theorem of Algebra discussed previously, are useful in this regard.

As mentioned previously in this competency, a polynomial of degree n may be factored as follows:

$$P(x) = (x - c_n)(x - c_{n-1})\ldots(x - c_2)(x - c_1)$$

The **Factor Theorem** states that a polynomial $P(x)$ has a factor $(x - a)$ if and only if $P(a) = 0$. Thus, the constants $c_n, c_{n-1}..c_1$ in the expression given above are the roots of the polynomial equation $P(x) = 0$.

If a single root c is known, then the polynomial can be simplified (that is, it can be reduced by one degree) using division.

$$Q(x) = \frac{P(x)}{x - c}$$

Here, if $P(x)$ has degree n, then $Q(x)$ has degree $n - 1$. If some roots are known, the task of finding the remainder of the roots can be simplified by performing the division represented above. As each successive root is found, the degree of the polynomial can be reduced to further simplify finding the remainder of the roots.

In some cases, dividing a polynomial by $(x - c)$ is simple, but generally speaking it is a complicated process. The process can be simplified using **synthetic division**, however.

To perform synthetic division of a polynomial $P(x)$ by $(x - c)$ to get a new polynomial $Q(x)$, first draw an upside-down division symbol as shown below, using the coefficients of $P(x)$ and the root c.

$$P(x) = a_n x^n + a_{n-1} x^{n-1} + \ldots + a_2 x^2 + a_1 x + a_0$$

$$Q(x) = \frac{P(x)}{x - c}$$

$$c \,\bigg|\, a_n \quad a_{n-1} \quad a_{n-2} \quad \ldots$$

The first step of synthetic division is to carry the first term, a_{n-1}.

$$
\begin{array}{c|cccc}
c & a_n & a_{n-1} & a_{n-2} & \cdots \\
& & & & \\
\hline
& a_n & & &
\end{array}
$$

Each successive step involves multiplying c by the previously carried term and then placing the result under the next term. Then add the two results to get the next carry value.

$$
\begin{array}{c|cccc}
c & a_n & a_{n-1} & a_{n-2} & \cdots \\
& & ca_n & & \\
\hline
& a_n & & &
\end{array}
$$

$$
\begin{array}{c|cccc}
c & a_n & a_{n-1} & a_{n-2} & \cdots \\
& & ca_n & & \\
\hline
& a_n & (a_{n-1} - ca_n) & &
\end{array}
$$

The process should be repeated until the last carry term is found. The result should be zero. (If the final carry value is non-zero, then c is not a root. This can be a useful test of whether a particular value is a root, especially for polynomials of high degrees.) The result of the division is the set of new coefficients for the quotient.

$$Q(x) = a_n x^{n-1} + (a_{n-1} - ca_n) x^{n-2} + \ldots$$

Example: Divide $x^4 - 7x^2 - 6x$ by $(x+2)$. Find the roots of the polynomial.

Use synthetic division. Notice that even the terms with coefficient zero must be included. (In other words, first write the polynomial as $x^4 + 0x^3 - 7x^2 - 6x + 0$.

$$
\begin{array}{c|ccccc}
-2 & 1 & 0 & -7 & -6 & 0 \\
& & & & & \\
\hline
& & & & &
\end{array}
$$

Perform the division.

$$-2 \,\big|\, 1 \quad 0 \quad -7 \quad -6 \quad 0$$
$$\underline{}$$
$$ 1$$

$$-2 \,\big|\, 1 \quad 0 \quad -7 \quad -6 \quad 0$$
$$ \underline{ -2}$$
$$ 1 \quad -2$$

$$-2 \,\big|\, 1 \quad 0 \quad -7 \quad -6 \quad 0$$
$$ \underline{ -2 \quad 4}$$
$$ 1 \quad -2 \quad -3$$

$$-2 \,\big|\, 1 \quad 0 \quad -7 \quad -6 \quad 0$$
$$ \underline{ -2 \quad 4 \quad 6}$$
$$ 1 \quad -2 \quad -3 \quad 0$$

$$-2 \,\big|\, 1 \quad 0 \quad -7 \quad -6 \quad 0$$
$$ \underline{ -2 \quad 4 \quad 6 \quad 0}$$
$$ 1 \quad -2 \quad -3 \quad 0 \quad 0$$

Thus, –2 is indeed a root of the polynomial. The result is then

$$\frac{x^4 - 7x^2 - 6x}{x+2} = x^3 - 2x^2 - 3x$$

Note that the remainder of the roots of this polynomial can be found much more easily than if the original polynomial was analyzed as is. Factor the result further.

$$x^3 - 2x^2 - 3x = x(x^2 - 2x - 3) = x(x-3)(x+1)$$

Thus, the roots are –2, –1, 0 and 3.

According to the **Complex Conjugate Root Theorem**, for a polynomial $P(x)$ with real coefficients, if $P(x)$ has a complex root z, then it must also have a complex root $\bar{z}$. (The bar notation indicates complex conjugate. Thus, if $z = a + bi$, then $\bar{z} = a - bi$.)

The **Rational Root Theorem**, also known as the Rational Zero Theorem, allows determination of all possible rational roots (or zeros) of a polynomial equation with integer coefficients. (A root is a value of x such that $P(x) = 0$.) Every rational root of $P(x)$ can be written as $x = \frac{p}{q}$, where p is an integer factor of the constant term a_0 and q is an integer factor of the leading coefficient a_n.

Example: Find the rational roots of $P(x) = 3x^3 - 7x^2 + 3x - 2$.

By the Rational Root Theorem, the roots must be of the form

$$x = \mp \frac{1, 2}{1, 3}$$

The candidates are then

$$x = \pm 1, \pm \frac{1}{3}, \pm \frac{2}{3}, \pm 2$$

Test each possibility. The only result that works is $x = 2$. (Note that the Rational Root Theorem does not guarantee that each potential rational number that includes factors of the leading and constant terms is a root. The theorem only states that roots will include these factors.)

In cases where a polynomial is highly complicated or involves constants that do not permit methods such as factoring, a numerical approach may be appropriate. Newton's method is one possible approach to solving a polynomial equation numerically. At other times it may require a graphical approach whereby the behavior of the function is examined on a visual plot. When using Newton's method, graphing the function can be helpful for estimating the locations of the real roots (if any).

To solve an **equation with rational expressions**, set the expression equal to zero (which leads to the elimination of the denominator) and solve, as with simple polynomials.

$$r(x) = 0 = \frac{p(x)}{q(x)}$$

$$p(x) = 0$$

Note, however, that solutions to $p(x) = 0$ may lead to undefined values for $r(x)$ (that is, values for which $q(x) = 0$), and must be checked prior to acceptance. This difficulty can be alleviated to some extent by factoring $p(x)$ and $q(x)$ and eliminating common factors.

Example: Find the solutions for $\dfrac{12}{2x^2 - 4x} + \dfrac{13}{5} = \dfrac{9}{x-2}$

Factor and rearrange the equation as follows, then solve for x.

$$\frac{12}{2x(x-2)} - \frac{9}{x-2} = -\frac{13}{5}$$

$$\frac{12}{2x(x-2)} - \frac{9(2x)}{2x(x-2)} = \frac{-18x+12}{2x(x-2)} = -\frac{13}{5}$$

$$-18x + 12 = -\frac{13}{5} \cdot 2x(x-2) = -\frac{26}{5}x^2 + \frac{52}{5}x$$

$$\frac{26}{5}x^2 - \frac{52}{5}x - 18x + 12 = 0$$

$$0 = \frac{26}{5}x^2 - \frac{142}{5}x + 12 = 26x^2 - 142x + 60 = 13x^2 - 71x + 30$$

The solutions for x can be found by factoring the above expression.

$$13x^2 - 71x + 30 = (x-5)(13x-6) = 0$$

Thus, $x = 5$ or $x = \dfrac{6}{13}$. These solutions can be confirmed by substitution into the original equation.

Solving a **radical equation** analytically typically involves either the use of logarithms or exponents. When the function is expressed in terms of exponents, simply follow the rules of exponents. (For instance, if two factors in a product have the same base, simply add the exponents to find the product.) A simple solution process for a radical equation is as follows.

1. Isolate the radical term on one side of the equation.
2. Raise both **sides** of the equation to the inverse of the exponent for the radical term. Combine any like terms.
3. If there is another radical still in the equation, repeat steps 1 and 2. Repeat as necessary to eliminate all radicals.
4. Solve the resulting equation.
5. Check all solutions in the original equation and discard any extraneous solutions.

Example: Find the roots of the equation $f(x) = 3\sqrt{x} - x$

Set $f(x)$ equal to zero and rearrange the results. Also, express each term using exponents instead of radicals.

$$f(x) = 0 = 3x^{1/2} - x$$
$$3x^{1/2} = x$$

Square both sides of the equation.

$$\left[3x^{1/2}\right]^2 = x^2 = 9x$$

Rearrange and factor the result.

$$x^2 - 9x = x(x-9) = 0$$

Thus, the roots of $f(x)$ are at $x = 0$ and $x = 9$. Test each result to be sure:

$$f(0) = 3\sqrt{0} - 0 = 0 \qquad f(9) = 3\sqrt{9} - 9 = 3(3) - 9 = 0$$

Example: Solve the equation $\sqrt{2x+1}+7=x$.

First, isolate the radical, then square both sides of the equation.

$$\sqrt{2x+1} = x-7$$
$$2x+1 = (x-7)^2 = x^2 - 14x + 49$$

Simplify and solve by factoring.

$$x^2 - 14x - 2x + 49 - 1 = 0$$
$$x^2 - 16x + 48 = 0$$
$$(x-12)(x-4) = 0$$

The solutions are $x = 12$ and $x = 4$. Check both solutions in the original expression.

$$\sqrt{2(12)+1}+7 = \sqrt{25}+7 = 5+7 = 12$$
$$\sqrt{2(4)+1}+7 = \sqrt{9}+7 = 3+7 = 10 \neq 4$$

Only $x = 12$ satisfies the original expression. This is therefore the only solution to the equation.

Some equations may be impossible or impractical to solve analytically. In such cases, a numerical approach, such as Newton's method, may be required.

Absolute value equations are, essentially, two functions that can be expressed as a single piecewise continuous function using absolute value notation. To solve an absolute value equation, follow these steps:

1. Isolate the absolute value expression on one side of the equation.
2. Split the absolute value equation into two separate standard equations. For one equation, set the expression in the absolute value equal to the expression on the other side of the original equality. For the second equation, set the expression in the absolute value equal to the negation of the expression on the other side of the original equality.
3. Solve each new equation.
4. Check each answer by substituting them into the original equation. Discard any extraneous solutions.

Example: Solve the following equation for x: $|2x-5|+1=12$.

Isolate the absolute value and then split the result into two equivalent equations.

$$|2x-5|=11$$
$$2x-5=11 \qquad\qquad 2x-5=-11$$

Solve each equation.

$$2x=16 \qquad\qquad 2x=-6$$
$$x=8 \qquad\qquad x=-3$$

Check both solutions in the original expression.

$$|2(8)-5|+1=11+1=12 \qquad |2(-3)-5|+1=11+1=12$$

Both solutions are valid in this case.

Example: Plot the system of inequalities given below:

$$x+2y\leq 3$$
$$-|x|+y>2$$

For each equation, find the equation of the line (or absolute value function) that bounds the solution areas.

$$2y\leq -x+3 \qquad\qquad y>|x|+2$$
$$y\leq -\frac{1}{2}x+\frac{3}{2}$$

Plot both functions as follows, using a solid line for the linear equation (since the line is included in the solution set) and a dashed line for the absolute value equation (since the line is not included in the solution set). Then shade the solution regions appropriately.

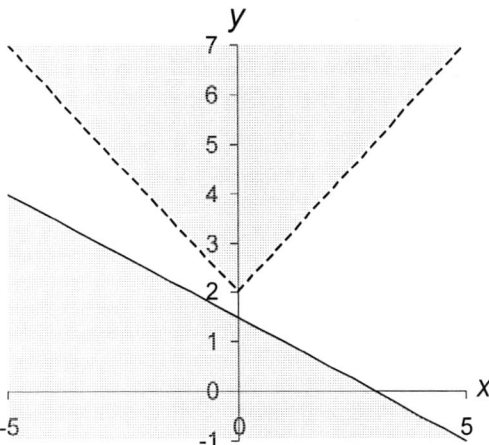

Note that there is no overlap of the solution regions for these two functions. As a result, the solution set is simply the null set (that is, there are no solutions to this system of equations).

Using a Graphing Calculator with Polynomial Functions

The following examples are illustrated using the basic functions of a Texas Instruments TI-type graphing calculator. Although the specifics may not apply to all graphing calculators, the general approach still should.

Solve $x^2 + 2x - 8$ for its roots by graphing on a calculator:

1. Press "Y=" and enter X^2+2X-8 for Y1.
2. Press "GRAPH".
3. Identify the x-intercepts of the parabola. The x-intercepts are -4 and 2.
4. Notice the parabola at x= -1- this is the line of the axis of symmetry and the vertex is at the bottom of the line; it is located halfway between the 2 x-intercepts.

Using a Graphing Calculator with Rational Functions

Graph the functions *f, g, f+g* for *x* for the interval [-*p, p*]:

1. Press the GRAPH key and then choose *y(x)=* by pressing the F1 key.
2. If needed, press the F4 key until only *y1=* appears on the screen.
3. Type in X^2, then press the ENTER key.
4. Type in COSX and press the ENTER key.

5. Type in y1+y2 and press the ENTER key. NOTE: Both *y*'s must be lower case.
6. The range settings for the graph above are [-*p, p*] × [-1, 10].
7. Note that the graphs of the functions are plotted in the order in which they are entered above. You can plot only one or two of the graphs by using the SELCT key on the **y(x)=** screen.

Using a Graphing Calculator with Absolute Value Functions

Solve $2|x-3| = 4$ using a table on the graphing calculator:

1. Enter the left side of the equation in the "Y=" editor. Press the "MATH" key and use the NUM menu for ABS(.
2. Use the defaults for "2nd" "WINDOW" (TBLSET) and then press "2nd" "GRAPH" TABLE to see the values of the equation above when $x = 0, 1, 2, 3, 4...$
3. Note that Y1=4 when $x = 1$ and when $x = 5$.

For more examples of equations and inequalities solved graphically and otherwise see the other skills in **Competency 007** as well as those in **Competency 006**.

The beginning teacher models situations using polynomial, rational, radical, absolute value, and piecewise functions and solves problems using a variety of methods, including technology.

In Competency 006 we modeled and solved problems using linear and quadratic functions. Some situations require higher-order polynomials. Other problems can be modeled using rational or radical functions. The rational and radical equations can usually be simplified to yield a polynomial equation. Some examples are given below.

Example: A cubic container is modified so that its length is increased by 4 inches and its width is shortened by 2 inches. The height of the container remains unchanged. If the volume of the container is 16 cubic inches, what is it height?

Let the side of the original cube be x inches.
The volume of the modified container is given by

$$x(x + 4)(x - 2) = 16$$

Distributing and rearranging we get

$$x(x^2 + 2x - 8) = 16$$
$$\Rightarrow x^3 + 2x^2 - 8x - 16 = 0$$

The third order polynomial equation above can be grouped and factored as follows:

$$x^2(x+2) - 8(x+2) = 0$$
$$\Rightarrow (x+2)(x^2 - 8) = 0$$

The solutions to the equation are, therefore, $x = -2, \pm 2\sqrt{2}$.
Since the height of the box must be a positive number, we choose the positive solution. Thus the height is $2\sqrt{2}$ inches.

Example: Elly Mae can feed the animals in 15 minutes. Jethro can feed them in 10 minutes. How long will it take them if they work together?

If Elly Mae can feed the animals in 15 minutes, then she could feed 1/15 of them in 1 minute, 2/15 of them in 2 minutes, $x/15$ of them in x minutes. In the same fashion Jethro could feed $x/10$ of them in x minutes. Together they complete 1 job. The equation is:

$$\frac{x}{15} + \frac{x}{10} = 1$$

Multiply each term by the LCD of 30:

$$2x + 3x = 30$$
$$x = 6 \text{ minutes}$$

Example: A salesman drove 480 miles from Pittsburgh to Hartford. The next day he returned the same distance to Pittsburgh in half an hour less time than his original trip took, because he increased his average speed by 4 mph. Find his original speed.

Since distance = rate × time, then time = distance/rate

original time − 1/2 hour = shorter return time

$$\frac{480}{x} - \frac{1}{2} = \frac{480}{x+4}$$

Multiplying by the LCD of 2x(x + 4), the equation becomes:

$$480\big[2(x+4)\big] - 1\big[x(x+4)\big] = 480(2x)$$
$$960x + 3840 - x^2 - 4x = 960x$$
$$x^2 + 4x - 3840 = 0$$
$$(x+64)(x-60) = 0 \quad \text{Either } (x-60=0) \text{ or } (x+64=0) \text{ or both} = 0$$
$$x = 60 \qquad\qquad \text{60 mph is the original speed.}$$

This is the solution since the time
cannot be negative. Check your answer

$$x + 4 = 64$$
$$\frac{480}{60} - \frac{1}{2} = \frac{480}{64}$$
$$8 - \frac{1}{2} = 7\frac{1}{2}$$
$$7\frac{1}{2} = 7\frac{1}{2}$$

Example: For a cone of height h and radius r, the slant height is given by $s = \sqrt{r^2 + h^2}$. The lateral surface area is πrs and the area of the base is πr^2.

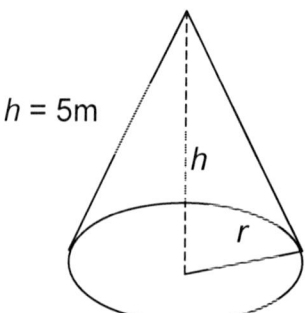

h = 5m

If the lateral surface area of a cone is twice that of its base and the height of the cone is 5m, find the radius.

The problem given may be modeled using the following radical equation:

$$\pi r\sqrt{25 + r^2} = 2\pi r^2$$

Canceling the common factor πr from both sides and squaring both sides we get

$$25 + r^2 = 4r^2$$
$$\Rightarrow 3r^2 = 25$$
$$\Rightarrow r = \sqrt{25/3} = 2.9$$

Thus, the radius of the cone is 2.9m.

Competency 008 The teacher understands exponential and logarithmic functions, analyzes their algebraic and graphical properties, and uses them to model and solve problems.

Exponential and logarithmic functions are closely related. This competency reviews the concepts and applications of exponential and logarithmic functions. The discussion covers the graphs and properties of these functions, solution of problems, and application of the functions to logarithmic scales and financial mathematics.

The beginning teacher recognizes and translates among various representations (e.g., written, numerical, tabular, graphical, algebraic) of exponential and logarithmic functions.

An **exponential function** is defined by the equation $y = ab^x$, where a is the starting value, b is the growth factor, and x is the exponent of the growth factor. For exponential functions, the **ratio** between successive y's or outputs are **constant**. In other words, each y or output, is a constant multiple of the previous y.

If $a > 0$ and b is between 0 and 1 the graph of the exponential function will be decreasing or decaying.

If $a > 0$ and b is greater than 1, the graph will be increasing or growing.

<u>Example:</u> Identify the pattern represented by $y = 100(0.5)^x$

x	y	ratio of change is a constant 50% increase indicated by multiplying by 1.5:
0	100	
1	150	$1.5(100) = 150$
2	225	$1.5(150) = 225$
3	337.5	$1.5(225) = 337.5$
4	506.25	$1.5(337.51) = 506.25$

Logarithmic functions of base *a* are of the basic form

$f(x) = \log_a x$, where $a > 0$ and not equal to 1.

Expressed verbally, the logarithm $f(x)$ of a number x is the exponent or power to which the base must be raised to equal x.

For example, 10 raised to the power 3 is 1000. Therefore, the base 10 logarithm of 1000 is 3.

The logarithmic function essentially transforms a geometrical progression into an arithmetic one. This is clear if one considers that the base 10 logarithm of $10, 10^2, 10^3, \ldots$ is 1, 2, 3, … and so on.

Graphing exponential and logarithmic functions involves finding a set of representative points, plotting these points on a graph and then connecting the points with appropriate curves. The domain of an exponential function includes all real numbers, but the domain of a logarithmic functions includes only the positive real numbers. The basic shapes of the exponential and logarithmic functions are illustrated below.

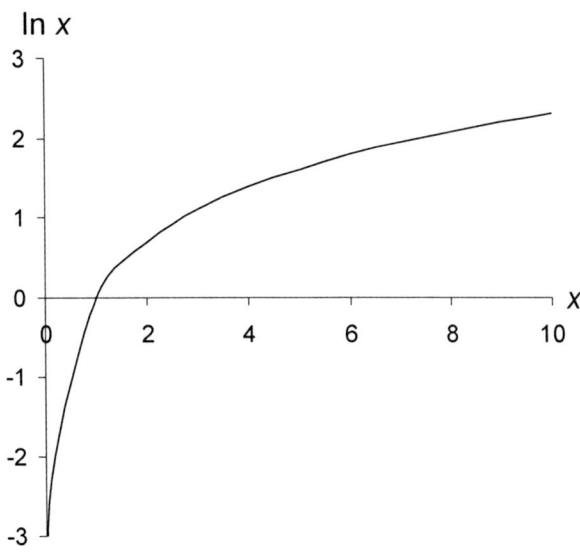

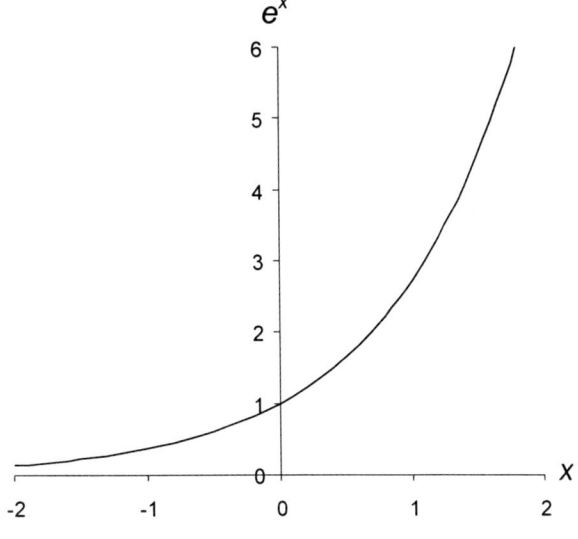

Note that the function e^x has an asymptote at $y = 0$ (the limit of the exponential function as x goes to negative infinity is zero), and the function ln x has an asymptote at $x = 0$ (the limit of the natural logarithmic function as x goes to zero from the right is negative infinity). The x-intercept of the logarithmic function, irrespective of base, is always (1,0) since any number raised to the power of 0 is equal to one. The y-intercept of e^x is (0,1) because any base raised to the power of 0 equals 1.

The beginning teacher recognizes and uses connections among significant characteristics (e.g., intercepts, asymptotes) of a function involving exponential or logarithmic expressions, the graph of the function, and the function's symbolic representation.

In order to understand the relationship between the graphical characteristics of an exponential or logarithmic function and its symbolic representation, we will look at several examples to see how a change in the algebraic expression is reflected on the graph. We will also see how knowing the intercepts and asymptotes aids the process of drawing a graph.

Example: Graph the function $f(x) = \log_2 (x + 1)$.

The domain of the function is all values of x such that $x + 1 > 0$. Thus, the domain of $f(x)$ is $x > -1$. The range of $f(x)$ is $(-\infty, +\infty)$.

The vertical asymptote of $f(x)$ is the value of x that satisfies the equation $x + 1 = 0$. Thus, the vertical asymptote is $x = -1$.

The x-intercept of $f(x)$ is the value of x that satisfies the equation $x + 1 = 1$ because $2^0 = 1$. Thus, the x-intercept of $f(x)$ is (0,0).

Finally, we find two additional values of $f(x)$, one between the vertical asymptote and the x-intercept and the other to the right of the x-intercept. $f(-0.5) = -1$ and $f(3) = 2$.

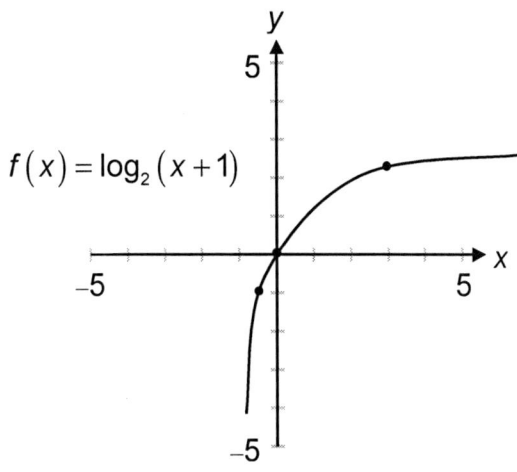

Example: Graph the function $f(x) = 2^x - 4$.

The domain of the function is the set of all real numbers and the range is $y > -4$. Because the base is greater than 1, the function is increasing. The y-intercept of f(x) is (0, –3). The x-intercept of f(x) is (2,0). The horizontal asymptote of f(x) is $y = -4$.

Finally, to construct the graph of f(x) we find two additional values for the function. For example, f(–2) = –3.75 and f(3) = 4.

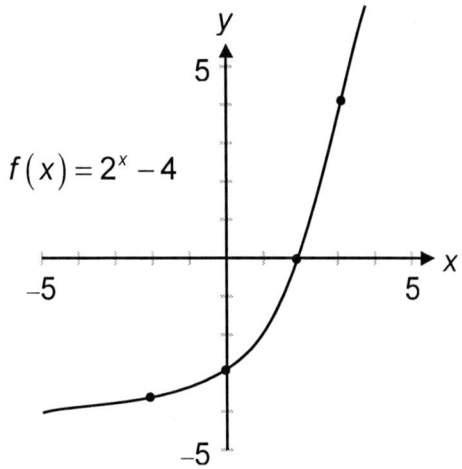

Note that the horizontal asymptote of any exponential function of the form $g(x) = a^x + b$ is $y = b$. Note also that the graph of such exponential functions is the graph of $h(x) = a^x$ shifted b units up or down. Finally, the graph of exponential functions of the form $g(x) = a^{(x+b)}$ is the graph of $h(x) = a^x$ shifted b units left or right.

Example: Sketch the graph of the function $f(x) = \ln(x^2 - 2)$.

In this case, the domain of the function is the set of values for which $x^2 - 2 > 0$, which requires that $x > \sqrt{2}$ or $x < -\sqrt{2}$. Note that the asymptotes are located at the x value for which the arguments of the natural logarithm is zero.

$$x^2 - 2 = 0$$
$$x^2 = 2$$
$$x = \pm\sqrt{2}$$

Also note that the function is symmetric about the y-axis. Some values for the function for $x > \sqrt{2}$ are shown in the table below.

x	f(x)
1.5	−1.39
1.75	0.06
2	0.69
3	1.95
4	2.64
5	3.14
6	3.53

The plot of the function is shown below, with the asymptotes displayed as dashed lines.

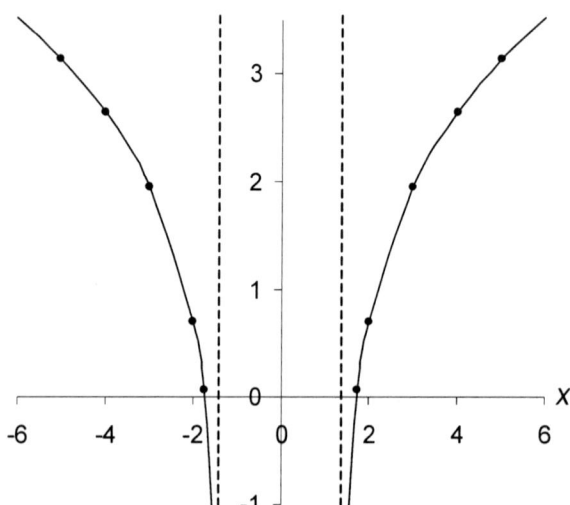

Example: Sketch the graph of the function $f(x) = e^{2x-1} + 1$.

The function f(x) has an asymptote at y = 1, since

$$\lim_{x \to -\infty} e^{2x-1} + 1 = e^{-2\infty-1} + 1 = 0 + 1 = 1$$

The y-intercept of the function is

$$f(0) = e^{2(0)-1} + 1 = e^{-1} + 1 = \frac{1}{e} + 1 \approx 1.368$$

The following table of values can be used to plot the function.

x	f(x)
−5	1.00
−4	1.00
−3	1.00
−2	1.01
−1	1.05
0	1.37
1	3.72
2	21.1

The graph of the function is shown below with the asymptote displayed as a dashed line.

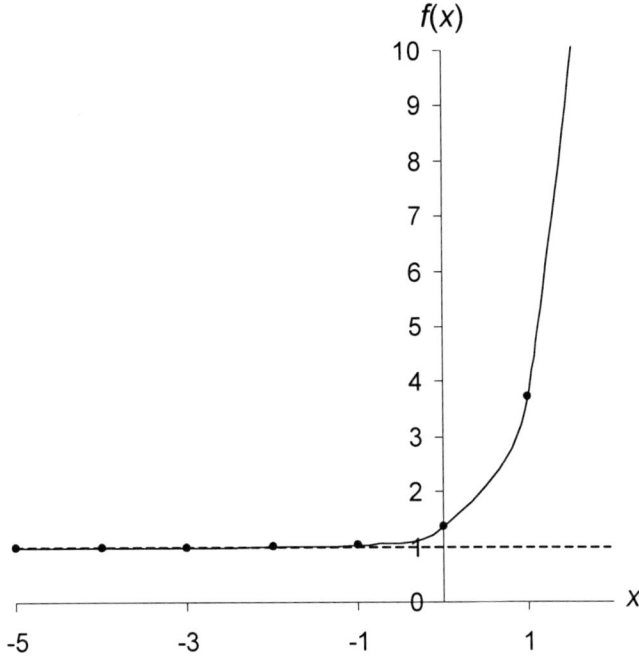

The beginning teacher understands the relationship between exponential and logarithmic functions and uses the laws and properties of exponents and logarithms to simplify expressions and solve problems.

Exponentials and **logarithms** are complementary. The general relationship for logarithmic and exponential functions is as follows.

$$y = \log_b x \quad \text{if and only if} \quad x = b^y$$

The relationship is as follows for the exponential base e and the natural logarithm (ln).

$$y = \ln x \quad \text{if and only if} \quad e^y = x$$

When changing common logarithms to exponential form,

$$y = \log_b x \quad \text{if and only if} \quad x = b^y$$

Natural logarithms can be changed to exponential form by using,

$$\log_e x = \ln x \quad \text{or} \quad \ln x = y \text{ can be written as } e^y = x$$

Example: Express in exponential form.

$\log_3 81 = 4$
$x = 81 \quad b = 3 \quad y = 4$ Identify values.
$81 = 3^4$ Rewrite in exponential form.

Example: Solve by writing in exponential form.

$\log_x 125 = 3$
$x^3 = 125$ Write in exponential form.
$x^3 = 5^3$ Write 125 in exponential form.
$x = 5$ Bases must be equal if exponents are equal.

Use a scientific calculator to solve:

Example: Find $\ln 72$.
$\ln 72 = 4.2767$ Use the $\ln x$ key to find natural logs.

Example: Find $\ln x = 4.2767$ Write in exponential form.
$e^{4.2767} = x$ Use the key (or 2nd ln x) to find x.
$x = 72.002439$ The small difference is due to rounding.

The following are properties that can be used to simplify expressions involving exponents.

Product Rule

1) $a^m \cdot a^n = a^{(m+n)}$

Example: $(3^4)(3^5) = 3^9$

2) $a^m \cdot a^m = (ab)^m$

Example: $(4^2)(5^2) = 20^2$

3) $(a^m)^n = a^{mn}$

Example: $(2^3)^2 = 2^6$

Quotient Rule

4) $\dfrac{a^m}{a^n} = a^{(m-n)}$

Example: $2^5 \div 2^3 = 2^2$

5) $a^{-m} = \dfrac{1}{a^m}$

Example: $2^{-2} = \dfrac{1}{2^2}$

Example: Simplify $\dfrac{3^5\left(3^{-2} + 3^{-3}\right)}{9}$

$\dfrac{3^5\left(3^{-2} + 3^{-3}\right)}{9} = \dfrac{3^5\left(3^{-2} + 3^{-3}\right)}{3^2} = 3^3\left(3^{-2} + 3^{-3}\right)$
$= 3^3 3^{-2} + 3^3 3^{-3} = 3^{3-2} + 3^{3-3} = 3 + 1 = 4$

Example: Simplify $\dfrac{3^2 \cdot 5^{-2} \cdot 2^5}{6^2 \cdot 5}$

$$\dfrac{3^2 \cdot 5^{-2} \cdot 2^5}{6^2 \cdot 5} = \dfrac{3^2 \cdot 5^{-2} \cdot 2^5}{3^2 \cdot 2^2 \cdot 5} = 5^{-2-1} \cdot 2^{5-2} = \dfrac{2^3}{5^3} = \dfrac{8}{125}$$

Note: Unless the negative sign is inside the parentheses and the exponent is outside the parentheses, the sign is not affected by the exponent.

Example:
$(^-2)^4 = (-2) \times (-2) \times (-2) \times (-2) = 16$
That is, -2 is multiplied by itself 4 times.
$^-2^4 = -(2 \times 2 \times 2 \times 2) = -16$
That is 2 is multiplied by itself 4 times and the answer is negated.

A radical may also be expressed using a rational exponent in the following way:

$$\sqrt[n]{a} = a^{\frac{1}{n}}$$

Example:
$$\sqrt{5} = 5^{\frac{1}{2}};\ \sqrt[5]{7} = 7^{\frac{1}{5}}$$

All the exponent laws discussed above also apply to rational exponents.

Example:
$$(\sqrt[5]{6})^3 = (6^{\frac{1}{5}})^3 = 6^{\frac{1}{5} \times 3} = 6^{\frac{3}{5}}$$

Example: Simplify $(-32)^{\frac{3}{5}} + 16^{\frac{3}{4}}$

$$(-32)^{\frac{3}{5}} + 16^{\frac{3}{4}} = (\sqrt[5]{-32})^3 + (\sqrt[4]{16})^3 = (-2)^3 + 2^3 = -8 + 8 = 0$$

The following properties of logarithms are helpful in solving equations.

Multiplication Property		$\log_b mn = \log_b m + \log_b n$
Quotient Property		$\log_b \dfrac{m}{n} = \log_b m - \log_b n$
Powers Property		$\log_b n^r = r \log_b n$
Equality Property		$\log_b n = \log_b m$ if and only if $n = m$.
Change of Base Formula		$\log_b n = \dfrac{\log n}{\log b}$
		$\log_b b^x = x$ and $b^{\log_b x} = x$

The beginning teacher uses a variety of representations and techniques (e.g., numerical methods, tables, graphs, analytic techniques, graphing calculators) to solve equations, inequalities, and systems involving exponential and logarithmic functions.

Solving equations involving exponentials or logarithms typically involves isolating the terms with the exponential or logarithmic function and using the inverse operation to "extract" the argument. For instance, given the following equation,

$$\ln f(x) = c$$

the function $f(x)$ can be determined by raising e to each side of the equation.

$$e^{\ln f(x)} = f(x) = e^c$$

Alternatively, if the function is in terms of an exponent e,

$$e^{f(x)} = c$$

solve by taking the natural logarithm of both sides.

$$\ln e^{f(x)} = f(x) = \ln c$$

Although these examples are in terms of e and the natural logarithm, the same logic applies to exponentials and logarithms involving different bases as well.

Example: Solve $\log_6(x-5) + \log_6 x = 2$.

$\log_6 x(x-5) = 2$ \hspace{1cm} Use product property.
$\log_6 x^2 - 5x = 2$ \hspace{1cm} Distribute.
$x^2 - 5x = 6^2$ \hspace{1cm} Write in exponential form.
$x^2 - 5x - 36 = 0$ \hspace{1cm} Solve quadratic equation.
$(x+4)(x-9) = 0$
$x = -4 \quad x = 9$

***Be sure to check results. Remember, x must be greater than zero in $\log x = y$.

$\log_6(x-5) + \log_6 x = 2$
$\log_6(-4-5) + \log_6(-4) = 2$ \hspace{1cm} Substitute the first answer
$\log_6(-9) + \log_6(-4) = 2$ \hspace{1cm} This is undefined (the argument is less than zero)

$\log_6(9-5) + \log_6 9 = 2$ \hspace{1cm} Substitute the second answer
$\log_6 4 + \log_6 9 = 2$
$\log_6(4)(9) = 2$ \hspace{1cm} Multiplication property
$\log_6 36 = 2$
$6^2 = 36$ \hspace{1cm} Write in exponential form.
$36 = 36$

Example: Find the roots of $f(x) = \ln(x^2 + 2) - 3$.

Set $f(x)$ equal to zero and simplify.

$f(x) = 0 = \ln(x^2 + 2) - 3$
$\ln(x^2 + 2) = 3$

Raise e to both sides of the equation and solve for x.

$$e^{\ln(x^2+2)} = x^2 + 2 = e^3$$
$$x^2 = e^3 - 2$$
$$x = \pm\sqrt{e^3 - 2} \approx \pm 4.252$$

Example: Find the roots of $f(x) = e^{-2x^2} - 1$.

Set f(x) equal to zero and simplify.

$$f(x) = 0 = e^{-2x^2} - 1$$
$$e^{-2x^2} = 1$$

Solve for x by taking the natural logarithm of both sides of the equation.

$$\ln e^{-2x^2} = -2x^2 = \ln 1 = 0$$

The solution is then x = 0.

Example: The water, w, in an open container is evaporating. The number of ounces remaining after h hours is shown in the table below:

Hours (h)	2	5	10	15	19	30
Water (w)	13	11	9	8.5	7.5	6.5

We construct a scatter plot for this data and find a logarithmic regression equation to model the data.

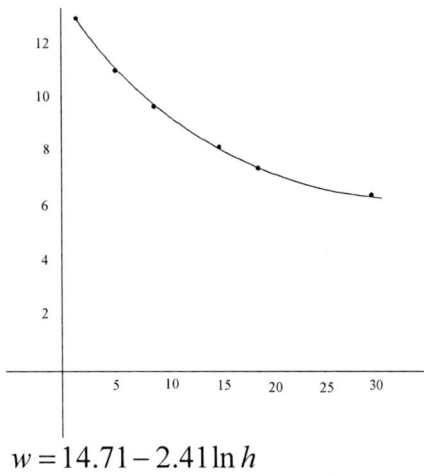

$$w = 14.71 - 2.41\ln h$$

Using this regression equation predict how many ounces of water are remaining in the container after 48 hours.

$$w = 14.71 - 2.41\ln(48)$$
$$w = 14.71 - 2.41(3.87)$$
$$w = 14.71 - 9.33$$
$$w = 5.4$$

Thus 5.4 ounces of water are remaining after 48 hours.

Using a graphing calculator with exponents: you can use the caret key ^ to evaluate powers. First enter the base, then ^ and then the exponent.

To use a table to evaluate 10^x and 10^{-x} for $x = 2, 3, 4, 5, 6$, by

1. Press "Y=" and enter "10^X" for Y1 and "10^(−)X" for Y2. Please note that the calculator key for a negative sign (−) is different from the key for subtraction -.
2. Press "2nd" "WINDOW" to choose the TABLE SETUP menu. Set the starting value, TBLStart to 0 and the set value, ⊔ Tbl to 1 so that the difference between each x-value in the table will be 1.
3. Press "2nd" "GRAPH" to view the table of values for 10^x and 10^{-x}.

The beginning teacher models and solves problems involving exponential growth and decay.

Exponential growth

A quantity which grows by a fixed percent at regular intervals (i.e., in proportion to the existing amount) demonstrates exponential growth. If a population has a constant birth rate through the years and is not affected by food or disease, it has exponential growth. The birth rate alone controls how fast the population grows exponentially.

Example: A population of a city is 20,000 and it increases at an annual rate of 20%. What will be the population of the city after 10 years?

The formula for the growth is $y = a(1 + r)^t$
where *a* is the initial amount, *r* is the growth rate, and *t* is the number of time intervals

Here,
> $a = 20000$
> $r = 20\% = 0.2$
> $t = 10$

Substituting the values,

$$\text{Population Growth} = y = 20000(1 + 0.2)^{10}$$
$$= 20000(1 + 0.2)^{10}$$
$$= 20000(1.2)^{10}$$
$$= 20000(6.19)$$
$$= 123800$$

So the population of the city after 10 years is 123,800.

What will be the population of the city after 50 years?

In 50 years,

$$\text{population} = 20000(1 + 0.2)^{50}$$
$$= 20000(1 + 0.2)^{50}$$
$$= 20000(1.2)^{50}$$
$$= 20000(9100.44)$$
$$= 182008800$$

The population after 50 years will be 182,008,800.

Another example of exponential growth is the growth of money through compound-interest. See later in this competency for examples.

Exponential Decay

Exponential decay is decrease by a fixed percent at regular intervals of time. Radioactive decay is an example of this.

Example: If 40 grams of Iodine has reduced to 20 grams in 6 days, what is the rate of decay?

The formula for exponential decay is $Q = ae^{rt}$ where Q is the amount of material at time t, a is the initial amount, r is the decay rate and t is the time period in days.

$Q = ae^{rt}$

$20 = 40e^{r(6)}$ Isolate e

$0.5 = e^{6r}$ Take ln of both sides

$\ln 0.5 = 6r$ Solve for r

$\dfrac{\ln 0.5}{6} = r$

$r = -0.1155$ Note: r is negative because it is decay

rate of decay is $0.1155 = 11.55\%$

Example: A 10-gram sample of Einsteinium-254 decays radioactively with a half-life of about 276 days. What is the remaining mass of Einsteinium-254 after 5 years (assume each year is 365 days).

Use the exponential decay formula derived above, where $m(t)$ is the mass of Einsteinium in the sample at time t.

$$m(t) = Ce^{kt}$$

The initial mass of Einsteinium-254 is 10 grams. Use this to find the value of C.

$$m(0) = 10g = Ce^{k(0)} = C$$

Thus, C is 10 grams. To find k, note that after 276 days, the amount of remaining Einsteinium-254 must be half the initial amount.

$$m(276d) = (10g)e^{k(276d)} = 5g$$

$$e^{k(276d)} = \dfrac{1}{2}$$

TEACHER CERTIFICATION STUDY GUIDE

Solve for k. Note that, to make the argument of the exponential dimensionless, the units of k should be inverse days.

$$\ln\left[e^{k(276d)}\right] = \ln\frac{1}{2}$$

$$(276d)k = \ln\frac{1}{2}$$

$$k = \frac{1}{276d}\ln\frac{1}{2}$$

$$k \approx -0.00251\frac{1}{d}$$

The complete expression for m is then the following, where t is in days.

$$m(t) \approx (10g)e^{-0.00251t}$$

Finally, calculate the amount of Einsteinium remaining after 5 years (1,825 days).

$$m(1825) \approx (10g)e^{-0.00251(1825)}$$

$$m(1825) \approx 0.102g$$

Thus, after 5 years, only about 0.102 grams of the initial sample remain.

The beginning teacher uses logarithmic scales (e.g., Richter, decibel) to describe phenomena and solve problems.

Many familiar measurement scales, including the Richter scale that measures the magnitude of an earthquake and the decibel scale used to measure the loudness of sound, are logarithmic: i.e., they transform a nonlinear scale that spans a wide range into a linear scale that is limited to a smaller range of numbers. Thus, phenomena that span a wide range of magnitudes can be conveniently represented and graphed using a narrow range of variable values. Values on the Richter or decibel scale do not have absolute units but are expressed in terms of a reference value.

On the Richter scale, the magnitude of an earthquake of intensity I is given by

$$R = \log_{10}\left(\frac{I}{I_0}\right)$$

where I_0 is the intensity of an earthquake of reference magnitude zero.

Example: What is the ratio of the intensities of two earthquakes of magnitudes 6.0 and 8.0 on the Richter scale?

The intensity of the earthquake of magnitude 6.0 is given by

$$I_1 = I_0 10^6$$

The intensity of the earthquake of magnitude 8.0 is given by

$$I_2 = I_0 10^8$$

Thus, the ratio of their intensities = $\dfrac{I_2}{I_1} = \dfrac{I_0 10^8}{I_0 10^6} = 100$.

Each unit on the Richter scale represents a factor 10. Thus, an earthquake that is 2 points higher on the scale is 100 times more powerful.

The decibel scale used to measure the loudness of sound is a similar relative logarithmic scale that uses the typical human hearing threshold intensity as the reference level.

$$\text{Loudness (dB)} = 10\log_{10}\left(\frac{I}{I_0}\right)$$

The factor of 10 transforms the original unit Bel into decibel, a unit 10 times smaller and of a more convenient size, since the ear can detect a change of 1 decibel.

Example: The threshold of human hearing has been measured at 10^{-12} Watts/m^2. Using this as the reference level, what is the loudness of a voice with a power of 6.4×10^{-6} Watts/m^2 in decibels?

$$\text{Loudness} = 10\log\left(\frac{6.4 \times 10^{-6}}{10^{-12}}\right) dB = 10\log(6.4 \times 10^{6}) dB = 68 dB$$

The beginning teacher uses exponential and logarithmic functions to model and solve problems involving the mathematics of finance (e.g., compound interest).

In finance, the **value of a sum of money** with compounded interest increases at a rate proportional to the original value. We use an exponential function to determine the growth of an investment accumulating compounded interest. The formula for calculating the value of an investment after a given compounding period is

$$A(t) = A_0(1 + \frac{i}{n})^{nt}.$$

A_0 is the principle, the original value of the investment. The rate of interest is i, the time in years is t, and the number of times the interest is compounded per year is n.

We can solve the compound interest formula for any of the variables by utilizing the properties of exponents and logarithms.

Example: Determine how long it will take $100 to amount to $1000 at 8% interest compounded 4 times annually.

In this problem we are given the principle ($A_0 = 100$), the final value ($A(t) = 1000$), the interest rate ($i = .08$), and the number of compounding periods per year ($n = 4$). Thus, we solve the compound interest formula for t. Solving for t involves the use of logarithms, the inverse function of exponents. To simplify calculations, we use the natural logarithm, ln.

$$A(t) = A_0(1 + \frac{i}{n})^{nt}$$

$$\frac{A(t)}{A_0} = (1 + \frac{i}{n})^{nt}$$

$$\ln \frac{A(t)}{A_0} = \ln(1 + \frac{i}{n})^{nt} \qquad \text{Take the ln of both sides.}$$

$$\ln \frac{A(t)}{A_0} = (nt)\ln(1 + \frac{i}{n})$$

Use the properties of logarithms with exponents.

$$\ln \frac{1000}{100} = (4t)\ln\left(1 + \frac{0.08}{4}\right)$$

$$t = \frac{\ln 10}{4(\ln 1.02)} = 29.07 \text{ years} \qquad \text{Substitute and solve for time } (t).$$

Example: Find the principle (A_0) that yields $500 with an interest rate of 7.5% compounded semiannually for 20 years.

In this problem $A(t) = 500$, the interest rate (i) is 0.075, $n = 2$, and $t = 20$. To find the principle value, we solve for A_0.

$$A(t) = A_0(1 + \frac{i}{n})^{nt}$$

$$500 = A_0(1 + \frac{0.075}{2})^{2(20)}$$

$$A_0 = \frac{500}{1.0375^{40}} = \$114.67 \qquad \text{Substitute and solve for } A_0.$$

Example: How long will it take $2,000 to triple if it is invested at 15% compounded continuously?

TEACHER CERTIFICATION STUDY GUIDE

Exponential growth formula for continuously compounding interest is

$$A = Pe^{rt}$$

where A is the amount of money in the account, P is Principal amount invested, r stands for rate of interest, t is the period in years, and e is the base of the natural log (an irrational number with value 2.71828…).

Substituting the given values in the formula:

$A = Pe^{rt}$
$6000 = 2000e^{0.15t}$
$3 = e^{0.15t}$
$\ln 3 = \ln e^{0.15t}$
$\ln 3 = 0.15t (\ln e)$
$\ln 3 = 0.15t$
$1.098 = 0.15t$
$t = \dfrac{1.098}{0.15} = 7.3$

Therefore, the amount of $2,000 is tripled in 7.3 years.

Example: Calculate the balance after 20 years for a savings account with an initial balance of $1,000 and an annual interest rate of 3%. Assume that the interest is compounded continuously.

In this case, the exponential growth formula can be used, where $A(t)$ is the amount of money in the account after t years. The equation is expressed using the standard "Pert" form.

$$A(t) = Pe^{rt}$$

The initial balance ($t = 0$) is $1,000.

$A(0) = \$1,000 = Pe^{r(0)} = P$
$A(t) = \$1,000 e^{rt}$

MATHEMATICS 8-12

The annual interest rate is 3%, or 0.03. Thus, after 1 year, the balance in the account must be 3% higher than the initial balance, or $1,030.

$$A(1) = \$1,030 = \$1,000e^{r(1)}$$
$$e^r = \frac{1030}{1000} = 1.03$$
$$\ln(e^r) = \ln 1.03$$
$$r = \ln 1.03 \approx 0.0296$$
$$A(t) = \$1,000e^{0.0296t}$$

The balance after 20 years is then

$$A(20) = \$1,000e^{0.0296(20)} = \$1807.60$$

The balance in the account after 20 years is then $1,807.60.

The beginning teacher uses the exponential function to model situations and solve problems in which the rate of change of a quantity is proportional to the current amount of the quantity [i.e., f '(x) = k f(x)].

Earlier portions of this competency presented examples of exponential growth and decay where the growth/decay happens at fixed intervals (e.g., compounding several times year) or continuously (radioactive decay or continuous compounding). The second type of problem corresponds to a situation where the rate of change of a quantity at any point in time is proportional to the current amount of the quantity. The differential equation describing this situation is

$$\frac{dy}{dx} = ky$$

To solve by applying separation of variables, move the occurrence of y to the left side of the equation and move dx to the right side.

$$\frac{1}{y}dy = k\,dx$$

Thus, the equation is now in the appropriate form:

$$A(y)dy = B(x)dx$$

Integrate both sides and simplify.

$$\int \frac{1}{y} dy = \int k\, dx + C$$
$$\ln y = kx + C$$
$$e^{\ln y} = e^{kx+C}$$
$$y = e^C e^{kx}$$

Note that e^C is simply a constant, so redefine the value C.

$$y = Ce^{kx}$$

This is the familiar solution that has been used to model and solve continuously varying problems in previous skills.

TEACHER CERTIFICATION STUDY GUIDE

Competency 009 The teacher understands trigonometric and circular functions, analyzes their algebraic and graphical properties, and uses them to model and solve problems.

Trigonometry can be used to systematically describe the relationship between circles, circular functions, right triangles, and trigonometric functions. This discussion reviews trigonometry in light of the unit circle, trigonometric functions and their inverses, and the solution of various problems involving these various concepts.

The beginning teacher analyzes the relationships among the unit circle in the coordinate plane, circular functions, and the trigonometric functions.

Trigonometric functions are typically defined in terms of right triangles: each trigonometric function corresponds to a ratio of certain sides of the triangle with respect to a particular angle. Thus, given the generic right triangle diagram below (not necessarily drawn to scale), the following functions can be specified.

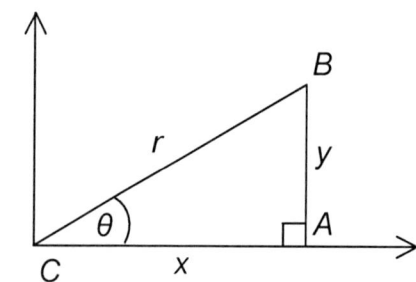

$$\sin\theta = \frac{y}{r} \qquad \csc\theta = \frac{r}{y}$$

$$\cos\theta = \frac{x}{r} \qquad \sec\theta = \frac{r}{x}$$

$$\tan\theta = \frac{y}{x} \qquad \cot\theta = \frac{x}{y}$$

With the aid of the **unit circle**, however, we see that trigonometric functions may be defined as circular functions or functions with values that repeat as the independent variable traverses a domain defined by successive turns of this hypothetical circle.

The unit circle is defined on the x-y plane and has a radius of 1.

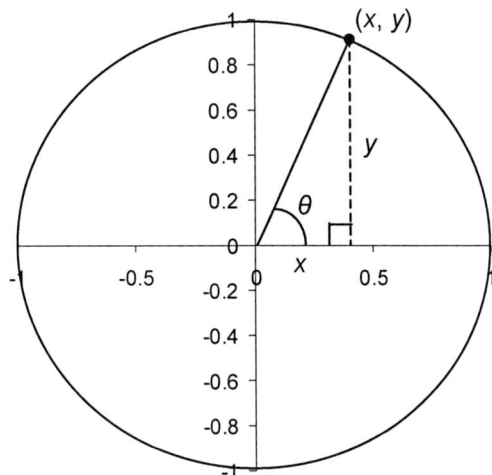

Notice that any given radius forms a right triangle with legs having lengths equal to the position of the point on the circle (x, y). Since the radius is equal to 1, the values of x and y are the following:

$x = \cos\theta$
$y = \sin\theta$

All the properties of trigonometric relationships for right triangles also apply in this case as well.

The argument of a trigonometric function is an angle that is typically expressed in either degrees or radians. A **degree** constitutes an angle corresponding to a sector that is 1/360th of a circle. Therefore, a circle has 360 degrees. A **radian**, on the other hand, is the angle corresponding to a sector of a circle where the arc length of the sector is equal to the radius of the circle. In the case of the unit circle (a circle of radius 1), the circumference is 2π. Thus, there are 2π radians in a circle. Conversion between degrees and radians is a simple matter of using the ratio between the total degrees in a circle and the total radians in a circle.

$$(\text{degrees}) = \frac{180}{\pi} \times (\text{radians})$$

$$(\text{radians}) = \frac{\pi}{180} \times (\text{degrees})$$

The beginning teacher recognizes and translates among various representations (e.g., written, numerical, tabular, graphical, algebraic) of trigonometric functions and their inverses.

The trigonometric functions sine, cosine and tangent (and their reciprocals) are **periodic functions**. The values of periodic functions repeat on regular intervals. The period, amplitude and phase shift are critical properties of periodic functions that can be determined by observation of the graph or by detailed study of the functions themselves.

The **period** is the smallest domain containing the complete cycle of the function. For example, the period of a sine or cosine function is the distance between the adjacent peaks or troughs of the graph. The **amplitude** of a function is half the distance between the maximum and minimum values of the function. The **phase shift** is the amount of horizontal displacement of a function from a given reference position.

Below is a generic sinusoidal graph with the period and amplitude labeled.

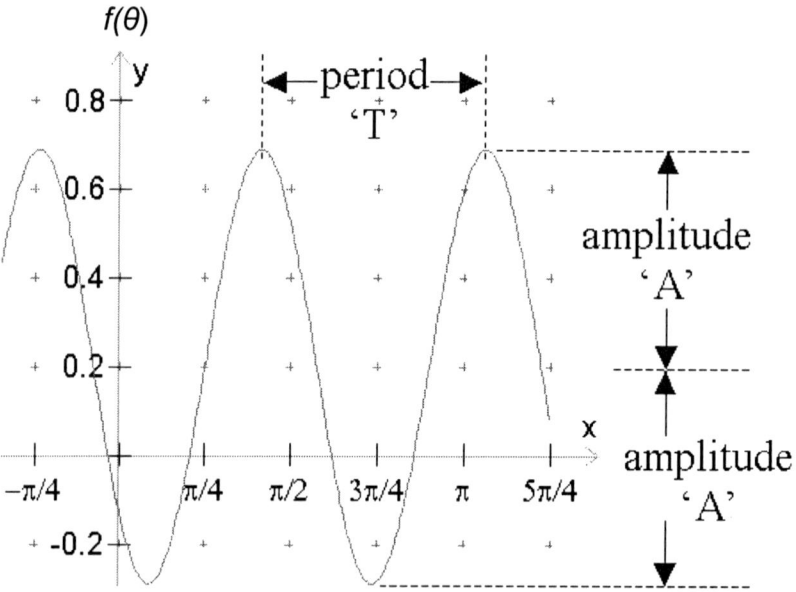

The period and amplitude for the three basic trigonometric functions are provided in the table below.

Function	Period (radians)	Amplitude
$\sin \theta$	2π	1
$\cos \theta$	2π	1
$\tan \theta$	π	Undefined

Below are the graphs of the basic trigonometric functions, (a) $y = \sin x$; (b) $y = \cos x$; and (c) $y = \tan x$.

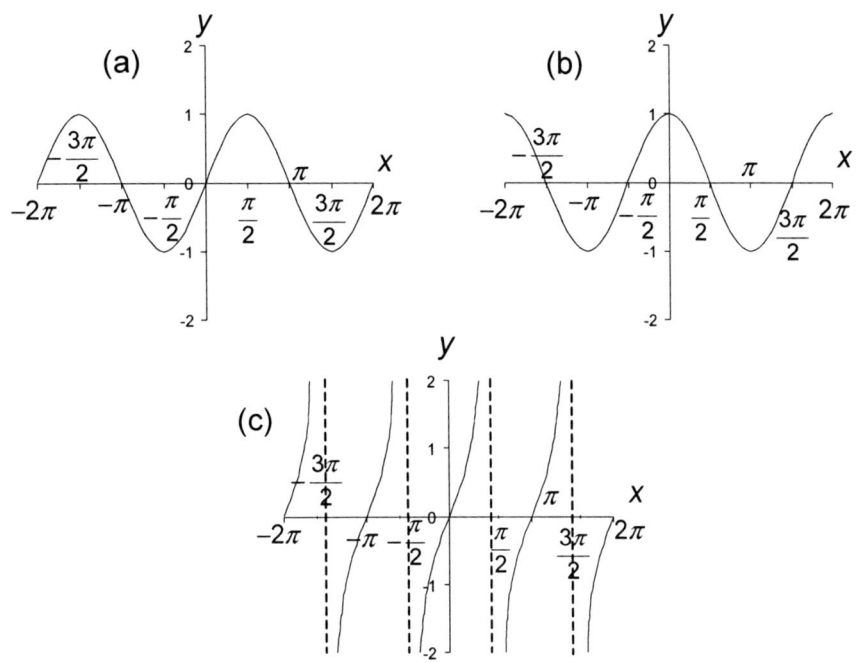

Note that the graph of the tangent function has asymptotes at $x = \dfrac{2n-1}{2}\pi$, where $n = 0, \pm 1, \pm 2, \pm 3, \ldots$.

The graphs of the reciprocal trigonometric functions are shown below, with (a) $y = \csc x$; (b) $y = \sec x$; and (c) $y = \cot x$.

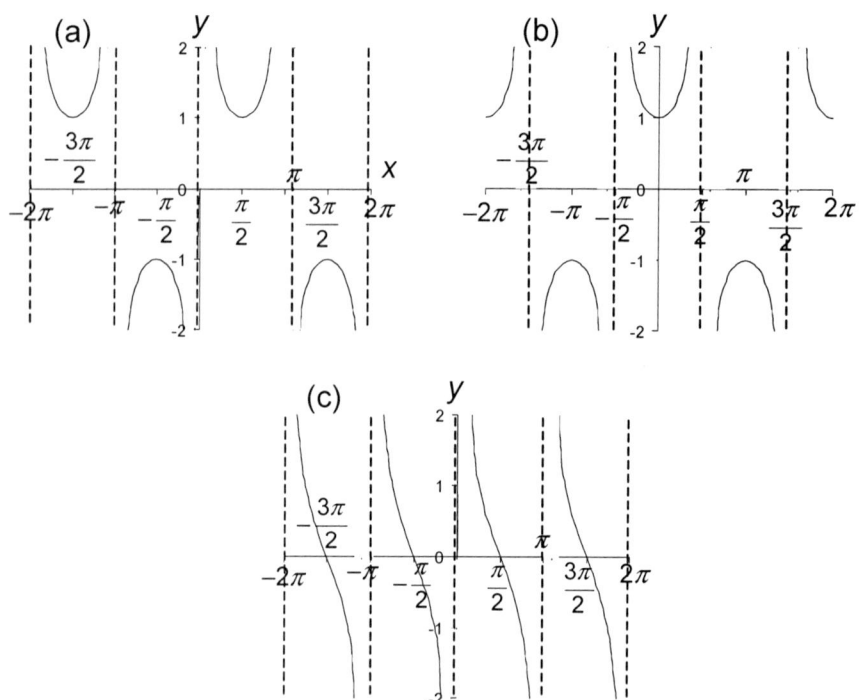

The phase and amplitude for the three reciprocal trigonometric functions are provided in the table below.

Function	Period (radians)	Amplitude
csc θ	2π	Undefined
sec θ	2π	Undefined
cot θ	π	Undefined

Inverse Trigonometric Functions

The inverse sine function of x is written as arcsin x or $\sin^{-1} x$ and is the angle for which the sine is x; i.e., sin(arcsin x) = x. Since the sine function is periodic, many values of arcsin x correspond to a particular x. In order to define arcsin as a function, therefore, its range needs to be restricted.

The function **y = arcsin x** has a domain [-1,1] and range $\left[-\frac{\pi}{2}, \frac{\pi}{2}\right]$.

In some books, a restricted inverse function is denoted by a capitalized beginning letter such as in Sin^{-1} or arctan. The arcsin function is shown below.

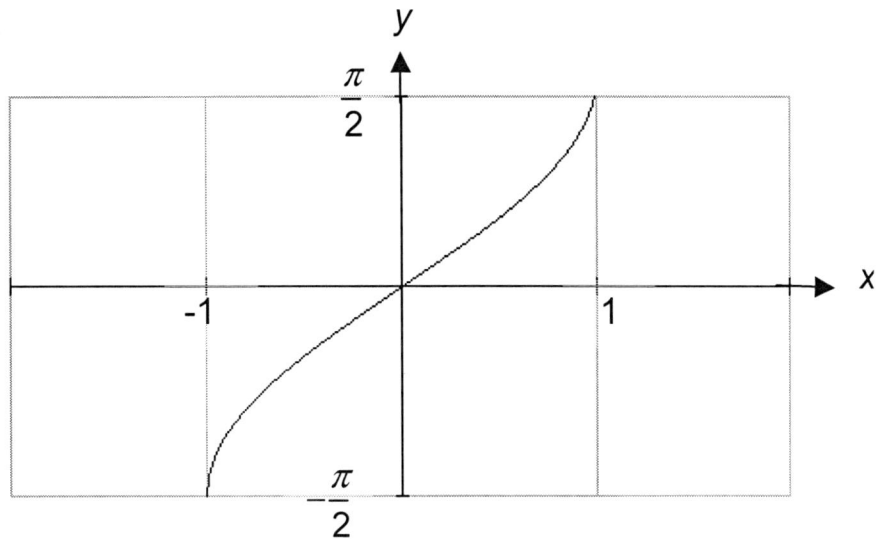

The inverse cosine and tangent functions are defined in the same way: cos(arccos x) = x; tan(arctan x) = x.

The function **y = arccos x** has a domain [-1,1] and range $[0,\pi]$. The graph of this function is shown below.

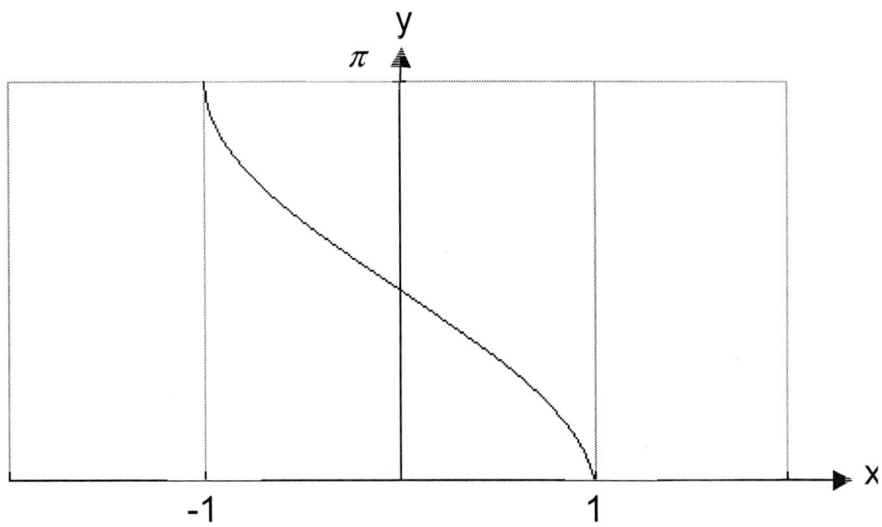

The function **y = arctan x** has a domain $[-\infty,+\infty]$ and range $\left[-\frac{\pi}{2},\frac{\pi}{2}\right]$. The plot of the function is shown below.

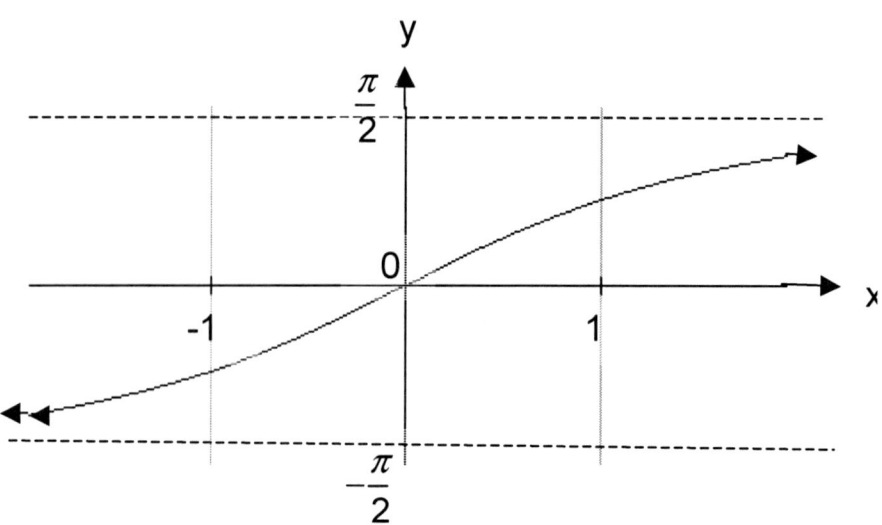

Example: Evaluate the following: (i) sin⁻¹(0) and (ii) arccos(-1)

(i) sin(sin⁻¹(0)) = 0.
The value of the inverse sine function must lie in the range $\left[-\frac{\pi}{2},\frac{\pi}{2}\right]$. Since 0 is the only argument in the range $\left[-\frac{\pi}{2},\frac{\pi}{2}\right]$ for which the sine function is zero, sin⁻¹(0) = 0.

(ii) $\cos(\arccos(-1)) = -1$
The value of the inverse cosine function must lie in the range $[0,\pi]$. π is the only argument for which the cosine function is equal to -1 in the range $[0,\pi]$.
Hence, $\arccos(-1) = \pi$.

TEACHER CERTIFICATION STUDY GUIDE

The beginning teacher recognizes and uses connections among significant properties (e.g., zeros, axes of symmetry, local extrema) and characteristics (e.g., amplitude, frequency, phase shift) of a trigonometric function, the graph of the function, and the function's symbolic representation.

Graphing a trigonometric function by hand typically requires a calculator for determining the value of the function for various angles. Nevertheless, simple functions can often be graphed by simply determining the amplitude, period and phase shift. Once these parameters are known, the graph can be sketched approximately. The amplitude of a simple sine or cosine function is simply the multiplicative constant (or function) associated with the trigonometric function. Thus, 2cos x, for instance, has an amplitude of 2. The phase shift is typically just a constant added to the argument of the function. For instance, sin(x + 1) includes a phase shift of 1. A positive phase shift constant indicates that the graph of the function is shifted to the left; a negative phase shift indicates that the graph is shifted to the right.

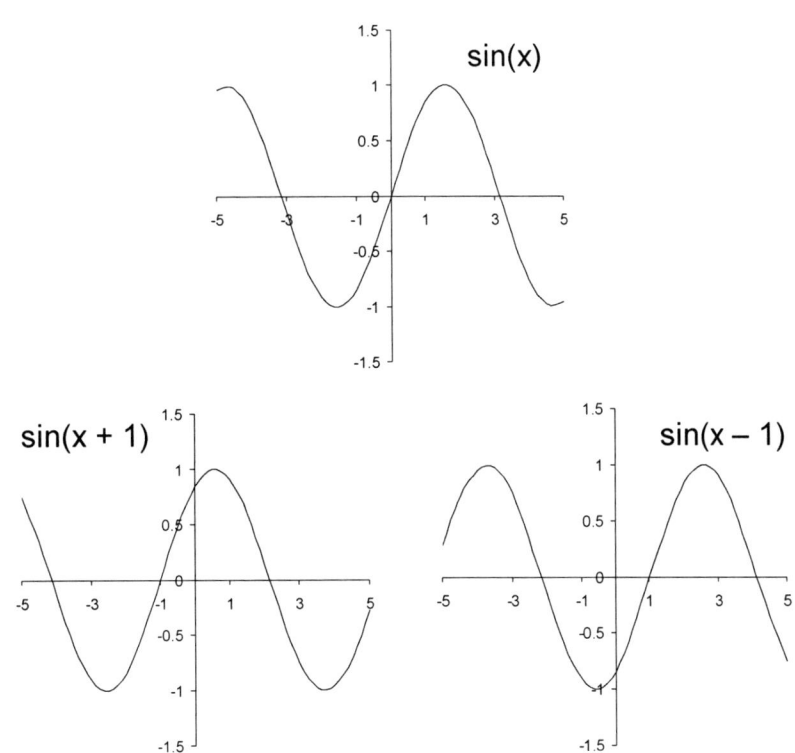

Example: Sketch the graph of the function $f(x) = 4\sin\left(2x + \dfrac{\pi}{2}\right)$.

Notice, first, that the amplitude of the function is 4. Since there is no constant term added to the sine function, the function is centered on the *x*-axis. Find crucial points on the graph by setting *f* equal to zero and solving for *x* to find the roots.

$$f(x) = 0 = 4\sin\left(2x + \dfrac{\pi}{2}\right)$$

$$\sin\left(2x + \dfrac{\pi}{2}\right) = 0$$

$$2x + \dfrac{\pi}{2} = n\pi$$

In the above expression, *n* is an integer.

$$2x = \left(n - \dfrac{1}{2}\right)\pi$$

$$x = \left(n - \dfrac{1}{2}\right)\dfrac{\pi}{2}$$

So, the roots of the function are at

$$x = \pm\dfrac{\pi}{4}, \pm\dfrac{3\pi}{4}, \pm\dfrac{5\pi}{4}, \ldots$$

The maxima and minima of the function are halfway between successive roots. Determine the location of a maximum by testing the function. Try *x* = 0.

$$f(0) = 4\sin\left(2[0] + \dfrac{\pi}{2}\right) = 4\sin\left(\dfrac{\pi}{2}\right) = 4$$

Thus, *f* is maximized at *x* = 4. The function can then be sketched.

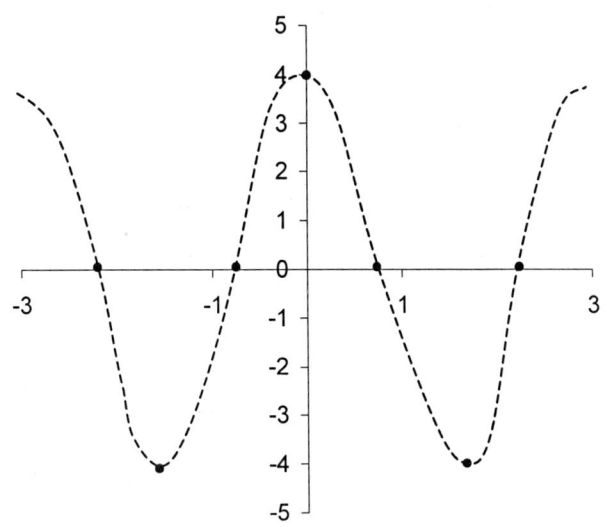

The beginning teacher understands the relationships between trigonometric functions and their inverses and uses these relationships to solve problems.

There are a range of **trigonometric identities** or relationships between different trigonometric functions. These include reciprocal and Pythagorean identities listed below.

Reciprocal identities:

$$\sin x = \frac{1}{\csc x} \qquad \sin x \csc x = 1 \qquad \csc x = \frac{1}{\sin x}$$

$$\cos x = \frac{1}{\sec x} \qquad \cos x \sec x = 1 \qquad \sec x = \frac{1}{\cos x}$$

$$\tan x = \frac{1}{\cot x} \qquad \tan x \cot x = 1 \qquad \cot x = \frac{1}{\tan x}$$

$$\tan x = \frac{\sin x}{\cos x} \qquad \qquad \cot x = \frac{\cos x}{\sin x}$$

Pythagorean identities:

$$\sin^2 x + \cos^2 x = 1 \qquad 1 + \tan^2 x = \sec^2 x \qquad 1 + \cot^2 x = \csc^2 x$$

Example: Prove that $\sin^2 x + \cos^2 x = 1$.

Use the definitions of the sine and cosine functions from right triangle trigonometry.

$$\left(\frac{y}{r}\right)^2 + \left(\frac{x}{r}\right)^2 = 1$$

$$\frac{x^2 + y^2}{r^2} = 1$$

But the numerator of the above fraction, by the Pythagorean theorem, is simply r^2. Then:

$$\frac{r^2}{r^2} = 1$$

The identity has been proven.

Example: Prove that $\cot x + \tan x = \csc x \sec x$.

Use the reciprocal identities to convert the left side of the equation to sines and cosines. Then combine terms using a common denominator.

$$\frac{\cos x}{\sin x} + \frac{\sin x}{\cos x}$$

$$\frac{\cos^2 x}{\sin x \cos x} + \frac{\sin^2 x}{\sin x \cos x} = \frac{\sin^2 x + \cos^2 x}{\sin x \cos x} = \frac{1}{\sin x \cos x}$$

Finally, convert the expression using the reciprocal identities.

$$\frac{1}{\sin x \cos x} = \csc x \sec x$$

The identity is then proven.

The following are some **identities for the inverse trigonometric functions**:

$$\csc^{-1}(x) = \sin^{-1}(1/x) \text{ for } |x| \geq 1$$

$$\sec^{-1}(x) = \cos^{-1}(1/x) \text{ for } |x| \geq 1$$

$$\cot^{-1}(x) = \begin{cases} \tan^{-1}(1/x) & \text{for } x > 0 \\ \tan^{-1}(1/x) + \pi & \text{for } x < 0 \\ \pi/2 & \text{for } x = 0 \end{cases}$$

$$\sin^{-1} x = \cos^{-1}(\sqrt{1-x^2}) \qquad \cos^{-1} x = \sin^{-1}(\sqrt{1-x^2})$$

$$\tan^{-1} x = \cos^{-1}\left(\frac{1}{\sqrt{1+x^2}}\right) \qquad \cos^{-1} x = \tan^{-1}\left(\frac{\sqrt{1-x^2}}{x}\right)$$

$$\tan^{-1} x = \sin^{-1}\left(\frac{x}{\sqrt{1+x^2}}\right) \qquad \sin^{-1} x = \tan^{-1}\left(\frac{x}{\sqrt{1-x^2}}\right)$$

Example: Simplify the expression $\cos(\arcsin x) + \sin(\arccos x)$

$$\arcsin x = \arccos(\sqrt{1-x^2}) \quad \text{identity}$$
$$\Rightarrow \cos(\arcsin x) = \sqrt{1-x^2}$$
$$\arccos x = \arcsin(\sqrt{1-x^2}) \quad \text{identity}$$
$$\Rightarrow \sin(\arccos x) = \sqrt{1-x^2}$$

Hence, $\cos(\arcsin x) + \sin(\arccos x) = \sqrt{1-x^2} + \sqrt{1-x^2} = 2\sqrt{1-x^2}$

Example: Using the identities given above, prove the identity

$$\sin^{-1} x + \cos^{-1} x = \frac{\pi}{2}$$

Since $\sin\left(\dfrac{\pi}{2}\right) = 1$, the identity may be proven by showing that

$$\sin\left(\sin^{-1}x + \cos^{-1}x\right) = 1$$

$$\sin\left(\sin^{-1}x + \cos^{-1}x\right) = \sin(\sin^{-1}x)\cos(\cos^{-1}x)$$
$$+ \cos(\sin^{-1}x)\sin(\cos^{-1}x) \quad \text{sine sum formula}$$
$$= x \cdot x + \sqrt{1-x^2}\sqrt{1-x^2} \quad \text{inverse identities}$$
$$= x^2 + 1 - x^2 = 1$$

Other similar identities include the following:

$$\tan^{-1}x + \cot^{-1}x = \pi/2$$
$$\sec^{-1}x + \csc^{-1}x = \pi/2$$

The beginning teacher uses trigonometric identities to simplify expressions and solve equations.

Unlike trigonometric identities that are true for all values of the defined variable, trigonometric equations and inequalities are true for some, but not all, of the values of the variable. Most often, trigonometric equations are solved for values between 0 and 360 degrees or 0 and 2π radians. For inequalities, the solution is often a set of intervals, since trigonometric functions are periodic. Solving trigonometric problems is largely the same as solving algebraic equations. Care must be taken, however, due to the periodic nature of trigonometric functions. This often yields multiple (or an infinite number of) solutions.

Trigonometric identities, including sum and difference formulas, are often indispensable in the problem-solving process. These identities allow many complicated functions to be simplified to forms that are more easily managed algebraically.
Some algebraic operation, such as squaring both sides of an equation, will yield extraneous answers. Avoid incorrect solutions by remembering to check all solutions to be sure they satisfy the original equation.

Example: Solve the following equation for x: $\cos x = 1 - \sin x$, where $0° \leq x \leq 360°$.

Start by squaring both sides of the equation.

$$\cos^2 x = (1-\sin x)^2 = 1 - 2\sin x + \sin^2 x$$

Substitute using the Pythagorean identity to replace the cosine term.

$$1 - \sin^2 x = 1 - 2\sin x + \sin^2 x$$

Simplify the results.

$$2\sin^2 x - 2\sin x = 0$$
$$\sin x (\sin x - 1) = 0$$

There are two possible solutions to the equation:

$$\sin x = 0 \quad \text{and} \quad \sin x = 1$$
$$x = 0°, 180° \qquad\qquad x = 90°$$

Thus, the apparent solutions to the problem are $x = 0°$, $90°$ and $180°$. By checking each solution, however, it is found that $x = 180°$ is not a legitimate solution and must be discarded. The actual solutions to the equation are thus $x = 0°$ and $90°$.

Example: Solve the following equation: $\cos^2 x = \sin^2 x$ for $0 \le x \le 2\pi$.

First, use the Pythagorean identity to convert either the cosine or sine term.

$$\cos^2 x = 1 - \cos^2 x$$

Simplify the results.

$$2\cos^2 x = 1$$
$$\cos^2 x = \frac{1}{2}$$
$$\cos x = \pm \frac{1}{\sqrt{2}}$$

Familiarity with the properties of trigonometric functions should lead to the realization that this corresponds to odd integer multiples of $\frac{\pi}{4}$. Alternatively, a calculator can be used to calculate the inverse function. (A detailed review of inverse trigonometric functions is provided earlier in this discussion.)

$$x = \arccos\left(\pm\frac{1}{\sqrt{2}}\right)$$

In either case, the solution is the following:

$$x = \frac{\pi}{4}, \frac{3\pi}{4}, \frac{5\pi}{4}, \frac{7\pi}{4}$$

Example: Solve for x: $\sin x \geq 0$.

Solving a trigonometric inequality involves the same general process as is involved in solving any other inequality. In this case, however, the set of solutions includes an infinite number of intervals, rather than a single interval as is the case for some non-periodic functions. First, replace the inequality symbol with an equal sign. Solve using the inverse function.

$$\sin x = 0$$
$$\arcsin[\sin x] = \arcsin[0]$$
$$x = \arcsin[0]$$

The solutions for x are the following.

$$x = n\pi \qquad n = 0, \pm 1, \pm 2, \pm 3, \ldots$$

These solutions are the points at which the sine function crosses the x-axis. Thus, some set of intervals bounded by these solutions is the set of solutions for the inequality. It is apparent that the sine function is greater than zero for x between 0 and π, and negative for x between π and 2π. This pattern then repeats. Thus, sin x is greater than zero between $2n\pi$ and $(2n + 1)\pi$ for n = 0, ±1, ±2, ±3,….

$$\sin x \geq 0 \text{ for } 2n\pi \leq x \leq (2n+1)\pi \text{ where } n = 0, \pm 1, \pm 2, \pm 3, \ldots$$

Note that the endpoints of the intervals are included in the solution set. The validity of this solution can be confirmed by looking at a graph of sin *x*.

The beginning teacher models and solves a variety of problems (e.g., analyzing periodic phenomena) using trigonometric functions.

Since trigonometric functions, as seen earlier, can be related to the properties of a circle, they have a cyclical (or **periodic**) behavior that is suited to modeling periodic phenomena. For instance, the height of a point marked on a wheel, tracked as the wheel rolls, can be modeled using trigonometric functions. The key to modeling such periodic phenomena is identification of the amplitude, period and phase (if necessary) of the phenomenon. This information allows expression of some parameter of the phenomenon as an equation involving a trigonometric function.

Example: Write an equation for the height of sea waves whose crests pass every 5 seconds and whose peaks are 10 feet above their troughs.

In this case, phase information is not needed since there is no fixed position to form a point of comparison (or origin). Thus, either a sine or cosine function can be used. The amplitude of the waves are half the distance between the peaks and troughs, or 5 feet. The period is simply the inverse of the time between peaks: 0.2 sec^{-1}. The equation for the height *h* of the waves with respect to time *t* is then the following:

$$h = 5\sin(0.2t) \text{ feet}$$

where *t* is measured in seconds.

Example: Find an equation to model the length of a 6-foot man's shadow with respect to the angle of the sun in the sky measured relative to the plane of the ground at the man's feet. Assume that the sun is directly overhead at midday.

It is helpful to first draw a diagram of the situation. Since it is assumed that the sun is directly overhead at midday, the situation can be drawn in a single plane.

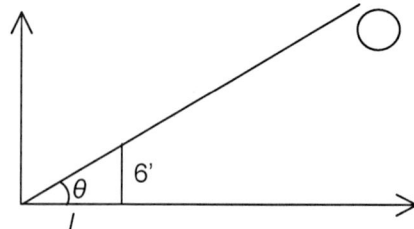

The length l of the shadow can be calculated using the tangent function for the angle θ of the sun in the sky. Use the definition of the tangent to find the function.

$$\tan\theta = \frac{6'}{l}$$

The resulting function can then be written in the following manner:

$$l(\theta) = \frac{6'}{\tan\theta}$$

To check the result, try some simple cases. For instance, when the sun is overhead, $\theta = 90°$.

$$l(90°) = \frac{6'}{\tan 90°} = \frac{6'}{\infty} = 0'$$

Thus, there is no shadow cast when the sun is overhead. Another example is when the sun approaches the horizon. In this case, the shadow should get very long. Try $\theta = 1°$.

$$l(1°) = \frac{6'}{\tan 1°} = \frac{6'}{0.175} = 342.9'$$

This answer is at least somewhat intuitive. These checks provide confidence in the equation.

The beginning teacher uses graphing calculators to analyze and solve problems involving trigonometric functions.

Graphing calculators are handheld devices with a display area that allows users to view the results of mathematical computations graphically as well as in numerical and symbolic terms. Online and computer-based graphing calculators are available for those who do not have access to a handheld one.

Most graphing calculators can evaluate and display graphs of mathematical expressions containing square root, exponential, logarithm and standard trigonometric functions. For example, to view the graph of the function sin(x) in most calculators, simply type in y = sin(x) and select GRAPH.

In order to solve an equation with a graphing calculator, both sides of the equation must be expressed in the form y = f(x). For instance, to plot the equation 3 cosx − 2 sinx = 1, one must plot the two graphs y1=3 cosx − 2 sinx and y2 = 1. The intersection points provide the solution of the equation. When the graph is displayed, the user can zoom in on any area of the graph. The calculator also displays the coordinates of selected points. One must be careful to note that the calculator may not display exact values. For instance, the exact solution (1, 2) may be displayed as (0.9999995, 1.9999995).

Note: When using a graphing calculator to evaluate trigonometric functions, one must be careful to **set the mode correctly to radians or degrees**.

Competency 010 The teacher understands and solves problems using differential and integral calculus.

This competency reviews limits, the relationship of limits and continuity as well as the fundamental concepts of differential and integral calculus. The modeling and solving of word problems (including through the use of technology) using differential and integral calculus is also discussed.

The beginning teacher understands the concept of limit and the relationship between limits and continuity.

A **limit** is the value that a function approaches as a variable of the function approaches a certain value. Thus, the limit L of a function $f(x)$ as x approaches some value c is written as follows:

$$\lim_{x \to c} f(x) = L$$

This concept can be understood through graphical illustrations, as with the following example. The graph shows a plot of $y = 2x$. A limit in this case might be the limit of y as x approaches 3. Graphically, this involves tracing the plot of the function to the point at $x = 3$.

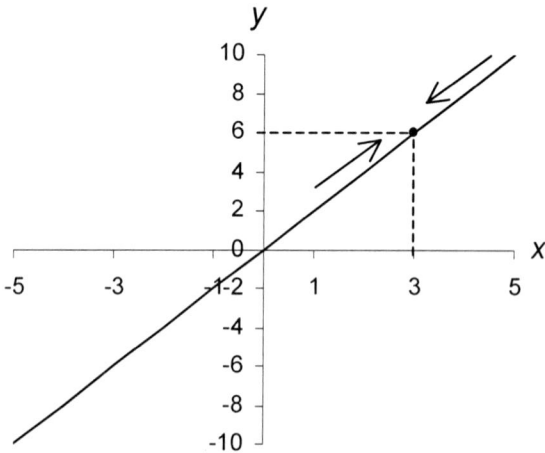

The limit in this case is clearly 6. The limit can be considered either from the right or left (that is, considering x decreasing toward 3 or x increasing toward 3). In the simple example above, the limit is the same from either direction, but, in some cases, the limit may be different from different directions or the function may not even exist on one side of the limit.

When evaluating a limit, especially when dealing with complicated functions, the following two steps should be taken.

1. Factor the expression completely and cancel all common factors in fractions.

2. Substitute the number to which the variable is approaching. In most cases this produces the value of the limit.

For simple functions, evaluating the limit may simply involve calculating $f(c)$ (according to the limit definition given above), but, for other functions, a more subtle approach may be required. In some instances, such as where the limit of the function differs depending on whether c is approached from the left or right, or where the limit of the function is positive or negative infinity, the limit of $f(x)$ at $x = c$ does not exist.

Evaluation of limits often deals with infinity. If, after simplification, the limit involves positive or negative infinity in the numerator (even if the fraction is simply the numerator divided by unity), then the limit does not exist (sometimes it is stated that "the limit is infinity", but this is not necessarily meaningful). If the denominator includes positive or negative infinity, then the limit is zero (since as the denominator of a fraction gets arbitrarily large, the fraction approaches zero).

Example: Evaluate the following limit: $\lim\limits_{x \to \infty} \dfrac{2x^2}{x^5}$.

$\lim\limits_{x \to \infty} \dfrac{2}{x^3}$ Cancel common factors.

$\dfrac{2}{\infty^3}$ Substitute.

0 Calculate limit.

As mentioned previously, simplification of the function can, in some cases, eliminate the need to deal with infinity.

Example: Evaluate the following limit: $\lim_{x \to {}^-3} \dfrac{x^2 + 5x + 6}{x + 3} + 4x$.

$\lim_{x \to {}^-3} \dfrac{(x+3)(x+2)}{(x+3)} + 4x$ Factor the numerator.

$\lim_{x \to {}^-3} (x+2) + 4x$ Cancel common factors.

$({}^-3 + 2) + 4({}^-3)$ Substitute ${}^-3$ for x.

${}^-1 + {}^-12$ Simplify.

${}^-13$

If this example is approached using solely substitution, the denominator becomes infinity and the limit becomes 4(–3) = –12. Simplification avoids such errors and leads to the correct answer in this case.

The above approach to the evaluation of limits sometimes results in an answer of either $\dfrac{0}{0}$ or $\dfrac{\infty}{\infty}$. In such cases, use **L'Hopital's rule** to find the limit. L'Hopital's rule states that the limit in such cases can be determined by taking the derivative of the numerator of the function and the derivative of the denominator of the function, and then calculating the limit of the resulting quotient.

Example: Evaluate the following limit: $\lim_{x \to \infty} \dfrac{3x - 1}{x^2 + 2x + 3}$.

$\dfrac{3\infty - 1}{\infty^2 + 2\infty + 3}$ Substitute ∞ for x.

$\dfrac{\infty}{\infty}$ The result is not defined.

$\lim_{x \to \infty} \dfrac{3}{2x + 2}$ Use L'Hopital's rule

$\dfrac{3}{2\infty + 2}$ Substitute.

$\dfrac{3}{\infty}$ Simplify.

0 Evaluate limit.

Example: Evaluate the following limit: $\lim\limits_{x \to 1} \dfrac{\ln x}{x-1}$.

$\dfrac{\ln 1}{1-1}$ Substitute 1 for x.

$\dfrac{0}{0}$ Simplify.

$\lim\limits_{x \to 1} \dfrac{\left(\dfrac{1}{x}\right)}{1} = \lim\limits_{x \to 1} \dfrac{1}{x}$ Use L'Hopital's rule

$\dfrac{1}{1} = 1$ Evaluate limit.

Thus, L'Hopital's rule can be used to find the limit in cases where direct substitution yields an undefined result.

Continuity

The continuity of a function is easily understood graphically as the absence of any missing points or any breaks in the plot of the function. A more rigorous definition can be formulated, however. A function $f(x)$ is continuous at a point c if all of the following apply:

1. The function $f(x)$ is defined at $x = c$.
2. The limit $\lim\limits_{x \to c} f(x)$ exists.
3. The limit can be found by substitution: $\lim\limits_{x \to c} f(x) = f(c)$.

A function can then be called **continuous** for an open interval (a, b) if the above definition applies to the function for every point c in the interval. The function is also continuous at the points $x = a$ and $x = b$ if $\lim\limits_{x \to a^+} f(x) = f(a)$ and $\lim\limits_{x \to b^-} f(x) = f(b)$ both exist. (The +/− notation simply signifies approaching the limiting value from either the right or left, respectively.) If both of these conditions apply, then the function is continuous for the closed interval $[a, b]$.

Example: Determine whether the function $\ln x$ is continuous over the closed interval $[0, 1]$.

One approach is to look at a plot of the function.

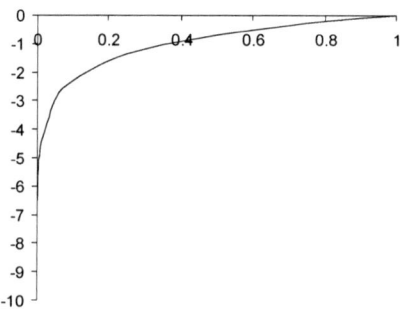

The limit of the function as x approaches zero from the right does not exist. Algebraically, this can be seen by noting that the equation $\ln x = L$ is the same as the equation $e^L = x$. For $x = 0$, the only possible value of L that satisfies this equation is $-\infty$. The limit does not exist, therefore, and the function is not continuous on the closed interval [0, 1].

Example: Determine if the function shown in the graph is continuous at $x = 2$.

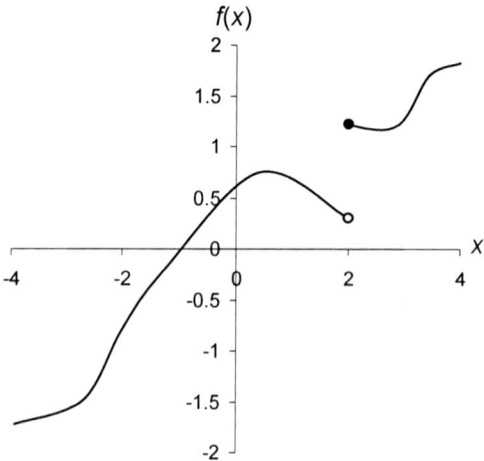

By inspection, it can be seen that the limits of the function as x approaches 2 from the right and left are not equal.

$$\lim_{x \to 2^+} f(x) \neq \lim_{x \to 2^-} f(x)$$

As a result, the function does not meet all of the criteria for continuity at $x = 2$ and is therefore discontinuous.

Example: Determine whether the following function is continuous at

$x = 1$: $f(x) = \begin{cases} x^2 & x \neq 1 \\ 0 & x = 1 \end{cases}$.

The plot of this piecewise function is shown below.

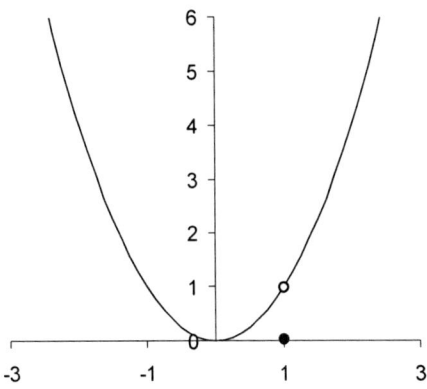

In this case, the function is defined at $x = 1$, and the limits of $f(x)$ as x approaches 1 from both the right and from the left are both equal to 1. Nevertheless, it is not the case that $\lim_{x \to 1} f(x) = f(1)$, since $\lim_{x \to 1} f(x) = 1$ and $f(1) = 0$. Thus, the function is not continuous at $x = 1$.

The beginning teacher relates the concept of average rate of change to the slope of the secant line and the concept of instantaneous rate of change to the slope of the tangent line.

The rate of change of a function is equivalent to its slope (change in the vertical direction for a given change in the horizontal direction). For nonlinear functions, the slope is continuously changing over the domain. The rate of change of a function can be found as an average rate of change over some portion of the domain (a difference quotient), or it can be found at a particular domain value (derivative).

The **difference quotient** is the average rate of change over an interval. For a function f, the **difference quotient** is represented by the formula:

$$\frac{f(x+h) - f(x)}{h}.$$

This formula computes the **slope of the secant line** through two points on the graph of f. These are the points with x coordinates x and x + h.

Example: Find the difference quotient for the function $f(x) = 2x^2 + 3x - 5$.

Use the difference quotient formula and simplify the results.

$$\frac{f(x+h) - f(x)}{h} = \frac{2(x+h)^2 + 3(x+h) - 5 - (2x^2 + 3x - 5)}{h}$$

$$= \frac{2(x^2 + 2hx + h^2) + 3x + 3h - 5 - 2x^2 - 3x + 5}{h}$$

$$= \frac{2x^2 + 4hx + 2h^2 + 3x + 3h - 5 - 2x^2 - 3x + 5}{h}$$

$$= \frac{4hx + 2h^2 + 3h}{h}$$

$$= 4x + 2h + 3$$

The **derivative** is the **slope of a line tangent to a graph** f(x) at x, and is usually denoted $f'(x)$. This is also referred to as the instantaneous rate of change. The derivative of f(x) at x = a is found by taking the limit of the average rates of change (computed by the difference quotient) as h approaches 0.

$$f'(a) = \lim_{h \to 0} \frac{f(a+h) - f(a)}{h}$$

Pick a point (for instance, at x = –3) on the graph of a function and draw a tangent line at that point. Find the derivative of the function and substitute the value x = –3. This result will be the slope of the tangent line.

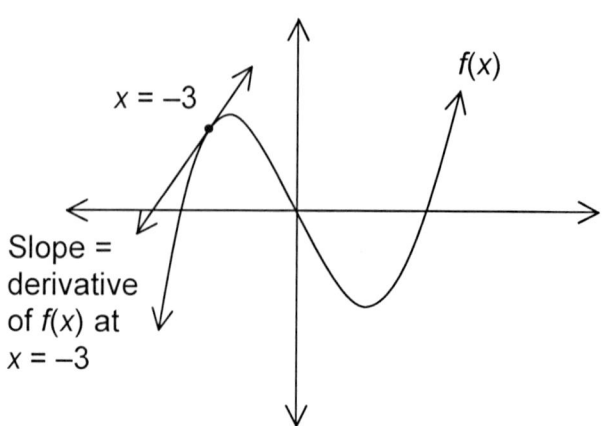

Slope = derivative of f(x) at x = –3

Example: Suppose a company's annual profit (in millions of dollars) is represented by the above function, $f(x) = 2x^2 + 3x - 5$, and x represents the number of years in the interval. Compute the rate at which the annual profit was changing over a period of 2 years.

$$f'(a) = \lim_{h \to 0} \frac{f(a+h) - f(a)}{h}$$
$$= f'(2) = \lim_{h \to 0} \frac{f(2+h) - f(2)}{h}$$

Using the difference quotient we computed previously, $4x + 2h + 3$, yields

$$f'(2) = \lim_{h \to 0}(4(2) + 2h + 3)$$
$$= 8 + 3$$

The annual profit for the company has increased at the average rate of $11 million per year over the two-year period.

Rules of Differentiation

The following properties of the derivative allow for differentiation of a wide range of functions (although the process of differentiation may be more or less difficult, depending on the complexity of the function). For illustration, consider two arbitrary functions, *f(x)* and *g(x)*, and an arbitrary constant, *c*.

Rule for multiplicative constants:

$$\frac{d}{dx}(cf) = cf'$$

Sum and difference rules:

$$\frac{d}{dx}(f+g) = f' + g'$$
$$\frac{d}{dx}(f-g) = f' - g'$$

Product rule:

$$\frac{d}{dx}(fg) = fg' + gf'$$

Quotient rule:

$$\frac{d}{dx}\left(\frac{f}{g}\right) = \frac{gf' - fg'}{g^2}$$

Another useful rule is the **chain rule**. The chain rule, as expressed below, allows differentiation of composite functions. The variable u can be an independent variable or it can be a function (of x). Note that, in the rule below, the differential elements du in the numerator of the first factor and the denominator of the second factor can otherwise cancel, making the right side of the equation identical to the left side.

$$\frac{df}{dx} = \frac{df}{du}\frac{du}{dx}$$

Example: Find the first derivative of the function $y = 5x^4$.

$$\frac{dy}{dx} = (5)(4)x^{4-1}$$
$$\frac{dy}{dx} = 20x^3$$

Example: Find y' where $y = \dfrac{1}{4x^3}$.

First, rewrite the function using a negative exponent, then apply the differentiation rule.

$$y' = \frac{1}{4}x^{-3}$$

$$y' = \frac{1}{4}(-3)x^{-3-1}$$

$$y' = -\frac{3}{4}x^{-4} = -\frac{3}{4x^4}$$

Example: Find the first derivative of $y = 3\sqrt{x^5}$.

Rewrite using $\sqrt[z]{x^n} = x^{n/z}$, then take the derivative.

$$y = 3x^{5/2}$$

$$\frac{dy}{dx} = (3)\left(\frac{5}{2}\right)x^{5/2-1}$$

$$\frac{dy}{dx} = \left(\frac{15}{2}\right)x^{3/2}$$

$$\frac{dy}{dx} = 7.5\sqrt{x^3} = 7.5x\sqrt{x}$$

Summary of differentiation rules for transcendental functions

$\dfrac{d}{dx}\sin x = \cos x$ $\dfrac{d}{dx}\csc x = -\csc x \cot x$

$\dfrac{d}{dx}\cos x = -\sin x$ $\dfrac{d}{dx}\sec x = \sec x \tan x$

$\dfrac{d}{dx}\tan x = \sec^2 x$ $\dfrac{d}{dx}\cot x = -\csc^2 x$

$\dfrac{d}{dx}\arcsin x = \dfrac{1}{\sqrt{1-x^2}}$ $\dfrac{d}{dx}\text{arc csc } x = -\dfrac{1}{|x|\sqrt{x^2-1}}$

$\dfrac{d}{dx}\arccos x = -\dfrac{1}{\sqrt{1-x^2}}$ $\dfrac{d}{dx}\text{arc sec } x = \dfrac{1}{|x|\sqrt{x^2-1}}$

$\dfrac{d}{dx}\arctan x = \dfrac{1}{1+x^2}$ $\dfrac{d}{dx}\text{arc cot } x = -\dfrac{1}{1+x^2}$

$\dfrac{d}{dx}\ln x = \dfrac{1}{x}$ $\dfrac{d}{dx}e^x = e^x$

Example: Find the derivative of the function $y = 4e^{x^2} \sin x$.

Apply the appropriate rules (product and chain rules) to the function.

$$\frac{dy}{dx} = 4\left(\sin x \frac{d}{dx}e^{x^2} + e^{x^2}\frac{d}{dx}\sin x\right)$$

$$\frac{dy}{dx} = 4\left(\sin x \left[2xe^{x^2}\right] + e^{x^2} \cos x\right)$$

$$\frac{dy}{dx} = 8xe^{x^2}\sin x + 4e^{x^2}\cos x$$

Example: Find the derivative of the function $y = \dfrac{5}{e^{\sin x}}$.

Rewrite the function with a negative exponent and use the chain rule.

$$y = 5e^{-\sin x}$$

$$\frac{dy}{dx} = 5\frac{d}{dx}e^{-\sin x}$$

$$\frac{dy}{dx} = 5e^{-\sin x}\left[-\cos x\right]$$

$$\frac{dy}{dx} = -5e^{-\sin x}\cos x = -\frac{5\cos x}{e^{\sin x}}$$

Using these properties of derivatives, the slopes (and therefore equations) of tangent lines can be found for a wide range of functions. The procedure simply involves finding the slope of the function at the given point using the derivative, then determining the equation of the line using point-slope form.

Example: Find the slope of the tangent line for the given function at the given point: $y = \dfrac{1}{x-2}$ at (3, 1).

Find the derivative of the function.

$$y' = \frac{d}{dx}(x-2)^{-1}$$

$$y' = (-1)(x-2)^{-2}(1) = -\frac{1}{(x-2)^2}$$

Evaluate the derivative at $x = 3$:

$$y' = -\frac{1}{(3-2)^2} = -1$$

Thus, the slope of the function at the point is -1.

Example: Find the points where the tangent to the curve $f(x) = 2x^2 + 3x$ is parallel to the line $y = 11x - 5$.

For the tangent line to be parallel to the given line, the only condition is that the slopes are equal. Thus, find the derivative of f, set the result equal to 11 and solve for x.

$$f'(x) = 4x + 3 = 11$$
$$4x = 8$$
$$x = 2$$

To find the y value of the point, simply substitute 2 into f.

$$f(2) = 2(2)^2 + 3(2) = 8 + 6 = 14$$

Thus, the tangent to f is parallel to $y = 11x - 5$ at the point $(2, 14)$ only.

Example: Find the equation of the tangent line to $f(x) = 2e^{x^2}$ at $x = -1$.

To find the tangent line, a point and a slope are needed. The x value of the point is given; the y value can be found by substituting $x = -1$ into f.

$$f(-1) = 2e^{(-1)^2} = 2e$$

Thus, the point is $(-1, 2e)$. The slope is found by substituting -1 into the derivative of f.

$$f'(x) = 2e^{x^2}(2x) = 4xe^{x^2}$$
$$f'(-1) = 4(-1)e^{(-1)^2} = -4e$$

Use the point-slope form of the line to determine the correct equation.

$$y - 2e = -4e(x - [-1])$$
$$y = 2e - 4ex - 4e = -4ex - 2e$$

Thus, the equation of the line tangent to f at $x = -1$ is $y = -4ex - 2e$.

The beginning teacher uses the first and second derivatives to analyze the graph of a function (e.g., local extrema, concavity, points of inflection).

Differential calculus can be a helpful tool in analyzing functions and the graphs of functions. Derivatives deal with the slope (or rate of change) of a function, and this information can be used to calculate the locations and values of extrema (maxima and minima) and inflection points, as well as to determine information concerning concavity.

The concept of extrema (maxima and minima) can be differentiated into local (or relative) and global (or absolute) extrema. For instance, consider the following function:

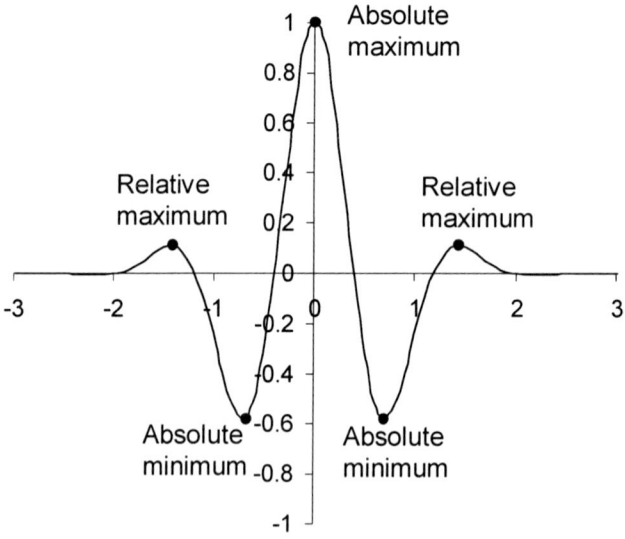

It is apparent that there are a number of peaks and valleys, each of which could, in some sense, be called a maximum or minimum. To allow greater clarity, local and global extrema can be specified. For instance, the peak at $x = 0$ is the maximum for the entire function. Additionally, the valleys at about $x = \pm 0.7$ both correspond to an (equivalent) minimum for the entire function. These are absolute extrema. On the other hand, the peaks at about $x = \pm 1.4$ are each a maximum for the function within a specific area; thus, they are relative maxima. Relative extrema are extreme values of the function over some limited interval. The points at which the derivative of a function is equal to zero are called **critical points** (the x values are called **critical numbers**).

By inspection of any graph, it is apparent that all extrema (where the function is continuous on either side of the maximum or minimum point) are located at points where the slope of the function is zero. This is to say that the derivative of the function at an extremum is zero. It is not necessarily the case, however, that all points where the derivative of the function is zero correspond to extrema. Consider the function $y = x^3$, whose graph is shown below.

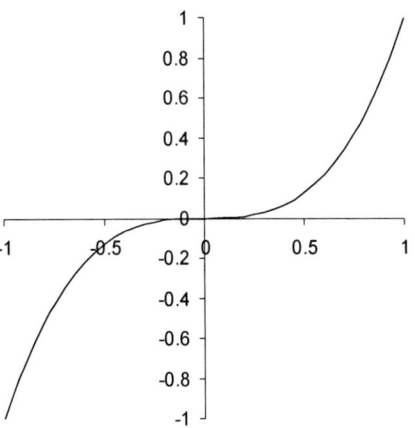

The derivative of y is $3x^2$, and the function is equal to zero only at $x = 0$. Nevertheless, the function y does not have an extremum at $x = 0$. The only cases where critical points correspond to extrema are when the derivative of the function actually crosses the x-axis. These cases correspond to the function having a positive slope on one side of the critical point and a negative slope on the other. This is a requirement for an extremum. (Notice that, for the plot of $y = x^3$, the function has a positive slope on both sides of the critical point.)

A positive slope on the left side of a critical point and a negative slope on the right side indicate that the critical point is a maximum. If the slope is negative on the left and positive on the right, then the critical point corresponds to a minimum. If the slopes on either side are both positive or both negative, then there is no extremum at the critical point.

Whether a critical point is an extremum can be determined using the second derivative f''. A critical point corresponds to the function f having zero slope. Thus, f' is zero at these points. As noted above, f has a maximum at the critical point only if f' crosses the x-axis. If f' is zero at a point but does not cross the x-axis, however, then that point is either a maximum or minimum of the function f', i.e., a critical number of f' ($f''=0$),. As a result, if the critical number of f is also a critical number of f' (i.e., $f''=0$), the critical point does not correspond to an extremum of f. The procedure for finding extrema for $f(x)$ is thus the following.

1. Calculate $f'(x)$.
2. Solve $f'(x) = 0$; the solutions of this equation are the critical numbers.
3. Calculate $f''(x)$.
4. Evaluate $f''(x)$ for each critical number c. If:
    a. $f''(c) = 0$, the critical point is not an extremum of f.
    b. $f''(c) > 0$, the critical point is a minimum of f.
    c. $f''(c) < 0$, the critical point is a maximum of f.

Example: Find the maxima and minima of $f(x) = 2x^4 - 4x^2$ on the closed interval [–2, 1].

First, differentiate the function and set the result equal to zero.

$$\frac{df}{dx} = 8x^3 - 8x = 0$$

Next, solve by factoring to find the critical numbers.

$$8x(x^2 - 1) = 0$$
$$x(x-1)(x+1) = 0$$

The solutions for this equation, which are also the critical numbers, are $x = -1$, 0 and 1. For each critical number, it is necessary to determine whether the point corresponds to a maximum, minimum or neither.

$$\frac{d^2f}{dx^2} = 24x^2 - 8$$

Test each critical point by substituting into the result above.

$$f''(-1) = 24(-1)^2 - 8 = 24 - 8 = 16 \rightarrow \text{minimum}$$
$$f''(0) = 24(0)^2 - 8 = -8 \rightarrow \text{maximum}$$
$$f''(1) = 24(1)^2 - 8 = 24 - 8 = 16 \rightarrow \text{minimum}$$

The critical numbers correspond to the minima (–1, –2) and (1, –2) and to the maximum (0, 0). The endpoint of the closed interval at x = –2 should also be tested to determine if it constitutes an extremum, as such may not be detectable using derivatives (the minimum at the endpoint x = 1 was detected, however). This endpoint corresponds to (–2, 16), which is the absolute maximum. Absolute minima exist at (–1, –2) and (1, –2), and a relative maximum exists at (0, 0).

The second derivative of a function can also be viewed in terms of concavity. The first derivative reveals whether a curve is increasing or decreasing (increasing or decreasing) from the left to the right. In much the same way, the second derivative relates whether the curve is concave up (slope increasing) or concave down (slope decreasing). Curves that are concave can be viewed as "collecting water"; curves that are concave down can be viewed as "dumping water."

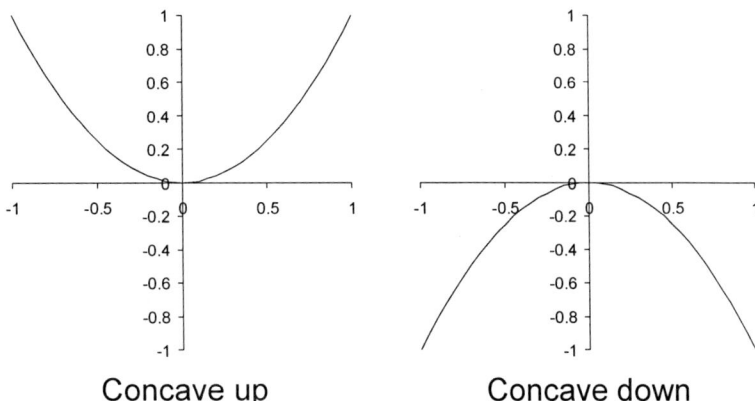

Concave up   Concave down

A **point of inflection** is a point where a curve changes from being concave up to concave down (or vice versa). To find these points, find the critical numbers of the first derivative of the function (that is, solve the equation for which the second derivative of the function is set equal to zero). A critical number coincides with an inflection point if the curve is concave up on one side of the value and concave down on the other. The critical number is the x coordinate of the inflection point. To get the y coordinate, plug the critical number into the **original** function.

Example: Find the inflection points of $f(x) = 2x - \tan x$ over the interval $-\frac{\pi}{2} < x < \frac{\pi}{2}$.

First, calculate the second derivative of f.

$$f''(x) = \frac{d^2 f(x)}{dx^2} = \frac{d}{dx}\left[\frac{d}{dx}(2x - \tan x)\right]$$

$$f''(x) = \frac{d}{dx}\left[2 - \sec^2 x\right] = -2\sec x \frac{d}{dx}\sec x$$

$$f''(x) = -2\sec x(\sec x \tan x) = -2\sec^2 x \tan x$$

Set the second derivative equal to zero and solve.

$$f''(x) = -2\sec^2 x \tan x = 0$$

The function is zero for either sec x = 0 or tan x = 0. Only tan x = 0, however, has real solutions. This means that the inflection points are at $x = n\pi$, where n = 0, 1, 2, etc. Within the given interval, however, the only solution is x = 0. Substituting this value into the original equation yields the following:

$$f(0) = 2(0) - \tan 0 = 0 - 0 = 0$$

Thus, the inflection point for this function on the interval $-\frac{\pi}{2} < x < \frac{\pi}{2}$ is (0, 0). The plot of the function is shown below, along with the associated inflection point. As hinted earlier, the inflection point can be seen graphically as the point at which the slope changes from an increasing value to a decreasing value (or vice versa).

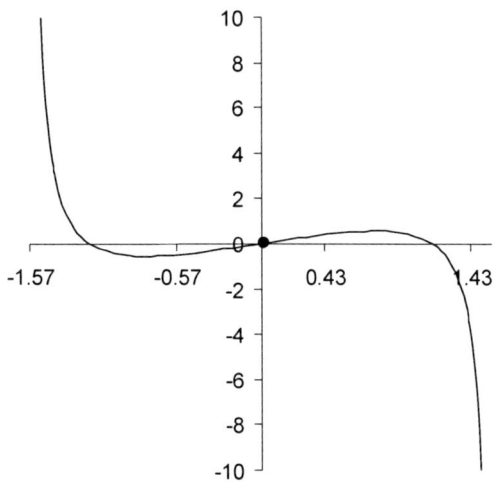

Example: Identify approximately the locations of the extrema (excluding the endpoints) and inflection points for the following graph.

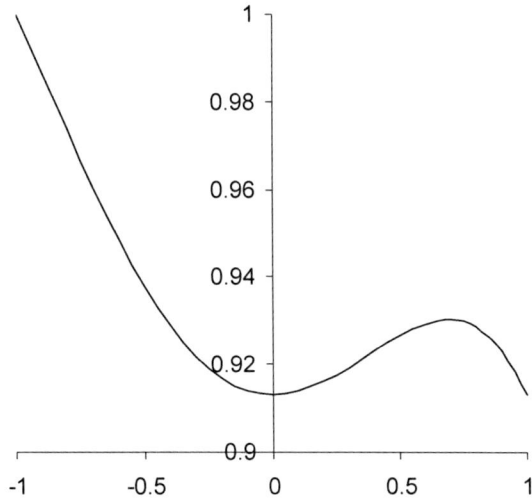

There are two obvious extrema in the graph: a minimum at about (0, .915) and a maximum at about (0.7, 0.93). These extrema are evidently relative extrema, since the function (at least apparently) has both larger and smaller values elsewhere. There is also an obvious concavity shift between the maximum and minimum. The inflection point is at about (0.35, 0.92). The extrema and inflection points are shown marked below.

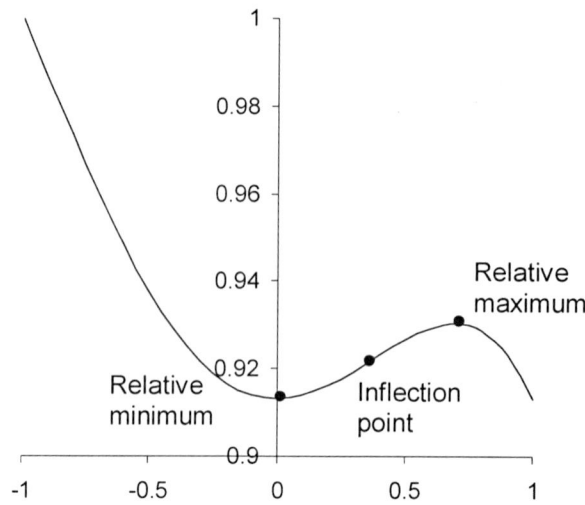

The beginning teacher understands and applies the fundamental theorem of calculus and the relationship between differentiation and integration.

The formal definition of an integral is based on the **Riemann sum**. A Riemann sum is the sum of the areas of a set rectangles that is used to approximate the area under the curve of a function. Given a function f defined over some closed interval $[a, b]$, the interval can be divided into a set of n arbitrary partitions, each of length Δx_i. Within the limits of each partition, some value $x = c_i$ can be chosen such that Δx_i and $f(c_i)$ define the width and height (respectively) of a rectangle. The sum of the aggregate of all the rectangles defined in this manner over the interval $[a, b]$ is the Riemann sum.

Consider, for example, the function $f(x) = x^2 + 1$ over the interval $[0, 1]$. The plot of the function is shown below.

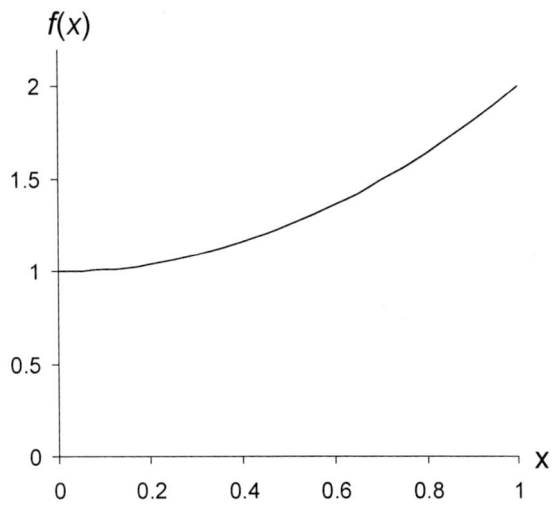

Partition the interval into segments of width 0.2 along the x-axis, and choose the function value $f(c_i)$ at the center of each interval. This function value is the height of the respective rectangle.

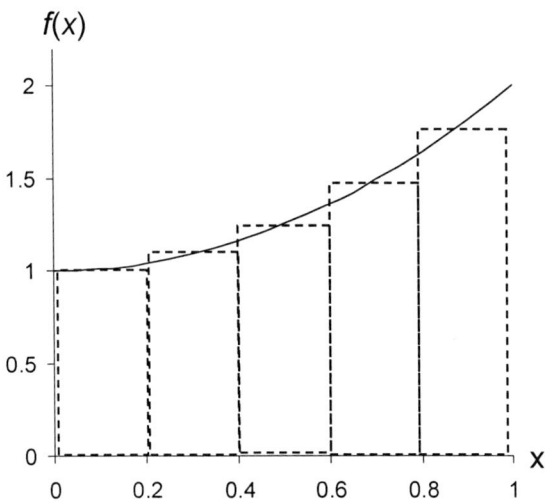

The Riemann sum for this case is expressed below.

$$\sum_{i=1}^{5} 0.2 f(0.2i - 0.1) = 1.33$$

This expression is the sum of the areas of all the rectangles shown above. This is an approximation of the area under the curve of the function (and a reasonably accurate one, as well—the actual area is $\frac{4}{3}$).

Generally, the Riemann sum for arbitrary partitioning and selection of the values c_i is the following:

$$\sum_{i=1}^{n} f(c_i) \Delta x_i$$

where c_i is within the closed interval defined by the partition Δx_i.

Definite Integrals

The **definite integral** is defined as the limit of the Riemann sum as the widths of the partitions Δx_i go to zero (and, consequently, n goes to infinity). Thus, the definite integral can be expressed mathematically as follows:

$$\int_a^b f(x)\,dx = \lim_{\Delta x_m \to 0} \sum_{i=1}^{n} f(c_i) \Delta x_i$$

where Δx_m is the width of the largest partition. If the partitioning of the interval is such that each partition has the same width, then the definition can be written as follows:

$$\int_a^b f(x)\,dx = \lim_{\Delta x \to 0} \sum_{i=1}^{n} f(c_i) \Delta x$$

Note that $n = \dfrac{b-a}{\Delta x}$ in this case.

The definite integral, therefore, is the area under the curve of $f(x)$ over the interval $[a, b]$. By taking the limit of the Riemann sum, the number of rectangles used to find the area under the curve becomes infinite and, therefore, the error in the result goes to zero since the width of each rectangle becomes infinitesimal.

The **Fundamental Theorem of Calculus** relates differentiation with definite integration, which is fundamentally defined in terms of the Riemann sum. According to the theorem, definite integration is the inverse of differentiation. The theorem is expressed below for the function $F(x)$, where $f(x) = F'(x)$ and where $f(x)$ is continuous on the interval [a, b].

$$\int_a^b f(x)\,dx = F(b) - F(a)$$

The function $F(x)$ is also called an **antiderivative** of $f(x)$, because the derivative of $F(x)$ is $f(x)$. Based on this theorem, it is clear that integrals can be evaluated without the need of finding the limit of a Riemann sum as long as the antiderivative of a function can be determined. Thus, the key to evaluating definite integrals is knowledge of how to find the antiderivative of a function.

The definite integral can thus be interpreted in terms of antiderivatives. Note that the definite integral for the interval [a, b] is the area under the curve $f(x)$ between $x = a$ and $x = b$. The antiderivative, $F(c)$, is then the cumulative area under $f(x)$ between $x = 0$ and $x = c$. Thus, the difference between the antiderivative evaluated at b and the antiderivative evaluated at a is the definite integral of $f(x)$ between a and b.

The example graph below shows a function $f(x)$. The antiderivative $F(x)$ evaluated at $x = 1$ is the solid shaded area; the antiderivative evaluated at $x = 2$ is the striped area. The difference $F(2) - F(1)$ is the difference between the two areas (the non-overlapping striped region).

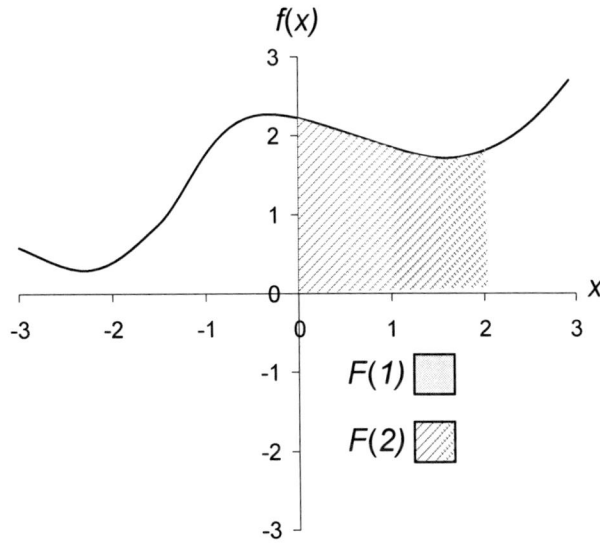

Properties of Integrals

The following summarizes some of the basic rules for integration in terms of some common functions.

Integration involving constants:

$$\int 0\, dx = C \qquad \int k\, dx = kx + C \qquad \int kf(x)\, dx = k\int f(x)\, dx$$

Integration of sums or differences of functions:

$$\int [f(x) + g(x)]\, dx = \int f(x)\, dx + \int g(x)\, dx$$
$$\int [f(x) - g(x)]\, dx = \int f(x)\, dx - \int g(x)\, dx$$

Power rule:

$$\int x^n\, dx = \frac{x^{n+1}}{n+1} + C \text{ for } n \neq 1$$

Trigonometric functions:

$$\int \sin x\, dx = -\cos x + C \qquad \int \cos x\, dx = \sin x + C$$
$$\int \tan x\, dx = \ln(\sec x) + C \qquad \int \cot x\, dx = \ln(\sin x) + C$$
$$\int \csc x\, dx = \ln\left(\tan \frac{x}{2}\right) + C \qquad \int \sec x\, dx = \ln\left(\tan\left[\frac{x}{2} + \frac{\pi}{4}\right]\right) + C$$
$$\int \sec^2 x\, dx = \tan x + C \qquad \int \sec x \tan x\, dx = \sec x + C$$
$$\int \csc^2 x\, dx = -\cot x + C \qquad \int \csc x \cot x\, dx = -\csc x + C$$

Logarithmic functions:

$$\int \ln x\, dx = x \ln x - x + C \qquad \int \frac{1}{x}\, dx = \ln x + C$$

Exponential functions:

$$\int e^x\, dx = e^x + C$$

The above rules are helpful for finding the antiderivatives of functions, but they are far from complete, since they do not permit in any obvious manner integration of composite functions or functions with arguments other than simply x. To this end, several helpful strategies can be applied.

If a function is expressed in terms of another function (that is, if it is a composite function), then a **change of variables** permits integration through conversion of the expression into a form similar to the rules given above. Consider a composite function $f(g(x))$ in the context of the following integral:

$$\int f(g(x))g'(x)dx$$

An example of such a composite function might be $f(x) = (x+1)^2$ or $f(x) = \sin(x^3)$, for instance. Assign $g(x)$ a new variable name u. Then differentiate $g(x)$ in terms of x and rearrange the differentials.

$$u = g(x)$$
$$\frac{dg(x)}{dx} = g'(x) = \frac{du}{dx}$$
$$du = g'(x)dx$$

Use this result in the indefinite integration of the composite function:

$$\int f(g(x))g'(u)dx = \int f(u)du$$

With this simple substitution (sometimes called a *u*-**substitution**), the integral can be made to look like one of the general forms. It is sometimes necessary to experiment with different *u* substitutions to find one that works (finding a *u* that allows complete elimination of *x* from the integral is not always trivial). Follow these steps for integration by substitution.

1. Select an appropriate value for *u* to perform the substitution.
2. Differentiate *u* as shown above.
3. Substitute *u* and *du* into the integral to eliminate *x*.
4. Evaluate the integral.
5. Substitute *g(x)* back into the result to get the antiderivative.

When dealing with definite integrals that require *u*-substitution, the only difference is that the limits of integration must be modified in accordance with the choice of *u*. Thus:

$$\int_a^b f(g(x))g'(u)\,dx = \int_{g(a)}^{g(b)} f(u)\,du$$

Example: Evaluate the following antiderivative:

$$\int 2x\left[\sin(x^2) + \cos(x^2)\right]dx.$$

First, split the integral into two parts.

$$\int 2x\left[\sin(x^2) + \cos(x^2)\right]dx = \int 2x\sin(x^2)\,dx + \int 2x\cos(x^2)\,dx$$

Next, select an appropriate value for *u*. In this case, choose $u = x^2$. Then:

$$du = 2x\,dx$$

Rewrite the integral in terms of *u*.

$$\int \sin(x^2)2x\,dx + \int \cos(x^2)2x\,dx = \int \sin u\,du + \int \cos u\,du$$

This result is in a form for which the antiderivative can be found easily.

$$\int \sin u\,du + \int \cos u\,du = -\cos u + \sin u + C$$

Substitute the definition of *u* back into the result to get the antiderivative in terms of *x*.

$$-\cos u + \sin u + C = -\cos(x^2) + \sin(x^2) + C$$

Example: Evaluate the following antiderivative: $\int e^{\sin x} \cos x\,dx$.

Try choosing $u = \cos x$.

$$du = -\sin x\,dx$$

Substitute into the integral.

$$\int e^{\sin x} \cos x \, dx = -\int \frac{u e^{\sin x}}{\sin x} du$$

Note that there is no apparent way to eliminate x from the integral. Thus, this choice of u should be abandoned. Instead, try $u = \sin x$.

$$du = \cos x \, dx$$

Substitute into the integral, as before.

$$\int e^{\sin x} \cos x \, dx = \int e^u \, du$$

This choice of u was successful. Evaluate the antiderivative and substitute the definition of u back into the result.

$$\int e^u \, du = e^u = e^{\sin x}$$

<u>Example</u>: Evaluate the following definite integral: $\int \frac{1}{x \ln x} dx$ over the interval $[e, e^e]$.

Substitute using $u = \ln x$. Then, $du = \frac{1}{x} dx$.

$$\int \frac{1}{x \ln x} dx = \int \frac{1}{xu} x \, du = \int \frac{1}{u} du$$

Evaluate the integral and apply the limits of integration, which, using $u = \ln x$, lead to the interval $[\ln e, \ln e^e] = [1, e]$.

$$\int_1^e \frac{1}{u} du = \ln u \Big|_1^e$$

$$\ln u \Big|_1^e = \ln e - \ln 1 = 1 - 0 = 1$$

Another useful technique for evaluating complicated integrals is **integration by parts**. Since this method is itself complicated, it should only be used as a last resort if simpler methods of integration are not successful. Integration by parts requires two substitutions. To remember the formula for integration by parts, it is helpful to remember that it is based on the product rule of differentiation for two functions, u and v.

$$\frac{d}{dx}(uv) = u\frac{dv}{dx} + v\frac{du}{dx}$$

Naturally, then, integrating this result should return the product *uv*.

$$\int \frac{d}{dx}(uv)\,dx = \int d(uv) = uv$$

$$\int \left[u\frac{dv}{dx} + v\frac{du}{dx} \right] dx = \int u\,dv + \int v\,du$$

Rewrite the equation and rearrange to get the formula for integration by parts:

$$uv = \int u\,dv + \int v\,du$$

$$\int u\,dv = uv - \int v\,du$$

Thus, by identifying substitution functions for *u* and *v*, a method for integration is available. Proper selection of *u* and *v* is crucial to making this technique work. Use the following steps to perform integration by parts.

1. Choose *dv* as the most complicated part of the integral that can be integrated by itself.
2. Choose *u* as the part of the integral that remains after the *dv* substitution is made. Preferably, the derivative of *u* should be simpler than *u*.
3. Integrate *dv* to get *v*.
4. Differentiate *u* to get *du*.
5. Rewrite the integral in the form $\int u\,dv = uv - \int v\,du$.
6. Integrate $\int v\,du$.
7. If you cannot integrate *v du*, go back to the first step and try a different set of substitutions.

<u>Example:</u> Find the antiderivative of the following function: $\int xe^{3x}\,dx$.

First, choose $dv = e^{3x}\,dx$ and $u = x$. Calculate *du* and *v*.

$$dv = e^{3x}\,dx \qquad\qquad u = x$$
$$\int dv = \int e^{3x}\,dx \qquad\qquad du = dx$$
$$v = \frac{e^{3x}}{3}$$

Substitute these results into the formula for integration by parts:

$$\int u \, dv = \int xe^{3x} \, dx = uv - \int v \, du = \frac{xe^{3x}}{3} - \int \frac{e^{3x}}{3} dx$$

The substitutions fit in this case, and the integration can now be performed easily.

$$\int xe^{3x} \, dx = \frac{xe^{3x}}{3} - \int \frac{e^{3x}}{3} dx = \frac{xe^{3x}}{3} - \frac{e^{3x}}{9} + C$$

$$\int xe^{3x} \, dx = \frac{e^{3x}}{3}\left(x - \frac{1}{3}\right) + C$$

This is the correct solution to the problem.

Example: Evaluate the following indefinite integral: $\int x \cos x \, dx$.
Try choosing $u = x$ and $dv = \cos x \, dx$.

$$du = dx \qquad \int dv = \int \cos x \, dx$$
$$v = \sin x$$

Substitute into the formula for integration by parts:

$$\int u \, dv = \int x \cos x \, dx = uv - \int v \, du = x \sin x - \int \sin x \, dx$$

The choices of *u* and *dv* work, so the integral can be evaluated to find the result.

$$\int x \cos x \, dx = x \sin x + \cos x + C$$

The beginning teacher models and solves a variety of problems (e.g., velocity, acceleration, optimization, related rates, work, center of mass) using differential and integral calculus.

Extreme value problems, also known as **max-min problems** or **optimization problems**, entail using the first derivative to find values that either maximize or minimize some quantity, such as area, profit or volume. The derivative is a critical tool in solving these types of problems. Follow these steps to solve an extreme value (optimization) problem.

1. Write an equation for the quantity to be maximized or minimized.
2. Use the other information in the problem to write secondary equations.
3. Use the secondary equations for substitutions, and rewrite the original equation in terms of only one variable.
4. Find the derivative of the primary equation (step 1) and the critical numbers of this derivative.
5. Substitute these critical numbers into the primary equation.

The value that produces either the largest or smallest result can be used to find the solution.

Example: A manufacturer wishes to construct an open box from a square piece of metal by cutting squares from each corner and folding up the sides. The metal is 12 feet on each side. What are the dimensions of the squares to be cut out such that the volume of the box is maximized?

First, draw a figure that represents the situation. Assume that the squares to be cut from the metal have sides of length x. Noting that the metal has sides of length 12 feet, this leaves 12 – 2x feet remaining on each side after the squares are cut out.

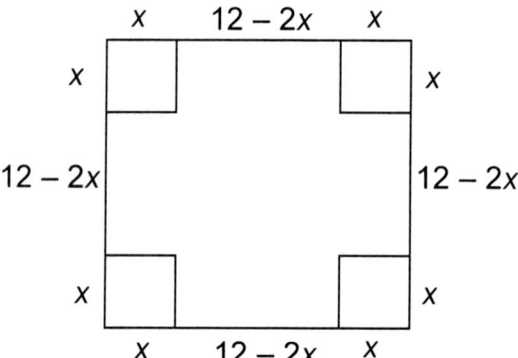

The volume V(x) of the box formed when the sides are folded up is the following:

$$V(x) = x(12-2x)^2$$

Simplify and take the first derivative of the result.

$$V(x) = x(144 - 48x + 4x^2) = 4x^3 - 48x^2 + 144x$$
$$V'(x) = 12x^2 - 96x + 144$$

Set the first derivative to zero and solve by factoring.

$$V'(x) = 12x^2 - 96x + 144 = 0$$
$$(x-6)(x-2) = 0$$

The solutions are then x = 2 feet and x = 6 feet. Note that, if x = 6 feet, the sides of the box become zero in width. This, therefore, is not a legitimate solution. Choose x = 2 feet as the solution that leads to the largest volume of the box.

Problems Involving Rectilinear Motion

If a particle (such as a car, bullet or other object) is moving along a line, then the position of the particle can be expressed as a function of time.

The rate of change of position with respect to time is the velocity of the object; thus, the first derivative of the distance function yields the velocity function for the particle. Substituting a value for time into this expression provides the instantaneous velocity of the particle at that time. The absolute value of the derivative is the speed (magnitude of the velocity) of the particle. A positive value for the velocity indicates that the particle is moving forward (that is, in the positive x direction); a negative value indicates the particle is moving backward (that is, in the negative x direction).

The acceleration of the particle is the rate of change of the velocity. The second derivative of the position function (which is also the first derivative of the velocity function) yields the acceleration function. If a value for time produces a positive acceleration, the particle's velocity is increasing; if it produces a negative value, the particle's velocity is decreasing. If the acceleration is zero, the particle is moving at a constant speed.

<u>Example:</u> The motion of a particle moving along a line is according to the equation $s(t) = 20 + 3t - 5t^2$, where s is in meters and t is in seconds. Find the position, velocity and acceleration of the particle at $t = 2$ seconds.
To find the position, simply use $t = 2$ in the given position function. Note that the initial position of the particle is s(0) = 20 meters.

$$s(2) = 20 + 3(2) - 5(2)^2$$
$$s(2) = 20 + 6 - 20 = 6\,m$$

To find the velocity of the particle, calculate the first derivative of $s(t)$ and then evaluate the result for $t = 2$ seconds.

$$s'(t) = v(t) = 3 - 10t$$
$$v(2) = 3 - 10(2) = 3 - 20 = -17 \text{ m/s}$$

Finally, for the acceleration of the particle, calculate the second derivative of $s(t)$ (also equal to the first derivative of $v(t)$) and evaluate for $t = 2$ seconds.

$$s''(t) = v'(t) = a(t) = -10 \text{ m/s}^2$$

Since the acceleration function $a(t)$ is a constant, the acceleration is always -10 m/s^2 (the velocity of the particle decreases every second by 10 meters per second).

Related rate problems

Some rate problems may involve functions with different parameters that are each dependent on time. In such a case, implicit differentiation may be required. Often times, related rate problems give certain rates in the description, thus eliminating the need to have specific functions of time for every parameter. Related rate problems are otherwise solved in the same manner as other similar problems.

Example: A spherical balloon is inflated such that its radius is increasing at a constant rate of 1 inch per second. What is the rate of increase of the volume of the balloon when the radius is 10 inches?

First, write the equation for the volume of a sphere in terms of the radius, r.

$$V(r) = \frac{4}{3}\pi r^3$$

Differentiate the function implicitly with respect to time, t, by using the chain rule.

$$\frac{dV(r)}{dt} = \frac{4}{3}\pi \frac{d}{dt}(r^3)$$
$$\frac{dV(r)}{dt} = \frac{4}{3}\pi (3r^2)\frac{dr}{dt} = 4\pi r^2 \frac{dr}{dt}$$

To find the solution to the problem, use the radius value $r = 10$ inches and the rate of increase of the radius $\frac{dr}{dt} = 1$ in/sec. Calculate the resulting rate of increase of the volume, $\frac{dV(r)}{dt}$.

$$\frac{dV(10)}{dt} = 4\pi(10\,\text{in})^2\, 1\,\text{in/sec} = 400\pi\,\text{in}^3/\text{sec} \approx 1257\,\text{in}^3/\text{sec}$$

The problem is thus solved.

Applications of Integrals

Taking the integral of a function and evaluating it over some interval on x provides the total **area under the curve** (or, more formally, the **area bounded by the curve and the x-axis**). Thus, the area of geometric figures can be determined when the figure can be cast as a function or set of functions in the coordinate plane. Remember, though, that regions above the x-axis have "positive" area and regions below the x-axis have "negative" area. It is necessary to account for these positive and negative values when finding the area under curves. The boundaries between positive and negative regions are delineated by the roots of the function. Follow these steps to find the total area under the curve.

1. Determine the interval or intervals on which the area under the curve is to be found. If portions of the function are negative, a given interval may need to be divided appropriately if all areas are to be considered positive.
2. Integrate the function.
3. Evaluate the integral once for each interval.
4. If any of the intervals evaluates to a negative number, reverse the sign (equivalently, take the absolute value of each integral).
5. Add the value of all the integral to get the area under the curve.

Example: Find the area under the following function on the given interval: $f(x) = \sin x$; $[0, 2\pi]$.

First, find the roots of the function on the interval.

$$f(x) = \sin x = 0$$
$$x = 0, \pi$$

The function sin x is positive over [0, π] (since $\sin\frac{\pi}{2} = 1$) and negative over [π, 2π] (since $\sin\frac{3\pi}{2} = -1$). Use these intervals for the integration to find the area A under the curve.

$$A = \int_0^{2\pi} |\sin x| dx = \left|\int_0^{\pi} \sin x\, dx\right| + \left|\int_{\pi}^{2\pi} \sin x\, dx\right|$$

$$A = \left|-\cos x\big|_0^{\pi}\right| + \left|-\cos x\big|_{\pi}^{2\pi}\right| = \left|-\cos\pi + \cos 0\right| + \left|-\cos 2\pi + \cos\pi\right|$$

$$A = |1+1| + |-1-1| = 2 + 2 = 4$$

Thus, the total area under the curve of $f(x) = \sin x$ on the interval $[0, 2\pi]$ is 4 square units.

Finding the **area between two curves** is similar to finding the area under one curve. The general process involves integrating the absolute value of the difference between the two functions over the interval of interest. In some instances, it is necessary to find the intervals over which the difference is positive and over which the difference is negative. For the former, the integral can simply be taken with no modifications; for the latter, however, the result of the integral must be negated. To find the points where the difference between the functions changes from positive to negative (or vice versa), simply set the functions equal to each other and solve. Take the absolute value of each portion of the integral (that is, each integral over a portion of the interval) and add all the parts. This yields the total area between the curves.

Example: Find the area of the regions bounded by the two functions on the indicated interval: $f(x) = x + 2$ and $g(x) = x^2$ on [−2, 3].

The integral of interest is the following:

$$\int_{-2}^{3} |f(x) - g(x)| dx$$

To eliminate the need to use the absolute value notation inside the integral, find the values for which $f(x) = g(x)$.

$$f(x) = x + 2 = g(x) = x^2$$
$$x^2 - x - 2 = 0 = (x-2)(x+1)$$

TEACHER CERTIFICATION STUDY GUIDE

The functions are then equal at $x = -1$ and $x = 2$. Perform the integration over the intervals defined by these values.

$$\int_{-2}^{3}|f(x)-g(x)|dx = \left|\int_{-2}^{-1}f(x)-g(x)dx\right| + \left|\int_{-1}^{2}f(x)-g(x)dx\right| + \left|\int_{2}^{3}f(x)-g(x)dx\right|$$

The antiderivative of $f(x)-g(x)$ is the following (ignoring the constant of integration).

$$\int[f(x)-g(x)]dx = \int(x+2-x^2)dx = \frac{x^2}{2}+2x-\frac{x^3}{3}$$

Evaluate over the intervals above.

$$\int_{-2}^{3}|f(x)-g(x)|dx = \left|\left[\frac{x^2}{2}+2x-\frac{x^3}{3}\right]_{-2}^{-1}\right| + \left|\left[\frac{x^2}{2}+2x-\frac{x^3}{3}\right]_{-1}^{2}\right| + \left|\left[\frac{x^2}{2}+2x-\frac{x^3}{3}\right]_{2}^{3}\right|$$

$$\left|\left[\frac{x^2}{2}+2x-\frac{x^3}{3}\right]_{-2}^{-1}\right| = \left|\left(\frac{1}{2}-2+\frac{1}{3}\right)-\left(\frac{4}{2}-4+\frac{8}{3}\right)\right| = \left|-\frac{7}{6}-\frac{2}{3}\right| = \frac{11}{6}$$

$$\left|\left[\frac{x^2}{2}+2x-\frac{x^3}{3}\right]_{-1}^{2}\right| = \left|\left(\frac{4}{2}+4-\frac{8}{3}\right)-\left(\frac{1}{2}-2+\frac{1}{3}\right)\right| = \left|\frac{10}{3}+\frac{7}{6}\right| = \frac{27}{6}$$

$$\left|\left[\frac{x^2}{2}+2x-\frac{x^3}{3}\right]_{2}^{3}\right| = \left|\left(\frac{9}{2}+6-\frac{27}{3}\right)-\left(\frac{4}{2}+4-\frac{8}{3}\right)\right| = \left|\frac{3}{2}-\frac{10}{3}\right| = \frac{11}{6}$$

The sum of these individual parts is $\frac{49}{6}$.

An area, bounded by a curve (or curves), that is revolved about a line is called a **solid of revolution**. To find the volume of such a solid, the **disc method** (called the **washer method** if the solid has an empty interior of some form) works in most instances. Imagine slicing through the solid perpendicular to the line of revolution. The cross section should resemble either a disc or a washer. The washer method involves finding the sum of the volumes of all "washers" that compose the solid, using the following general formula:

$$V = \pi(r_1^2 - r_2^2)t$$

where V is the volume of the washer, r_1 and r_2 are the interior and exterior radii and t is the thickness of the washer.

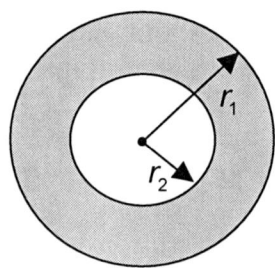

Depending on the situation, the radius is the distance from the line of revolution to the curve; or if there are two curves involved, the radius is the difference between the two functions. The thickness is dx if the line of revolution is parallel to the x-axis and dy if the line of revolution is parallel to the y-axis. The integral is then the following, where dV is the differential volume of a washer.

$$\int dV = \int \pi \left(r_1^2 - r_2^2 \right) dt$$
$$V = \pi \int \left(r_1^2 - r_2^2 \right) dt$$

It is assumed here that r_1 is the outer radius and r_2 is the inner radius. For the disc method, where only one radius is needed, $r_2 = 0$.

Example: Find the volume of the solid of revolution made by revolving $f(x) = 9 - x^2$ about the x axis on the interval $[0, 4]$.

This problem can be solved using the disc method. First, note that the radius is $9 - x^2$ and the thickness of the disc is dx. Write the appropriate integral as follows.

$$V = \pi \int_0^4 (9 - x^2)^2 \, dx$$

Next, expand the radius term and evaluate the integral.

$$V = \pi \int_0^4 \left(81 - 18x^2 + x^4 \right) dx$$
$$V = \pi \left[81x - \frac{18}{3}x^3 + \frac{1}{5}x^5 \right]_0^4$$
$$V = \pi \left[81(4) - \frac{18}{3}(4)^3 + \frac{1}{5}(4)^5 \right]$$
$$V = \pi [324 - 384 + 204.8] = 144.8\pi \approx 454.9$$

The volume is thus approximately 454.9 cubic units.

The **arc length** of a curve is another useful application of integration. The arc length is the distance traversed by a curve over a given interval. Geometrically, the distance d between two points (x_1, y_1) and (x_2, y_2) is given by the following formula.

$$d = \sqrt{(x_2 - x_1)^2 + (y_2 - y_1)^2}$$

If the points are only an infinitesimal distance apart (ds, which is the differential arc length), then the above expression can be written as follows in differential form.

$$ds = \sqrt{dx^2 + dy^2}$$

Factor out the dx term:

$$ds = \sqrt{1 + \left(\frac{dy}{dx}\right)^2} \, dx$$

But $\frac{dy}{dx}$ is simply the derivative of a function $y(x)$ (which can be expressed as $f(x)$ instead). Thus, the integral of the above expression over the interval $[a, b]$ yields the formula for the arc length.

$$\int ds = s = \int_a^b \sqrt{1 + [f'(x)]^2} \, dx$$

Example: Find the distance traversed by the function $f(x) = \ln(\cos x)$ on the interval $\left[-\frac{\pi}{4}, \frac{\pi}{4}\right]$.

Use the formula for arc length s, applying trigonometric identities as appropriate.

$$s = \int \sqrt{1 + \left[\frac{d}{dx}\ln(\cos x)\right]^2} = \int \sqrt{1 + \left[\frac{\sin x}{\cos x}\right]^2} \, dx$$

$$s = \int \sqrt{1 + [\tan x]^2} \, dx = \int \sqrt{1 + \tan^2 x} \, dx = \int \sqrt{\sec^2 x} \, dx$$

$$s = \int \sqrt{1 + [\tan x]^2} \, dx = \int \sqrt{1 + \tan^2 x} \, dx = \int \sec x \, dx$$

Evaluate the integral over the limits of integration.

$$s = \int_{-\pi/4}^{\pi/4} \sec x \, dx = \ln(\sec x + \tan x)\Big|_{-\pi/4}^{\pi/4}$$

$$s = \ln\left(\sec\frac{\pi}{4} + \tan\frac{\pi}{4}\right) - \ln\left(\sec\left[-\frac{\pi}{4}\right] + \tan\left[\frac{\pi}{4}\right]\right)$$

$$s = \ln(\sqrt{2} + 1) - \ln(\sqrt{2} - 1) \approx 1.763$$

The result is approximately 1.763 units.

Integral calculus, in addition to differential calculus, is a powerful tool for analysis of problems involving linear motion. The derivative of the position (or displacement) function is the velocity function, and the derivative of a velocity function is the acceleration function. As a result, the antiderivative of an acceleration function is a velocity function, and the antiderivative of the velocity function is a position (or displacement) function. Solving word problems of this type involve converting the information given into an appropriate integral expression. To find the constant of integration, use the conditions provided in the problem (such as an initial displacement, velocity or acceleration).

Example: A particle moves along the x-axis with acceleration $a(t) = 3t - 1 \frac{cm}{\sec^2}$. At time $t = 4$ seconds, the particle is moving to the left at 3 cm per second. Find the velocity of the particle at time $t = 2$ seconds.

Evaluate the antiderivative of the acceleration function $a(t)$ to get the velocity function $v(t)$ along with the unknown constant of integration C.

$$v(t) = \int a(t) \, dt = \int (3t - 1) \, dt$$

$$v(t) = \frac{3t^2}{2} - t + C$$

Use the condition that at time $t = 4$ seconds, the particle has a velocity of -3 cm/sec.

$$v(4) = \frac{3(4)^2}{2} - 4 + C = -3$$

$$\frac{48}{2} - 4 + C = -3$$

$$C = -3 + 4 - 24 = -23 \frac{cm}{sec}$$

Now evaluate $v(t)$ at time $t = 2$ seconds to get the solution to the problem.

$$v(t) = \frac{3t^2}{2} - t - 23 \frac{cm}{sec}$$

$$v(2) = \frac{3(2)^2}{2} - 2 - 23 \frac{cm}{sec} = 6 - 25 \frac{cm}{sec} = -19 \frac{cm}{sec}$$

Example: Find the displacement function of a particle whose acceleration is described by the equation $a(t) = 3\sin 2t$. Assume that the particle is initially motionless at the origin.

Find the antiderivative of the acceleration function $a(t)$ to get the velocity function $v(t)$.

$$v(t) = \int 3\sin 2t \, dt = -3\frac{1}{2}\cos 2t + C$$

Note that the initial velocity is zero; thus:

$$v(0) = 0 = -\frac{3}{2}\cos 2(0) + C = -\frac{3}{2} + C$$

$$C = \frac{3}{2}$$

$$v(t) = -\frac{3}{2}(1 - \cos 2t)$$

Find the antiderivative of the velocity function to get the displacement function $s(t)$.

$$s(t) = -\int \frac{3}{2}(1 - \cos 2t) \, dt = -\frac{3}{2}\left(t - \frac{1}{2}\sin 2t\right) + C'$$

The initial position is at the origin, so C' can be found.

$$s(0) = 0 = -\frac{3}{2}\left(0 - \frac{1}{2}\sin 2(0)\right) + C'$$

$$s(0) = 0 = -\frac{3}{2}(0-0) + C'$$

$$C' = 0$$

$$s(t) = -\frac{3}{2}\left(t - \frac{1}{2}\sin 2t\right)$$

This last result is the solution to the problem. A necessary (but not sufficient) check of the answer is to differentiate $s(t)$ twice and compare with $a(t)$.

$$s'(t) = \frac{3}{2}(1 - \cos 2t)$$

$$s''(t) = 3\sin 2t = a(t)$$

The beginning teacher analyzes how technology can be used to solve problems and illustrate concepts involving differential and integral calculus

Technology provides extremely powerful tools for math education, particularly in the form of aids to the visualization of mathematical concepts. A central concept in calculus is that of a function. **Graphing functions** using computer software or graphing calculators makes it easy for students to understand the connections between different representations of functions and the characteristics of different types of functions. Visual representations of functions also help to clarify the concepts of limit and continuity. In addition, graphing helps students understand differentiation in terms of slope of a tangent and integration as area under the curve.

Apart from the clarification of fundamental concepts, technology can also help in solving specific problems in calculus. Many graphing calculators implement **numerical integration and differentiation** algorithms. These algorithms can be used to solve problems that cannot be solved analytically and also to verify analytical solutions.

TEACHER CERTIFICATION STUDY GUIDE

DOMAIN III. **GEOMETRY AND MEASUREMENT**

Competency 011 The teacher understands measurement as a process.

This discussion reviews measurement in the context of unit analysis, various parameters of two- and three-dimensional figures, the Pythagorean theorem, and right triangle trigonometry. The section concludes with a review of Riemann sums as applied to the area under a curve.

The beginning teacher applies dimensional analysis to derive units and formulas in a variety of situations (e.g., rates of change of one variable with respect to another) and to find and evaluate solutions to problems.

There are many methods for converting measurements among various units within a system or between systems. One method is multiplication of the given measurement by a conversion factor. This conversion factor is the following ratio, which is always equal to unity.

$$\frac{\text{new units}}{\text{old units}} \quad \text{OR} \quad \frac{\text{what you want}}{\text{what you have}}$$

The fundamental feature of **unit analysis** or **dimensional analysis** is that conversion factors can be multiplied together and units can be cancelled in the same way as numerators and denominators of numerical fractions. The following examples help clarify this point.

Example: Convert 3 miles to yards.

Multiply the initial measurement by the conversion factor, cancel the mile units, and solve:

$$\frac{3 \text{ miles}}{1} \times \frac{1,760 \text{ yards}}{1 \text{ mile}} = 5,280 \text{ yards}$$

Example: It takes Cynthia 45 minutes to get ready each morning. How many hours does she spend getting ready each week?

Multiply the initial measurement by the conversion factors from minutes to hours and from days to weeks, then cancel the minute and day units and solve:

$$\frac{45\,\text{min}}{1\,\text{day}} \times \frac{1\,\text{hour}}{60\,\text{min}} \times \frac{7\,\text{days}}{1\,\text{week}} = 5.25\,\frac{\text{hours}}{\text{week}}$$

Conversion factors for different types of units are listed below:

Measurements of length (English system)

12 inches (in)	=	1 foot (ft)
3 feet (ft)	=	1 yard (yd)
1760 yards (yd)	=	1 mile (mi)

Measurements of length (metric system)

Kilometer (km)	=	1000 meters (m)
Hectometer (hm)	=	100 meters (m)
Decameter (dam)	=	10 meters (m)
Meter (m)	=	1 meter (m)
Decimeter (dm)	=	1/10 meter (m)
Centimeter (cm)	=	1/100 meter (m)
Millimeter (mm)	=	1/1000 meter (m)

Conversion of length from English to metric

1 inch	=	2.54 centimeters
1 foot	≈	30.48 centimeters
1 yard	≈	0.91 meters
1 mile	≈	1.61 kilometers

Measurements of weight (metric system)

kilogram (kg)	=	1000 grams (g)
gram (g)	=	1 gram (g)
milligram (mg)	=	1/1000 gram (g)

Conversion of weight from metric to English

28.35 grams (g)	=	1 ounce (oz)
16 ounces (oz)	=	1 pound (lb)
2000 pounds (lb)	=	1 ton (t) (short ton)
1.1 ton (t)	=	1 metric ton (t)

Conversion of weight from English to metric

1 ounce	≈	28.35 grams
1 pound	≈	0.454 kilogram
1.1 ton	=	1 metric ton

Measurement of volume (English system)

8 fluid ounces (oz)	=	1 cup (c)
2 cups (c)	=	1 pint (pt)
2 pints (pt)	=	1 quart (qt)
4 quarts (qt)	=	1 gallon (gal)

Measurement of volume (metric system)

Kiloliter (kl)	=	1000 liters (l)
Liter (l)	=	1 liter (l)
Milliliter (ml)	=	1/1000 liter (ml)

Conversion of volume from English to metric

1 teaspoon (tsp)	≈	5 milliliters
1 fluid ounce	≈	29.57 milliliters
1 cup	≈	0.24 liters
1 pint	≈	0.47 liters
1 quart	≈	0.95 liters
1 gallon	≈	3.8 liters

Note: (') represents feet and (") represents inches.

Example: Convert 8,750 meters to kilometers.

$$\frac{8,750 \text{ meters}}{1} \times \frac{1 \text{ kilometer}}{1,000 \text{ meters}} = 8.75 \text{ km}$$

Example: 4 mi. = _____ yd.

1760 yd = 1 mi
4 mi × 1760 yd/mi = 7040 yd

Square units can be derived from the basic units of length by squaring the equivalent measurements.

1 square foot (sq. ft. or ft^2) = 144 sq. in.
1 sq. yd. = 9 sq. ft.
1 sq. yd. = 1296 sq. in.

Note that conversion in each case is performed in the following manner.

$$1 \text{ ft}^2 = (1 \text{ ft})(1 \text{ ft}) = (12 \text{ in})(12 \text{ in}) = 144 \text{ in}^2$$

Example: 14 sq. yd. = _____ sq. ft.

$$14 \text{ yd}^2 = 14 (1 \text{ yd})(1 \text{ yd}) = 14 (3 \text{ ft})(3 \text{ ft}) = 126 \text{ ft}^2$$

Example: A car skidded 170 yards on an icy road before coming to a stop. How long is the skid distance in kilometers?

Since 1 yard ≈ 0.9 meters, multiply 170 yards by 0.9 meters/1 yard.

$$170 \text{ yd.} \times \frac{0.9 \text{ m}}{1 \text{ yd.}} = 153 \text{ m}$$

Since 1000 meters = 1 kilometer, multiply 153 meters by 1 kilometer/1000 meters.

$$153 \text{ m} \times \frac{1 \text{ km}}{1000 \text{ m}} = 0.153 \text{ km}$$

Example: The distance around a race course is exactly 1 mile, 17 feet, and $9\frac{1}{4}$ inches. Approximate this distance to the nearest tenth of a foot.

Convert the distance to feet.

$$1 \text{ mile} = 1760 \text{ yards} = 1760 \times 3 \text{ feet} = 5280 \text{ feet.}$$
$$9\frac{1}{4} \text{ in.} = \frac{37}{4} \text{ in.} \times \frac{1 \text{ ft.}}{12 \text{ in.}} = \frac{37}{48} \text{ ft.} \approx 0.77083 \text{ ft.}$$

So 1 mile, 17 ft. and $9\frac{1}{4}$ in. = 5280 ft. + 17 ft. + 0.77083 ft.
= 5297.77083 ft.

The answer rounds to 5297.8 feet.

Example: If the temperature is 90° F, what is it expressed in Celsius units?

To convert between Celsius (C) and Fahrenheit (F), use the following formula.

$$\frac{C}{5} = \frac{F - 32}{9}$$

If F = 90, then $C = 5\dfrac{(90-32)}{9} = \dfrac{5 \times 58}{9} = 32.2$.

Example: A map shows a scale of 1 inch = 2 miles. Convert this scale to a numerical ratio so that any unit system (such as metric) can be used to measure distances.

The scale is a ratio—1 inch: 2 miles. If either value is converted so that the two values have the same units, then this scale can be converted to a purely numerical ratio. To avoid fractions, convert miles to inches.

$$2\,\text{mi} = 2\,\text{mi} \times \dfrac{5,280\,\text{ft}}{1\,\text{mi}} \times \dfrac{12\,\text{in}}{1\,\text{ft}} = 126,720\,\text{in}$$

The ratio is then 1:126,720.

The beginning teacher applies formulas for perimeter, area, surface area, and volume of geometric figures and shapes (e.g., polygons, pyramids, prisms, cylinders, cones, spheres) to solve problems.

Some common parameters associated with geometric figures in two and three dimensions include perimeter, area, surface area, and volume. Such figures may include one or more polygons, circles, or three-dimensional shapes. For more on polygons and circles, see **Competency 013**.

The **perimeter** of any polygon is the sum of the lengths of the sides. The **area** of a polygon is the number of square units covered by the figure or the space that a figure occupies. In the area formulae below, *b* refers to the base and *h* to the height or altitude of a figure. For a trapezoid, *a* and *b* are the two parallel bases.

FIGURE	AREA FORMULA	PERIMETER FORMULA
Rectangle	LW	$2(L+W)$
Triangle	$\dfrac{1}{2}bh$	$a+b+c$
Parallelogram	bh	sum of lengths of sides
Trapezoid	$\dfrac{1}{2}h(a+b)$	sum of lengths of sides

Even though different figures have different area formulae, the formulae are connected and one can be easily converted from one to another. For instance, it is easy to see from the diagram below that the area of the triangle ABD is half that of rectangle EABD and the area of triangle ADC is half that of rectangle AFDC. Thus, the area of triangle ABC is half that of rectangle EFBC and is equal to

$$\frac{1}{2}BC \cdot EB = \frac{1}{2}BC \cdot AD$$

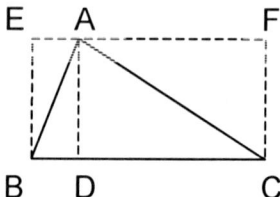

The area of a parallelogram may similarly be shown to be equal to that of an equivalent rectangle. Since triangles ACE and BDF are congruent in the diagram below (AE and BF are altitudes), parallelogram ABCD is equal in area to rectangle AEFB that has the same base (CD=EF) and height.

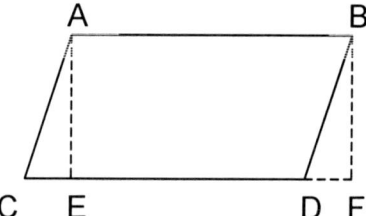

The area of a trapezoid is the sum of the areas of two triangles each of which has one of the parallel sides as a base. In the diagram below,

area of trapezoid ABCD = area of ABC + area of ACD

$$= \frac{1}{2}BC \cdot AE + \frac{1}{2}AD \cdot CF$$

$$= \frac{1}{2}AE(BC + AD) \quad \text{(Since AE = CF)}$$

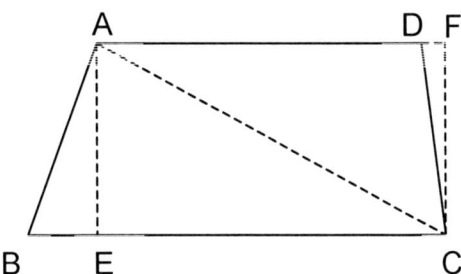

Example: A farmer has a piece of land shaped as shown below. He wishes to fence this land at an estimated cost of $25 per linear foot. What is the total cost of fencing this property to the nearest foot?

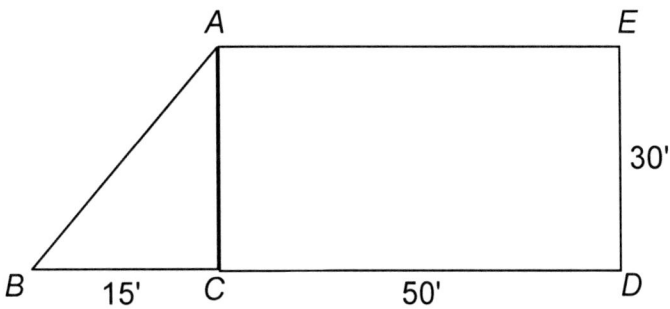

From the right triangle ABC, AC = 30 and BC = 15.

Since $(AB) = (AC)^2 + (BC)^2$, $(AB) = (30)^2 + (15)^2$. So, AB is $\sqrt{1{,}125}$ feet, or about 33.5 feet. To the nearest foot, AB = 34 feet. The perimeter of the piece of land is

$AB + BC + CD + DE + EA$ = 34 + 15 + 50 + 30 + 50 = 179 feet

The cost of fencing is $25 x 179 = $4,475.00

The area of any regular polygon having *n* sides can be expressed as a sum of the areas of *n* congruent triangles. If each side of the polygon is of length *a*, and the apothem (distance from center of polygon to one side) is *h*,

area of the polygon = $n \times \frac{1}{2} \times a \times h$ (*n* times the area of one triangle)

Since *n* x *a* is the perimeter of the polygon, we can also write

area of the polygon = $\frac{1}{2} \times \text{perimeter} \times \text{apothem}$

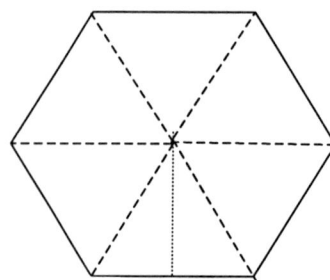

Three-Dimensional Figures

Three-dimensional figures require slightly more complicated mathematical manipulations to derive or apply such properties as surface area and volume. In some instances, two-dimensional concepts can be applied directly. In other instances, a more rigorous approach is needed.

To represent three-dimensional objects in a coordinate system, three coordinates are required. Thus, a point in three dimensions must be represented as (x, y, z), instead of simply (x, y) as is used in the two-dimensional representation.

The **volume** and **surface area** of three-dimensional figures can be derived most clearly (in some cases) using integral calculus. (See **Competency 010** for more information on integrals and on the washer method for finding the volume of a geometric figure.)

The surface area can be found by a similar integral that calculates the surface of revolution around the diameter, but there is a simpler method. Note that the differential change in volume of a sphere (dV) for a differential change in the radius (dr) is an infinitesimally thin spherical shell.

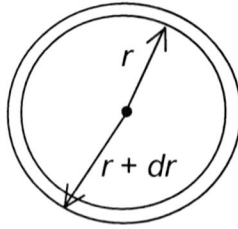

This infinitesimally thin shell is simply a surface with an area, but no volume. Find the derivative of the volume with respect to the radius to get the surface area. (Alternatively, this can be viewed as the differential volume of a thin spherical shell divided by the differential change in radius, which leads to an area.)

$$S = \frac{dV}{dr} = \frac{d}{dr}\frac{4}{3}\pi r^3$$
$$S = 4\pi r^2$$

The volume and surface area of a **right cone**, use an approach similar to that of the sphere. In this case, however, a line segment, rather than a semicircle, is revolved around the horizontal axis. The cone has a height h and a base radius r.

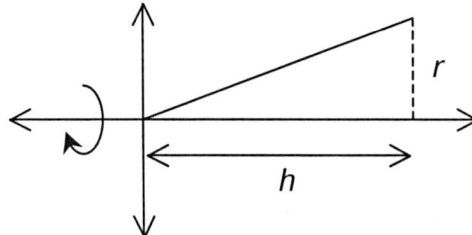

The function $f(x)$ that defines the line segment in this case is

$$f(x) = \frac{r}{h}x$$

from $x = 0$ to $x = h$.

$$V = \pi \int_0^h \left(\frac{r}{h}x\right)^2 dx$$

$$V = \pi \left(\frac{r}{h}\right)^2 \int_0^h x^2 dx = \pi \left(\frac{r}{h}\right)^2 \left[\frac{x^3}{3}\right]_0^h$$

$$V = \pi \left(\frac{r}{h}\right)^2 \frac{h^3}{3} = \frac{\pi r^2 h}{3}$$

To find the **lateral surface area** (which excludes the flat end of the cone), an integral must be used to find the surface of revolution. This integral uses $f(x)$ as follows (which is based on the arc length integral derived later in this section).

$$S = 2\pi \int_0^h f(x)\sqrt{1+\left[f'(x)\right]^2}\,dx$$

$$f'(x) = \frac{r}{h}$$

$$S = 2\pi\sqrt{1+\left(\frac{r}{h}\right)^2}\,\frac{r}{h}\int_0^h x\,dx = \frac{2\pi r}{h}\sqrt{1+\left(\frac{r}{h}\right)^2}\left[\frac{x^2}{2}\right]_0^h$$

$$S = \frac{2\pi r}{h}\sqrt{1+\left(\frac{r}{h}\right)^2}\left[\frac{h^2}{2}\right] = \pi rh\left(\frac{1}{h}\right)\sqrt{r^2+h^2}$$

$$S = \pi r\sqrt{r^2+h^2}$$

For right circular cylinders, the volume is simply the area of a cross section (a circle of radius r) multiplied by the height h of the cylinder. The lateral surface area (the surface area excluding the area on the ends of the figure) is simply the circumference of the circular cross section multiplied by the height h.

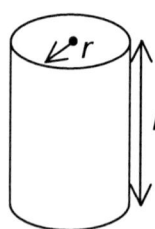

$$V = \pi r^2 h$$
$$S = 2\pi rh$$

The volumes and surface areas of these figures are summarized below.

Figure	Volume	Lateral Surface Area
Right Cylinder	$\pi r^2 h$	$2\pi rh + 2\pi r^2$
Right Cone	$\dfrac{\pi r^2 h}{3}$	$\pi r\sqrt{r^2+h^2} + \pi r^2$
Sphere	$\dfrac{4}{3}\pi r^3$	$4\pi r^2$

For figures such as pyramids and prisms, the volume and surface areas must be derived by breaking the figure into portions for which these values can be calculated easily. For instance, consider the following figure.

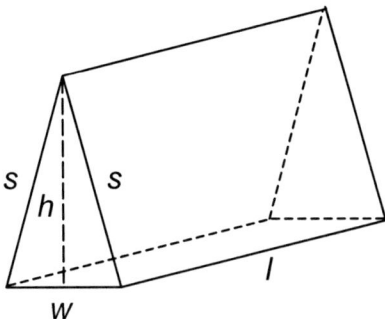

The volume of this figure can be found by calculating the area of the triangular cross section and then multiplying by l.

$$V = \frac{1}{2}hwl$$

The lateral surface area can be found by adding the areas of each side.

$$S = 2sl + lw$$

Similar reasoning applies to other figures composed of sides that are defined by triangles, quadrilaterals and other planar or linear elements.

Example: Find the height of a box whose volume is 120 cubic meters and the area of the base is 30 square meters.

$$V = Bh$$
$$120 = 30h$$
$$h = 4 \text{ meters}$$

Example: How much material is needed to make a basketball that has a diameter of 15 inches? How much air is needed to fill the basketball?

Draw and label a sketch:

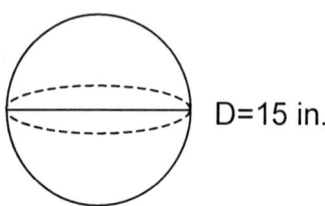

D=15 in.

Surface Area

$SA = 4\pi r^2$

$= 4\pi (7.5)^2$

$= 706.858 \text{ in}^2$

3. solve

Volume

$V = \dfrac{4}{3}\pi r^3$

$= \dfrac{4}{3}\pi (7.5)^3$

$= 1767.1459 \text{ in}^3$

1. write formula

2. substitute

Similar solids share the same shape but are not necessarily the same size. The ratio of any two corresponding measurements of similar solids is the scale factor. For example, the scale factor for two square pyramids, one with a side measuring 2 inches and the other with a side measuring 4 inches, is 2:4.

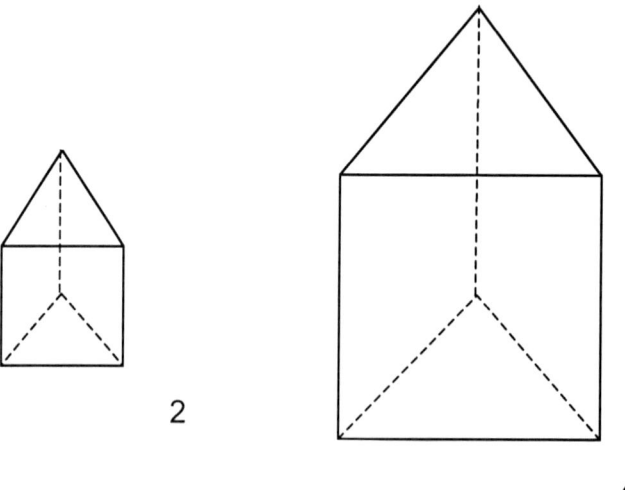

The base perimeter, the surface area, and the volume of similar solids are directly related to the scale factor. If the scale factor of two similar solids is a:b, then

> ratio of base perimeters = a:b
> ratio of areas = $a^2:b^2$
> ratio of volumes = $a^3:b^3$

Thus, for the above example,

> ratio of base perimeters = 2:4
> ratio of areas = $2^2:4^2$ = 4:16
> ratio of volumes = $2^3:4^3$ = 8:64

Example: What happens to the volume of a square pyramid when the length of the sides of the base are doubled?

> scale factor = a:b = 1:2
> ratio of volume = $1^3:2^3$ = 1:8 (The volume is increased 8 times.)

Example: Given the following measurements for two similar cylinders with a scale factor of 2:5 (cylinder A to cylinder B), determine the height, radius, and volume of each cylinder.

> cylinder A: $r = 2$
> cylinder B: $h = 10$

For cylinder A,

$$\frac{h_a}{10} = \frac{2}{5}$$
$$5h_a = 20 \quad \text{Solve for } h_a$$
$$h_a = 4$$

Volume of cylinder A = $\pi r^2 h = \pi(2)^2 4 = 16\pi$

For cylinder B,

$$\frac{2}{r_b} = \frac{2}{5}$$
$$2r_b = 10 \quad \text{Solve for } r_b$$
$$r_b = 5$$

Volume of cylinder B = $\pi r^2 h = \pi(5)^2 10 = 250\pi$

Example: A water glass of height 6" and an inner radius of 2" is filled to the top with water. How high would a conical glass with inner sides meeting at 45° have to be to hold the same amount of water?

This question requires comparing volumes. First, find the volume V_c of the cylindrical glass.

$$V_c = \pi r^2 h = \pi (2 \text{ in})^2 (6 \text{ in}) = 24\pi \text{ in}^3 \approx 75.4 \text{ in}^3$$

Next, write the volume of the cone in terms of the radius r. The relationship between the height h of the cone and the radius r is the following (since the sides meet at 45° at the apex):

$$\frac{r}{h} = \tan 22.5° \approx 0.414$$
$$r = 0.414h$$

The volume of a cone is

$$V = \frac{\pi r^2 h}{3}$$

In this case,

$$V \approx \frac{\pi h}{3}(0.414h)^2 \approx 0.057\pi h^3$$

After equating the volume of the conical glass with that of the cylindrical glass, the height of the cone that holds the same volume can be calculated.

$$75.4 \text{ in}^3 = 0.057\pi h^3$$
$$h^3 = \frac{75.4 \text{ in}^3}{0.057\pi} \approx 421.1 \text{ in}^3$$
$$h \approx 7.50 \text{ in}$$

Thus, the cone must be about 7.5 inches high.

The beginning teacher uses integral calculus to compute various measurements associated with curves and regions (e.g., area, arc length) in the plane and measurements associated with curves, surfaces, and regions in three-space.

The use of integral calculus in computing parameters associated with various figures and shapes in three-space is discussed at length in the preceding skill section.

The beginning teacher recognizes the effects on length, area, or volume when the linear dimensions of plane figures or solids are changed.

Some of the problems in the above skill section on formulas for geometric figures treat cases where dimensions of figures or solids are changed, resulting in changes to certain other parameters.

The beginning teacher applies the Pythagorean theorem, proportional reasoning, and right triangle trigonometry to solve measurement problems.

The Pythagorean Theorem

A **right triangle** is a triangle with one right angle. The side opposite the right angle is called the **hypotenuse**. The other two sides are the **legs**.

The Pythagorean theorem states that, for any right triangle, the square of the length of the hypotenuse is equal to the sum of the squares of the lengths of the legs. Symbolically, this is stated as:

$$c^2 = a^2 + b^2$$

Example: Given the right triangle below, find the missing side.

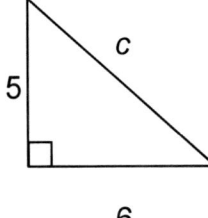

$c^2 = a^2 + b^2$ 1. write formula
$c^2 = 5^2 + 6^2$ 2. substitute known values
$c^2 = 61$ 3. take square root
$c = \sqrt{61}$ or 7.81   4. solve

The Converse of the Pythagorean Theorem states that if the square of one side of a triangle is equal to the sum of the squares of the other two sides, then the triangle is a right triangle.

Example: Given $\triangle XYZ$, with sides measuring 12, 16 and 20 cm. Is this a right triangle?

$$c^2 = a^2 + b^2$$
$$20^2 \ ? \ 12^2 + 16^2$$
$$400 \ ? \ 144 + 256$$
$$400 = 400$$

Yes, the triangle is a right triangle.

This theorem can be expanded to determine if triangles are obtuse or acute.

If the square of the longest side of a triangle is greater than the sum of the squares of the other two sides, then the triangle is an obtuse triangle. If the square of the longest side of a triangle is less than the sum of the squares of the other two sides, then the triangle is an acute triangle.

Example: Given $\triangle LMN$ with sides measuring 7, 12, and 14 inches. Is the triangle right, acute, or obtuse?

$$14^2 \ ? \ 7^2 + 12^2$$
$$196 \ ? \ 49 + 144$$
$$196 > 193$$

Therefore, the triangle is obtuse.

When an altitude is drawn to the hypotenuse of a right triangle, then the two triangles formed are similar to the original triangle and to each other.

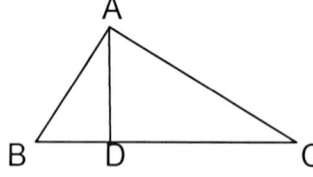

Given right triangle ABC with right angle at A, altitude AD drawn to hypotenuse BD at D, $\triangle ABC \sim \triangle ABD \sim \triangle ACD$.

If a, b and c are positive numbers such that $\dfrac{a}{b} = \dfrac{b}{c}$, then b is called the **geometric mean** between a and c.

The geometric mean is significant when the altitude is drawn to the hypotenuse of a right triangle.

The length of the altitude is the geometric mean between each segment of the hypotenuse. Also, each leg is the geometric mean between the hypotenuse and the segment of the hypotenuse that is adjacent to the leg.

Right Triangle Trigonometry

For a discussion of right triangle trigonometry, see **Competency 009**.

Proportional Reasoning

The use of such concepts as similarity (discussed above for geometric figures) is part of proportional reasoning. Often, it is helpful to find an unknown parameter in a problem through comparison to a known parameter. This approach can be applied to triangle similarity (discussed in **Competency 012**), for instance.

The beginning teacher relates the concept of area under a curve to the limit of a Riemann sum.

The formal definition of an integral (which can be interpreted as the area under a curve) is based on the **Riemann sum**, as discussed in **Competency 010**.

The use of the Riemann sum, leading to integral calculus, provides an intellectual foundation for understanding the use of integral calculus in deriving such parameters as volume and area for various geometric figures. Since integral calculus can handle general functions, the results obtained from this approach are not limited to cases where, for instance, a figure is a polygon or otherwise has a perimeter composed exclusively of finite-length line segments. The application of integral calculus to geometric figures is discussed earlier in this competency.

Competency 012   **The teacher understands geometries, in particular Euclidean geometry, as axiomatic systems.**

Euclidean geometry is the study of the properties of two-dimensional (planar) and three-dimensional (solid) figures. It (like other axiomatic systems) is based on the **undefined terms** (or concepts) and a set of self-evident statements or **axioms**. Starting from these building blocks, deductive reasoning is used to prove a set of propositions or **theorems**. Euclidean geometry is based on the undefined terms of **point, line** and **plane**, which, along with certain axioms, can be used to derive the properties of different geometric figures. The axioms and theorems and the process of formal proof provide a consistent logical framework that can be used to derive further results. The following review of geometric concepts provides a basic overview of proof relating to geometry, as well as the characteristics of various types of polygons, circles, compound shapes and three-dimensional figures.

The beginning teacher understands axiomatic systems and their components (e.g., undefined terms, defined terms, theorems, examples, counterexamples).

Deductive thinking is the process of arriving at a conclusion based on other statements that are all known to be true, such as theorems, axioms or postulates. Valid mathematical arguments are deductive in nature.

A **direct proof** demonstrates a proposition by beginning with the given information and showing that it leads to the proposition through logical steps. An **indirect proof** of a proposition can be carried out by demonstrating that the opposite of the proposition is untenable.

A proof of a geometrical proposition is typically presented in a format with two columns side by side. In a **two-column proof** of this type, the left column consists of the given information or statements that can be proved by deductive reasoning. The right column consists of the reasons used to justify each statement on the left. The right side should identify given information or state the theorems, postulates, definitions or algebraic properties used to show that the corresponding steps are valid.

The following **algebraic postulates** are frequently used as justifications for statements in two-column geometric proofs:

Addition Property:	If $a = b$ and $c = d$, then $a + c = b + d$.
Subtraction Property:	If $a = b$ and $c = d$, then $a - c = b - d$.
Multiplication Property:	If $a = b$, then $ac = bc$.
Division Property:	If $a = b$ and $c \neq 0$, then $a/c = b/c$.
Reflexive Property:	$a = a$
Symmetric Property:	If $a = b$, then $b = a$.
Transitive Property:	If $a = b$ and $b = c$, then $a = c$.
Distributive Property:	$a(b + c) = ab + ac$
Substitution Property:	If $a = b$, then b may be substituted for a in any other expression (a may also be substituted for b).

The beginning teacher uses properties of points, lines, planes, angles, lengths, and distances to solve problems.

A **point** is a dimensionless location and has no length, width or height.

A **line** connects a series of points and continues "straight" infinitely in two directions. Lines extend in one dimension. A line is defined by any two points that fall on the line; therefore a line may have multiple names.

A **line segment** is a portion of a line. A line segment is the shortest distance between two endpoints and is named using those end points. Line segments therefore have exactly two names (i.e., $\overline{AB}$ or $\overline{BA}$). Because line segments have two endpoints, they have a defined length or distance.

A **ray** is a portion of a line that has only one end point and continues infinitely in one direction. Rays are named using the endpoint as the first point and any other point on the ray as the second.

Note that the symbol for a line includes two arrows (indicating infinite extent in both directions), the symbol for a ray includes only one arrow (indicating that it has one end point) and the symbol for a line segment has no arrows (indicating two end points).

Example: Use the diagram below, calculate the length of $\overline{AB}$ given $\overline{AC}$ is 6 cm and $\overline{BC}$ is twice as long as $\overline{AB}$.

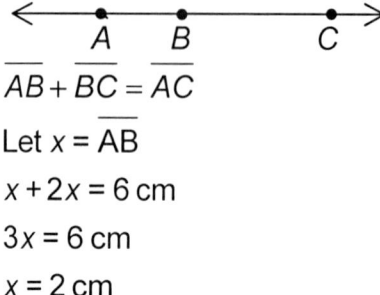

$\overline{AB} + \overline{BC} = \overline{AC}$

Let $x = \overline{AB}$

$x + 2x = 6$ cm

$3x = 6$ cm

$x = 2$ cm

A **plane** is a flat surface defined by three points. Planes extend indefinitely in two dimensions. A common example of a plane is x-y plane used in the Cartesian coordinate system.

In geometry, the point, line, and plane are key concepts and can be discussed in relation to each other.

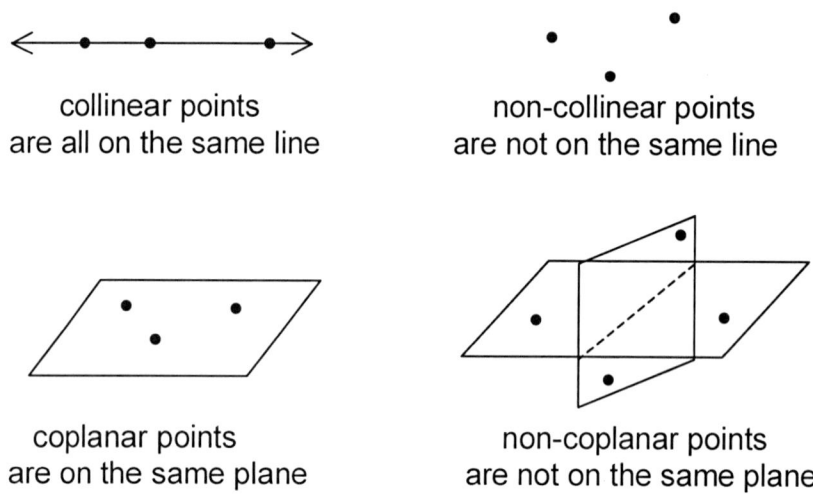

collinear points
are all on the same line

non-collinear points
are not on the same line

coplanar points
are on the same plane

non-coplanar points
are not on the same plane

Problems throughout this competency illustrate the use of these various geometric elements in the solution of problems.

The beginning teacher applies the properties of parallel and perpendicular lines to solve problems.

Parallel lines in two dimensions can be sufficiently defined as lines that do not intersect. In three dimensions, however, this definition is insufficient. **Parallel lines** in three dimensions are defined as lines for which every pair of nearest points on the lines has a fixed distance.

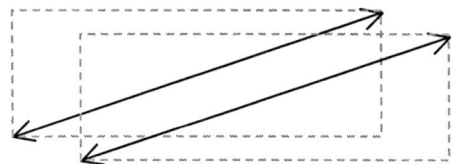

Lines in three dimensions that do not intersect and are not parallel are called **skew lines**. Parallel lines are coplanar, skew lines are not.

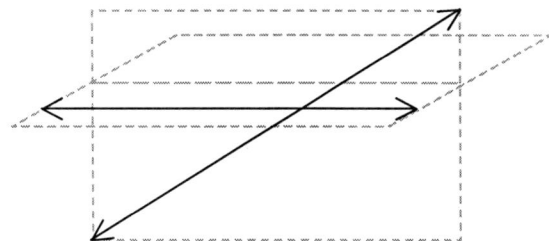

Two planes intersect on a single line. If two planes do not intersect, then they are parallel. Parallel and non-parallel planes are shown in the diagram below.

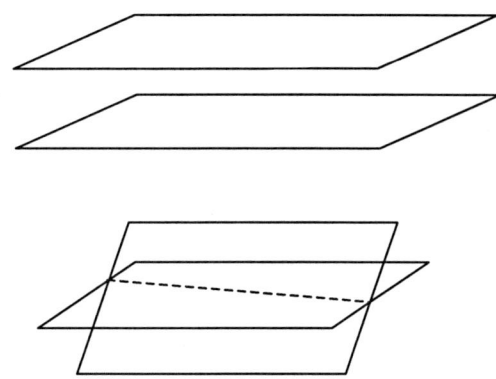

Parallelism between two planes may also be defined in the same way as parallel lines: the distance between any pair of nearest points (one point on each plane) is constant.

Perpendicularity of lines and planes in three dimensions is largely similar to that of two dimensions. Two lines are **perpendicular**, in two or three dimensions, if they intersect at a point and form 90° angles between them. Consequently, perpendicular lines are always coplanar.

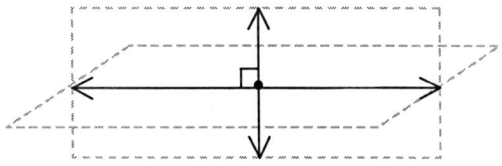

Notice that, for any line and coincident point on that line, there are an infinite number of perpendicular lines to the line through that point. In two dimensions, there is only one.

Two planes are perpendicular if they intersect and the angles formed between them are 90°. For any given plane and line on that plane, there is only one perpendicular plane.

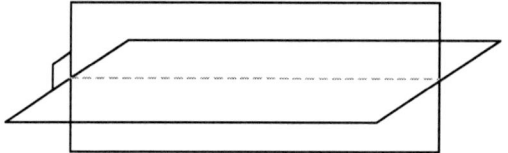

Properties of Parallel Lines

The **Parallel Postulate** in Euclidean planar geometry states that if a line *l* is crossed by two other lines *m* and *n* (where the crossings are not at the same point on *l*), then *m* and *n* intersect on the side of *l* where the sum of the interior angles α and β is less than 180°. This scenario is illustrated below.

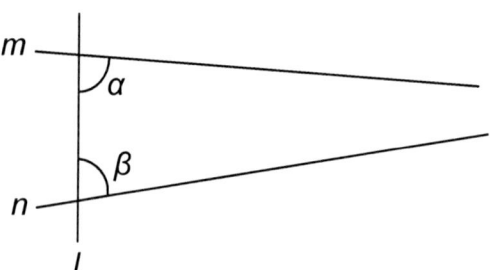

Based on this definition, a number of implications and equivalent formulations can be derived. First, note that the lines *m* and *n* intersect on the right-hand side of *l* above only if $\alpha + \beta < 180°$. This implies that if α and β are both 90° and, therefore, $\alpha + \beta = 180°$, then the lines do not intersect on either side. This is illustrated below.

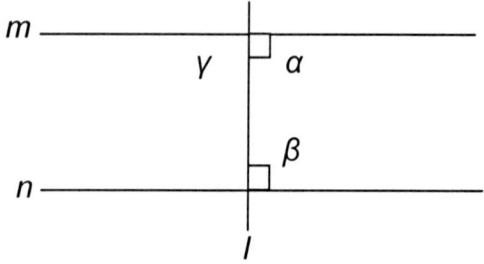

The supplementary angles formed by the intersection of *l* and *m* (and the intersection of *l* and *n*) must sum to 180°:

$$\alpha + \gamma = 180° \qquad \beta + \delta = 180°$$

Since these sums are both equal to 180°, the lines *m* and *n* do not intersect on either side of *l*. That is to say, these lines are **parallel**.

Let the non-intersecting lines *m* and *n* used in the above discussion remain parallel, but adjust *l* such that the interior angles are no longer right angles.

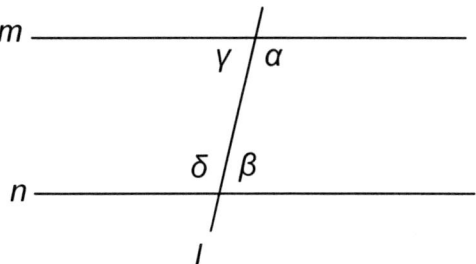

The Parallel Postulate still applies, and it is therefore still the case that $\alpha + \beta = 180°$ and $\gamma + \delta = 180°$. Combined with the fact that $\alpha + \gamma = 180°$ and $\beta + \delta = 180°$, the **Alternate Interior Angle Theorem** can be justified. This theorem states that if two parallel lines are cut by a transversal, the alternate interior angles are congruent.

By manipulating the four relations based on the above diagram, the relationships between alternate interior angles (γ and β form one set of alternate interior angles, and α and δ form the other) can be established.

$$\alpha = 180° - \beta$$
$$\alpha + \gamma = 180° = 180° - \beta + \gamma$$
$$-\beta + \gamma = 0$$
$$\gamma = \beta$$

By the same reasoning,

$$\gamma = 180° - \delta$$
$$\beta + \delta = 180° = \beta + 180° - \delta$$
$$\beta = \delta$$

One of the consequences of the Parallel Postulate, in addition the Alternate Interior Angle Theorem, is that **corresponding angles** are equal. If two parallel lines are cut by a transversal line, then the corresponding angles are equal. The diagram below illustrates one set of corresponding angles (α and β) for the parallel lines *m* and *n* cut by *l*.

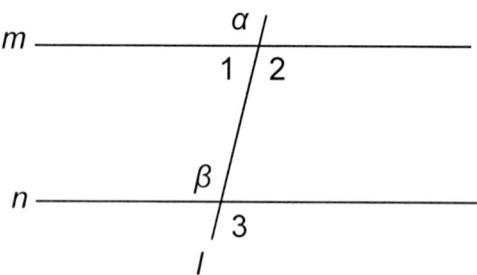

That α and β are equal can be proven as follows.

$\angle \beta = \angle 2$	Alternate Interior Angle Theorem
$\angle 1 + \angle 2 = 180°$	Supplementary angles
$\angle 2 = 180° - \angle 1$	
$\angle 1 + \angle \alpha = 180°$	Supplementary angles
$\angle \alpha = 180° - \angle 1$	
$\angle 2 = 180° - \angle 1 = \angle \alpha$	
$\angle 2 = \angle \alpha$	
$\angle \beta = \angle 2 = \angle \alpha$	
$\angle \beta = \angle \alpha$	

Thus, it has been proven that corresponding angles are equal. Note, also, that the above proof also demonstrates that vertical angles are equal ($\angle 2 = \angle \alpha$). Thus, opposite angles formed by the intersection of two lines (called **vertical angles**) are equal. Furthermore, **alternate exterior angles** (angles α and 1 in the diagram above) are also equal.

$\angle \beta = \angle 3$	Vertical angles
$\angle \alpha = \angle 2$	Vertical angles
$\angle \beta = \angle 2$	Alternate Interior Angle Theorem
$\angle \alpha = \angle 2 = \angle \beta = \angle 3$	
$\angle \alpha = \angle 3$	

TEACHER CERTIFICATION STUDY GUIDE

The beginning teacher uses properties of congruence and similarity to explore geometric relationships, justify conjectures, and prove theorems.

Congruence

Congruent figures have the same size and shape; i.e., if one of the figures is superimposed on the other, the boundaries coincide exactly. Congruent line segments have the same length; congruent angles have equal measures. The symbol ≅ is used to indicate that two figures, line segments or angles are congruent.

The **reflexive, symmetric** and **transitive** properties described for algebraic equality relationships may also be applied to congruence. For instance, if $\angle A \cong \angle B$ and $\angle A \cong \angle D$, then $\angle B \cong \angle D$ (transitive property).

The polygons (pentagons) *ABCDE* and *VWXYZ* shown below are congruent since they are exactly the same size and shape.

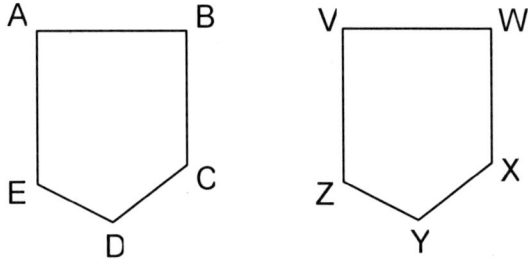

$ABCDE \cong VWXYZ$

Corresponding parts are congruent angles and congruent sides. For the polygons shown above:

corresponding angles	corresponding sides
$\angle A \leftrightarrow \angle V$	$AB \leftrightarrow VW$
$\angle B \leftrightarrow \angle W$	$BC \leftrightarrow WX$
$\angle C \leftrightarrow \angle X$	$CD \leftrightarrow XY$
$\angle D \leftrightarrow \angle Y$	$DE \leftrightarrow YZ$
$\angle E \leftrightarrow \angle Z$	$AE \leftrightarrow VZ$

Two triangles are congruent if each of the three angles and three sides of one triangle match up in a one-to-one fashion with congruent angles and sides of the second triangle. To see how the sides and angles match up, it is sometimes necessary to imagine rotating or reflecting one of the triangles so the two figures are oriented in the same position.

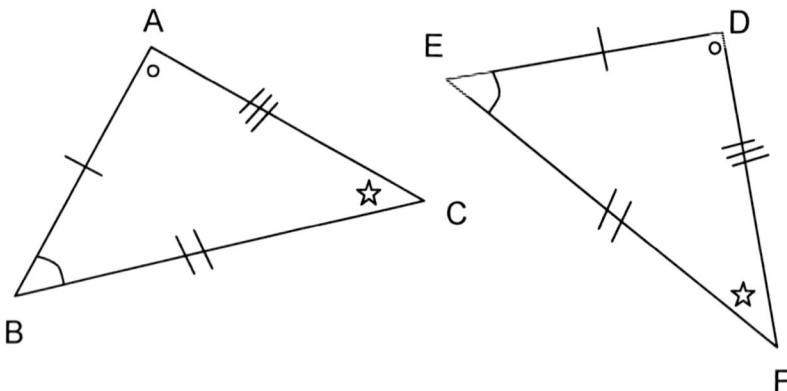

In the example above, the two triangles ABC and DEF are congruent if these 6 conditions are met:

1. ∠A ≅ ∠D   4. $\overline{AB} \cong \overline{DE}$
2. ∠B ≅ ∠E   5. $\overline{BC} \cong \overline{EF}$
3. ∠C ≅ ∠F   6. $\overline{AC} \cong \overline{DF}$

The congruent angles and segments "correspond" to each other.

It is not always necessary to demonstrate all of the above six conditions to prove that two triangles are congruent. There are several "shortcut" methods described below.

The **SAS Postulate** (side-angle-side) states that if two sides and the included angle of one triangle are congruent to two sides and the included angle of another triangle, then the two triangles are congruent.

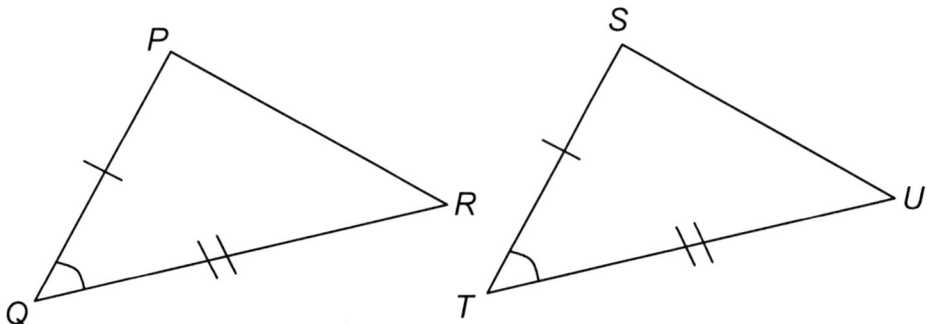

To see why this is true, imagine moving the triangle PQR (shown above) in such a way that the point P coincides with the point S, and line segment PQ coincides with line segment ST. Point Q will then coincide with T since PQ ≅ ST. Also, segment QR will coincide with TU, because ∠Q ≅ ∠T. Point R will coincide with U, because QR ≅ TU. Since P and S coincide and R and U coincide, line PR will coincide with SU because two lines cannot enclose a space. Thus the two triangles match perfectly point for point and are congruent.

Example: Are the following triangles congruent?

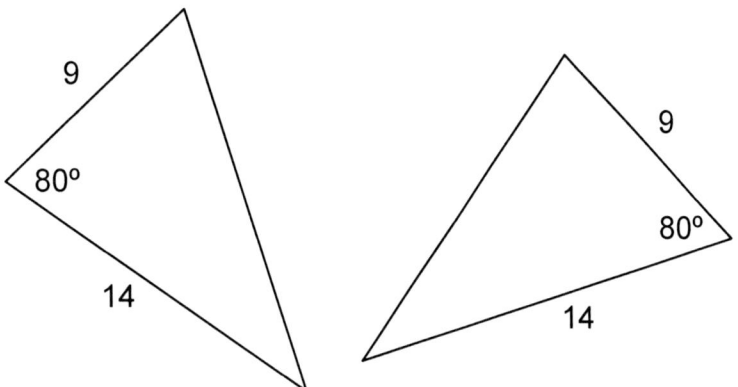

Each of the two triangles has a side that is 14 units and another that is 9 units. The angle included in the sides is 80° in both triangles. Therefore, the triangles are congruent by SAS.

The **SSS Postulate** (side-side-side) states that if three sides of one triangle are congruent to three sides of another triangle, then the two triangles are congruent.

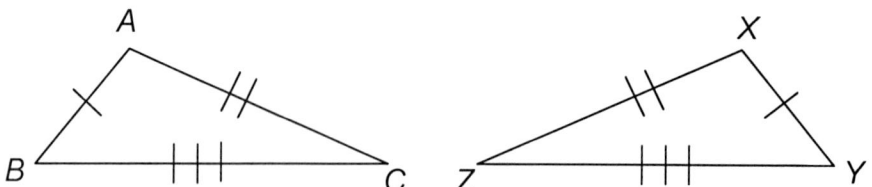

Since $AB \cong XY$, $BC \cong YZ$ and $AC \cong XZ$, then $\triangle ABC \cong \triangle XYZ$.

Example: Given isosceles triangle ABC with D being the midpoint of base AC, prove that the two triangles ABD and ADC are congruent.

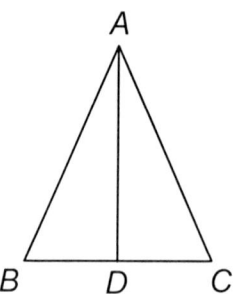

Proof:

1. Isosceles triangle ABC,
   D midpoint of base AC   Given
2. $AB \cong AC$ An isosceles triangle has two congruent sides
3. $BD \cong DC$ Midpoint divides a line into two equal parts
4. $AD \cong AD$ Reflexive property
5. $\triangle ABD \cong \triangle BCD$ SSS

The **ASA Postulate** (angle-side-angle) states that if two angles and the included side of one triangle are congruent to two angles and the included side of another triangle, the triangles are congruent.

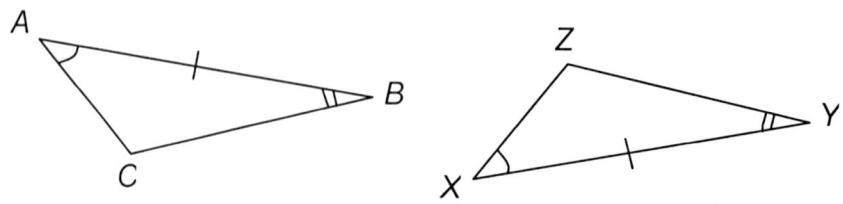

$\angle A \cong \angle X$, $\angle B \cong \angle Y$, $AB \cong XY$ then $\triangle ABC \cong \triangle XYZ$ by ASA

Example: Given two right triangles with one leg (*AB* and *KL*) of each measuring 6 cm and the adjacent angle 37°, prove the triangles are congruent.

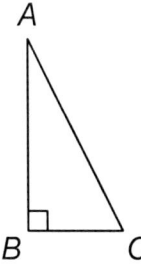
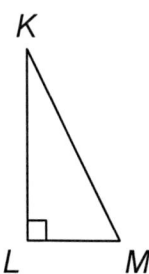

Proof:

1. Right △*ABC* and △*KLM* Given
   AB = *KL* = 6 cm
   ∠*A* = ∠*K* = 37°
2. *AB* ≅ *KL* Figures with the same
   ∠*A* ≅ ∠*K* measure are congruent
3. ∠*B* ≅ ∠*L* All right angles are congruent.
4. △*ABC* ≅ △*KLM* ASA

Example: What method could be used to prove that triangles ABC and ADE are congruent?

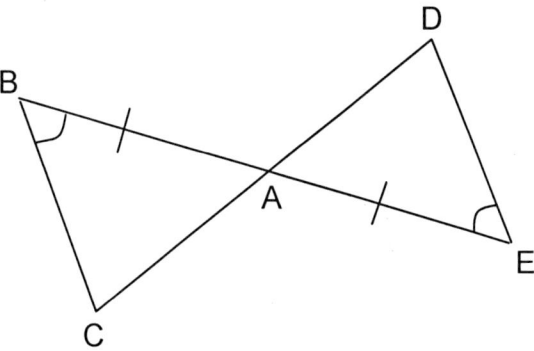

The sides *AB* and *AE* are given as congruent, as are ∠*BAC* and ∠*DAE*. ∠*BAC* and ∠*DAE* are vertical angles and are therefore congruent. Thus triangles △*ABC* and △*ADE* are congruent by the ASA postulate.

The **HL Theorem** (hypotenuse-leg) is a congruence shortcut that can only be used with right triangles. According to this theorem, if the hypotenuse and leg of one right triangle are congruent to the hypotenuse and leg of the other right triangle, then the two triangles are congruent.

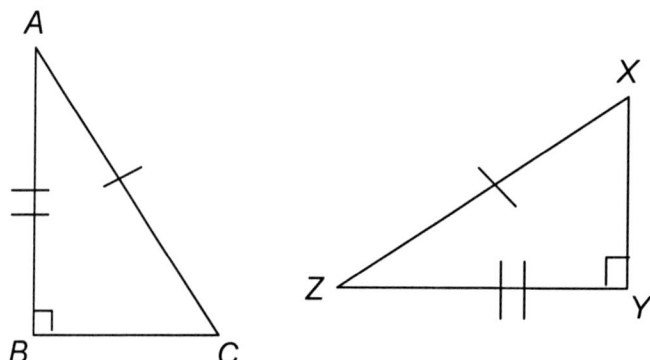

If ∠B and ∠Y are right angles and $AC \cong XZ$ (hypotenuse of each triangle), $AB \cong YZ$ (corresponding leg of each triangle), then $\triangle ABC \cong \triangle XYZ$ by HL.

Proof:

1. $\angle B \cong \angle Y$
   $AB \cong YZ$
   $AC \cong XZ$ Given

2. $BC = \sqrt{AC^2 - AB^2}$ Pythagorean theorem

3. $XY = \sqrt{XZ^2 - YZ^2}$ Pythagorean theorem

4. $XY = \sqrt{AC^2 - AB^2} = BC$ Substitution ($XZ \cong AC$, $YZ \cong AB$)

5. $\triangle ABC \cong \triangle XYZ$ SAS ($AB \cong YZ$, $\angle B \cong \angle Y$, $BC \cong XY$)

Similarity

Two figures that have the same shape are **similar**. To be the same shape, corresponding angles must be equal. Therefore, polygons are similar if and only if there is a one-to-one correspondence between their vertices such that the corresponding angles are congruent. For similar figures, the lengths of corresponding sides are proportional. The symbol ~ is used to indicate that two figures are similar.

The polygons ABCDE and VWXYZ shown below are similar.

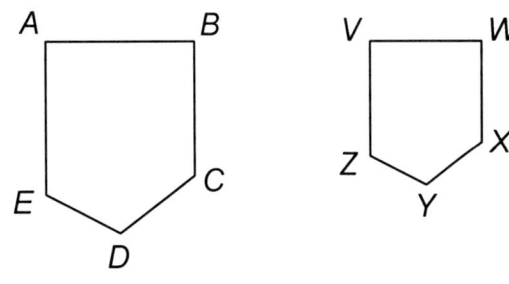

ABCDE ~ VWXYZ

Corresponding angles: $\angle A = \angle V$, $\angle B = \angle W$, $\angle C = \angle X$, $\angle D = \angle Y$, $\angle E = \angle Z$

Corresponding sides: $\dfrac{AB}{VW} = \dfrac{BC}{WX} = \dfrac{CD}{XY} = \dfrac{DE}{YZ} = \dfrac{AE}{VZ}$

Example: Given two similar quadrilaterals, find the lengths of sides x, y, and z.

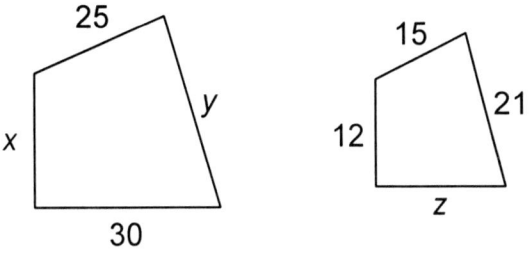

Since corresponding sides are proportional, 15/25 = 3/5, so the scale factor is 3/5.

$\dfrac{12}{x} = \dfrac{3}{5}$ $\dfrac{21}{y} = \dfrac{3}{5}$ $\dfrac{z}{30} = \dfrac{3}{5}$
$3x = 60$ $3y = 105$ $5z = 90$
$x = 20$ $y = 35$ $z = 18$

Just as for congruence, there are shortcut methods that can be used to prove similarity.

According to the **AA Similarity Postulate**, if two angles of one triangle are congruent to two angles of another triangle, then the triangles are similar. It is obvious that if two of the corresponding angles are congruent, the third set of corresponding angles must be congruent as well. Hence, showing AA is sufficient to prove that two triangles are similar.

The **SAS Similarity Theorem** states that, if an angle of one triangle is congruent to an angle of another triangle and the sides adjacent to those angles are in proportion, then the triangles are similar.

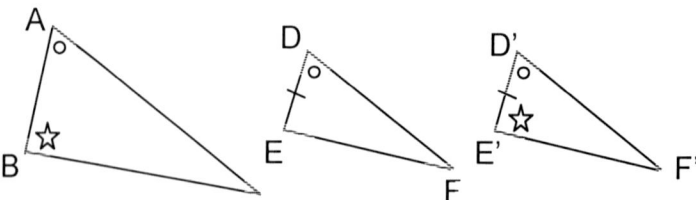

If $\angle A = \angle D$ and $\dfrac{AB}{DE} = \dfrac{AC}{DF}$, $\triangle ABC \sim \triangle DEF$.

Example: A graphic artist is designing a logo containing two triangles. The artist wants the triangles to be similar. Determine whether the artist has created similar triangles.

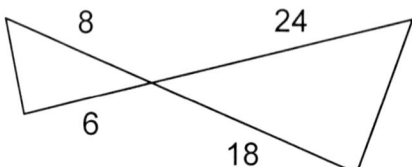

The sides are proportional $\dfrac{8}{24} = \dfrac{6}{18} = \dfrac{1}{3}$ and vertical angles are congruent. The two triangles are therefore similar by the SAS similarity theorem.

According to the **SSS Similarity Theorem**, if the sides of two triangles are in proportion, then the triangles are similar.

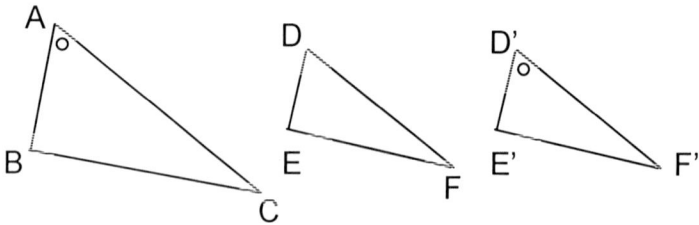

If $\dfrac{AB}{DE} = \dfrac{AC}{DF} = \dfrac{BC}{EF}$, $\triangle ABC \sim \triangle DEF$

Example: Tommy draws and cuts out 2 triangles for a school project. One of them has sides of 3, 6, and 9 inches. The other triangle has sides of 2, 4, and 6. Is there a relationship between the two triangles?

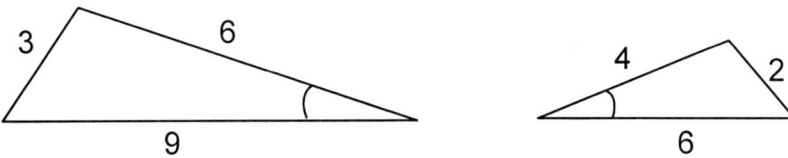

Determine the proportions of the corresponding sides.

$$\frac{2}{3} \qquad \frac{4}{6} = \frac{2}{3} \qquad \frac{6}{9} = \frac{2}{3}$$

The smaller triangle is 2/3 the size of the large triangle, therefore they are similar triangles by the SSS similarity theorem.

The beginning teacher describes and justifies geometric constructions made using compass and straightedge, reflection devices, and other appropriate technologies.

Classical construction refers to the use of a straightedge and compass for creating geometrical figures that match certain criteria. A construction consists of only segments, arcs, and points. Typical constructions includes the replication of line segments, angles or shapes, bisection of angles and lines, drawing lines that are parallel or perpendicular to a given line, as well as drawing different kinds of polygons and circles.

Duplication of line segments and angles

The easiest construction to make is to **duplicate a given line segment**. Given segment AB, construct a segment equal in length to segment AB by following these steps.

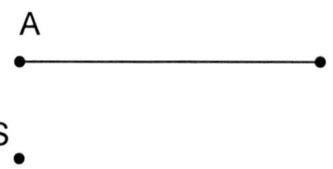

1. Place a point anywhere in the plane to anchor the duplicate segment. Call this point S.

2. Open the compass to match the length of segment AB. Keeping the compass rigid, swing an arc from S.

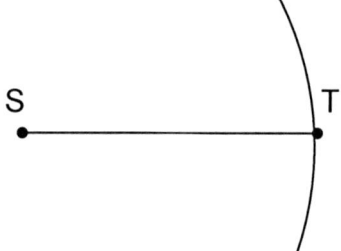

3. Draw a segment from S to any point on the arc. This segment will be the same length as AB.

To construct an angle congruent to a given angle:

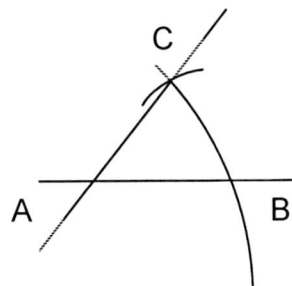

1. Given the angle A, draw an arc with any radius such that it intersects the sides of the angle at B and C.

2. On a working line w, select a point A' as the vertex of the angle to be drawn. With A' as the center and the same radius as the previous arc, draw an arc that intersects the line w in B'.

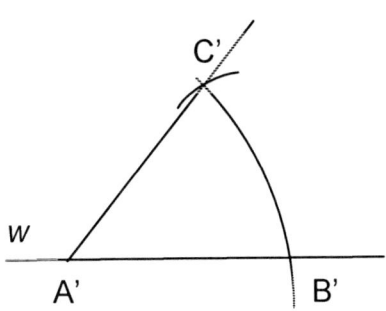

3. With B' as the center and radius equal to BC (measured off by placing ends of compass on B and C), draw a second arc that intersects the first arc at C'.

4. Join points A' and C'. The angle A' is congruent to the given angle A.

Construction of parallel lines

Angle duplication may be used to construct a line parallel to a given line AB through point W as shown below:

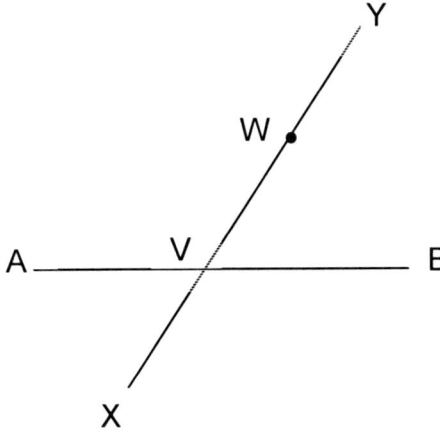

1. Draw a line XY through W intersecting AB in V.

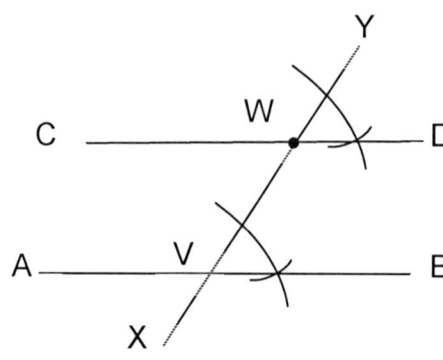

2. Construct angle YWD congruent to angle WVB. Then CD is the line parallel to AB.

Construction of perpendicular lines and bisectors

Given a line such as line l and a point P not on l, follow these steps to construct a **perpendicular line to l that passes through P**.

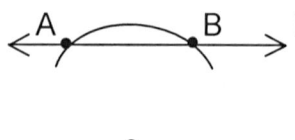

1. Swing an arc of any radius from P so that the arc intersects line l in two points A and B.

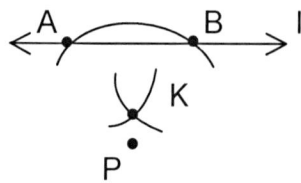

2. Open the compass to any length and swing two arcs of the same radius, one from A and the other from B. These two arcs will intersect at a new point K.

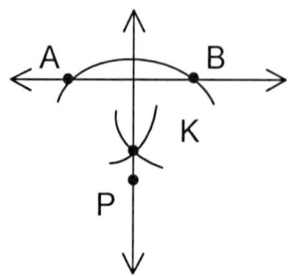

3. Connect K and P to form a line perpendicular to line l that passes through P.

Given a line segment with two endpoints such as A and B, follow these steps to **construct the line that both bisects and is perpendicular** to the line given segment.

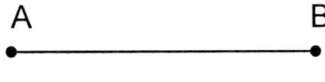

1. Swing an arc of any radius from point A. Swing another arc of the same radius from B. The arcs will intersect at two points. Label these points C and D.

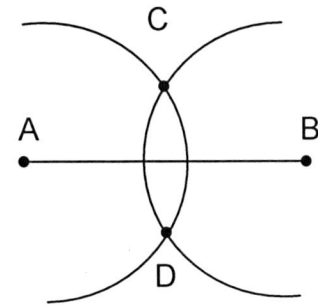

2. Connect C and D to form the perpendicular bisector of segment

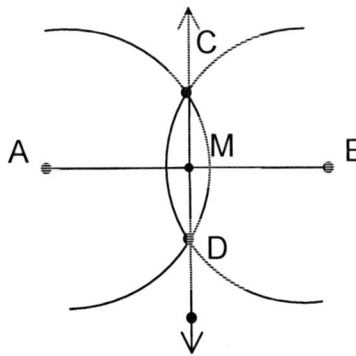

3. The point M where line CD and segment AB intersect is the midpoint of segment AB.

Construction of angle bisectors

To **bisect a given angle** such as angle *FUZ*, follow these steps.

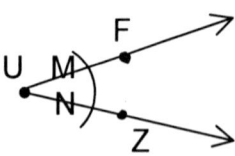

1. Swing an arc of any length with its center at point U. This arc will intersect rays *UF* and *UZ* at M and N.

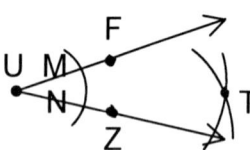

2. Open the compass to any length and swing one arc from point M and another arc of the same radius from point N. These arcs will intersect in the interior or angle *FUZ* at point T.

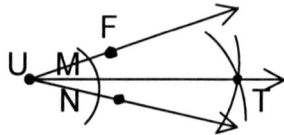

3. Connect U and T for the ray which bisects angle *FUZ*. Ray *UT* is the angle bisector of angle *FUZ*

The beginning teacher demonstrates an understanding of the use of appropriate software to explore attributes of geometric figures and to make and evaluate conjectures about geometric relationships.

Various software packages can be useful in analyzing and comparing the characteristics of geometric figures. For instance, simple drawing programs are one possibility; although these types of programs do not generally allow for sophisticated analyses, they do allow for simple manipulation of shapes, such as rotations, reflections, overlays, and other operations.

More sophisticated software packages may allow measurement of angles and lengths, construction of figures with specified attributes, automatic calculation of parameters such as area and volume, and other features.

A list of some available software packages can be found at http://mathforum.org/geometry/geometry.software.html.

The beginning teacher Compares and contrasts the axioms of Euclidean geometry with those of non-Euclidean geometry (i.e., hyperbolic and elliptic geometry).

Euclid wrote a set of 13 books around 330 B.C. called the Elements. He outlined 10 axioms and then deduced 465 theorems. Euclidean geometry is based on the undefined concept of the point, line and plane.

The fifth of Euclid's axioms (referred to as the parallel postulate) was not as readily accepted as the other nine axioms. Many mathematicians throughout the years have attempted to prove that this axiom is not necessary because it could be proved by the other nine. Among the many who attempted to prove this was Carl Friedrich Gauss; his work led to the development of hyperbolic geometry. Elliptical, spherical or Riemannian geometry were hypothesized by G.F. Berhard Riemann, who based his work on the theory of surfaces and used models as physical interpretations of the undefined terms that satisfy the axioms.

The variants of the Parallel Postulate that lead to non-Euclidean geometries are based on the number of possible unique lines that can be parallel to a line *l*, where the potential parallel lines must pass through a specific point not on *l*.

The chart below lists the fifth axiom (the Parallel Postulate) as it is given in each of the three geometries:

Euclidean Geometry	Spherical or Riemannian Geometry	Hyperbolic or Saddle Geometry
Through a point not on a line, there is no more than one line parallel to that line.	If l is any line and P is any point not on l, then there are no lines through P that are parallel to l.	If l is any line and P is any point not on l, then there exists at least two lines through P that are parallel to l.

Euclidean geometry is the study of flat, two-dimensional space. Non-Euclidean geometries involve **curved surfaces**, such as that of a sphere. These geometries have a direct connection to our experiences: for instant, the surface of the Earth is (roughly) spherical. Note the results of the parallel postulate for spherical geometry. (A line on a sphere is a so-called great circle, which is a circle of the same radius as the sphere.)

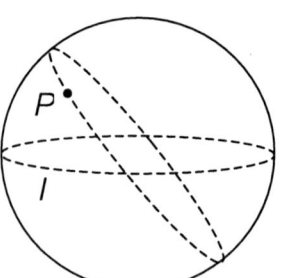

The line through point P is not parallel to line l. Furthermore, it is clear through observation of the figure that, since there is only one great circle through any given point, there are non-collinear parallel lines in spherical geometry. Thus, as is demonstrated by this example, varying the details of the Parallel Postulate leads to various types of non-Euclidean geometries.

Non-Euclidean geometries have application beyond just abstract mathematical theory. For instance, hyperbolic geometry is a central concept in Einstein's theory of relativity.

Competency 013 The teacher understands the results, uses, and applications of Euclidean geometry.

Various aspects of Euclidean geometry are covered in this section, including the properties of polygons and circles, cross sections and nets, and using various views of three-dimensional figures to facilitate accurate and complete representations of these figures.

The beginning teacher analyzes the properties of polygons and their components.

A polygon is a simple closed figure composed of line segments. Here we will consider only **convex polygons**, i.e. polygons for which the measure of each internal angle is less than 180°. Of the two polygons shown below, the one on the left is a convex polygon.

A **regular polygon** is one for which all sides are the same length and all interior angles are the same measure.

The sum of the measures of the **interior angles** of a polygon can be determined using the following formula, where n represents the number of angles in the polygon.

Sum of $\angle s = 180(n - 2)$

The measure of each angle of a regular polygon can be found by dividing the sum of the measures by the number of angles.

Measure of $\angle = \dfrac{180(n-2)}{n}$

Example: Find the measure of each angle of a regular octagon.
Since an octagon has eight sides, each angle equals:

$$\dfrac{180(8-2)}{8} = \dfrac{180(6)}{8} = 135°$$

The sum of the measures of the **exterior angles** of a polygon, taken one angle at each vertex, equals 360°.

The measure of each exterior angle of a regular polygon can be determined using the following formula, where *n* represents the number of angles in the polygon.

Measure of exterior ∠ of regular polygon
$$= 180 - \frac{180(n-2)}{n} = \frac{360}{n}$$

Example: Find the measure of the interior and exterior angles of a regular pentagon.

Since a pentagon has five sides, each exterior angle measures:

$$\frac{360}{5} = 72°$$

Since each exterior angle is supplementary to its interior angle, the interior angle measures 180 − 72 or 108°.

A **quadrilateral** is a polygon with four sides. The sum of the measures of the angles of a convex quadrilateral is 360°.

A **trapezoid** is a quadrilateral with exactly <u>one</u> pair of parallel sides. The two parallel sides of a trapezoid are called the bases, and the two non-parallel sides are called the legs. If the two legs are the same length, then the trapezoid is called isosceles.

The segment connecting the two midpoints of the legs is called the median. The median has the following two properties:
1. The median is parallel to the two bases.
2. The length of the median is equal to one-half the sum of the length of the two bases.

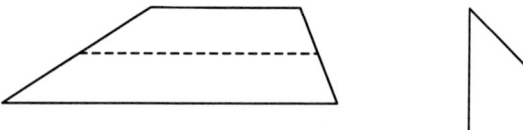

In an **isosceles trapezoid**, the non-parallel sides are congruent.

An isosceles trapezoid has the following properties:

1. The diagonals of an isosceles trapezoid are congruent.
2. The base angles of an isosceles trapezoid are congruent.

Example: An isosceles trapezoid has a diagonal of 10 and a base angle measure of 30°. Find the measure of the other 3 angles.

Based on the properties of trapezoids, the measure of the other base angle is 30° and the measure of the other diagonal is 10. The other two angles have a measure of

$$360 = 30(2) + 2x$$
$$x = 150°$$

The other two angles measure 150° each.

A **parallelogram** is a quadrilateral with <u>two</u> pairs of parallel sides and has the following properties:

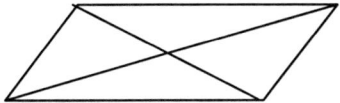

1. The diagonals bisect each other.
2. Each diagonal divides the parallelogram into two congruent triangles.
3. Both pairs of opposite sides are congruent.
4. Both pairs of opposite angles are congruent.
5. Two adjacent angles are supplementary.

Example: Find the measures of the other three angles of a parallelogram if one angle measures 38°.

Since opposite angles are equal, there are two angles measuring 38°. Since adjacent angles are supplementary, 180 – 38 = 142. Hence the other two angles measure 142° each.

Example: The measures of two adjacent angles of a parallelogram are 3x + 40 and x + 70. Find the measures of each angle.

$$2(3x + 40) + 2(x + 70) = 360$$
$$6x + 80 + 2x + 140 = 360$$
$$8x + 220 = 360$$
$$8x = 140$$
$$x = 17.5$$
$$3x + 40 = 92.5$$
$$x + 70 = 87.5$$

Thus the angles measure 92.5°, 92.5°, 87.5°, and 87.5°.

A **rectangle** is a parallelogram with a right angle. Since a rectangle is a special type of parallelogram, it exhibits all the properties of a parallelogram. All the angles of a rectangle are right angles because of congruent opposite angles. Additionally, the diagonals of a rectangle are congruent.

A **rhombus** is a parallelogram with all sides equal in length. A rhombus also has all the properties of a parallelogram. Additionally, its diagonals are perpendicular to each other and they bisect its angles.

A **square** is a rectangle with all sides equal in length. A **square** has all the properties of a rectangle <u>and</u> a rhombus.

Example: True or false?

All squares are rhombuses.	True
All parallelograms are rectangles.	False - <u>some</u> parallelograms are rectangles
All rectangles are parallelograms.	True
Some rhombuses are squares.	True
Some rectangles are trapezoids.	False - only <u>one</u> pair of parallel sides
All quadrilaterals are parallelograms.	False - some quadrilaterals are parallelograms
Some squares are rectangles.	False - all squares are rectangles
Some parallelograms are rhombuses.	True

Example: In rhombus *ABCD* side *AB* = 3*x* - 7 and side *CD* = *x* + 15. Find the length of each side.

Since all the sides are the same length, 3*x* – 7 = *x* + 15
2*x* = 22
x = 11

Since 3(11) – 7 = 25 and 11 + 15 = 25, each side measures 25 units.

The beginning teacher analyzes the properties of circles and the lines that intersect them.

The distance around a circle is the **circumference**. The ratio of the circumference to the diameter is represented by the Greek letter pi, where $\pi \approx 3.14$. The circumference of a circle is given by the formula $C = 2\pi r$ or $C = \pi d$ where *r* is the radius of the circle and *d* is the diameter. The **area** of a circle is given by the formula $A = \pi r^2$.

We can extend the area formula of a regular polygon to get the area of a circle by considering the fact that a circle is essentially a regular polygon with an infinite number of sides. The radius of a circle is equivalent to the apothem of a regular polygon. Thus, applying the area formula for a regular polygon to a circle we get

$$\frac{1}{2} \times perimeter \times apothem = \frac{1}{2} \times 2\pi r \times r = \pi r^2$$

If two circles have radii that are in a ratio of $a:b$, then the following ratios also apply to the circles:

1. The diameters are in the ratio $a:b$.
2. The circumferences are in the ratio $a:b$.
3. The areas are in the ratio $a^2 : b^2$, or the ratio of the areas is the square of the ratio of the radii.

If you draw two radii in a circle, the angle they form with the center as the vertex is a **central angle**. The piece of the circle "inside" the angle is an arc. Just like a central angle, an arc can have any degree measure from 0 to 360. The measure of an arc is equal to the measure of the central angle that forms the arc. Since a diameter forms a semicircle and the measure of a straight angle like a diameter is 180°, the measure of a semicircle is also 180°.

Given two points on a circle, the two points form two different arcs. Except in the case of semicircles, one of the two arcs will always be greater than 180° and the other will be less than 180°. The arc less than 180° is a **minor arc** and the arc greater than 180° is a **major arc**.

Example: If $m\angle BAD = 45°$, what is the measure of the major arc BD?

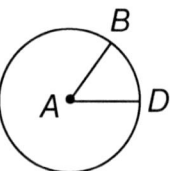

The minor arc BD is the same as $m\angle BAD = 45°$. Since the sum of the minor and major arcs formed by two points on a circle always add to 360°, the major arc BD must be 360° – 45° = 315°.

Example: If $\overline{AC}$ is a diameter of the circle below, what is the measure of $\angle BDC$?

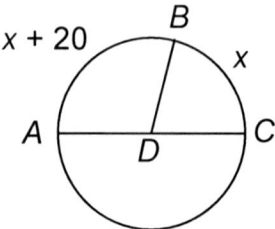

Since the diameter forms a semicircle ABC with a measure of 180°, the following expression applies.

$$(x + 20) + x = 180°$$

Solving for x yields

$$2x + 20 = 180°$$
$$2x = 160°$$
$$x = 80°$$

Finally, since the measure of an arc is the same as the measure of the central angle,

$$\angle BDC = x = 80°$$

Although an arc has a measure associated with the degree measure of the corresponding central angle, it also has a length that is a fraction of the circumference of the circle. For each central angle and its associated arc, there is a sector of the circle that resembles a pie piece. The area of such a sector is a fraction of the area of the circle. The fractions used for the area of a sector and length of its associated arc are both equal to the ratio of the central angle to 360°.

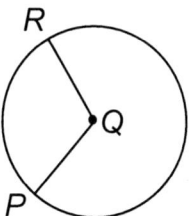

$$\frac{\angle PQR}{360°} = \frac{\text{length of arc RP}}{\text{circumference of circle}} = \frac{\text{area of sector PQR}}{\text{area of circle}}$$

Example: Circle A has a radius of 4 cm. What is the length of arc ED?

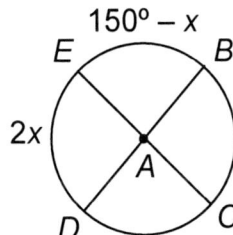

Because arcs BE and ED form a semicircle, arc BED is 180°. Use this to write an equation and solve for x.

$$(150° - x) + 2x = 180°$$
$$x = 30°$$

The angle corresponding to arc ED is thus 2x = 60°. The ratio of this arc to that of the entire circle (360°) must be the same as the ratio of the arc length (labeled L) to the circumference of the circle, as shown below.

$$\frac{60°}{360°} = \frac{L}{2\pi(4\text{cm})}$$

$$L = \frac{1}{6}2\pi(4\text{cm}) = \frac{4}{3}\pi \text{ cm} \approx 4.19\text{cm}$$

Example: The radius of circle M is 3 cm. The length of arc PF is 2π cm. What is the area of sector MPF?

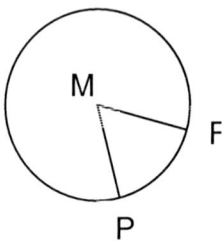

The circumference of the circle is 2π(3 cm), making the circumference equal to 6π cm. The total area of the circle is π(3 cm)², or 9π cm². The ratio of the arc length PF to the circumference of the circle must be the same as the ratio of the area of sector MPF (labeled A) to the total area of the circle. Thus,

$$\frac{A}{9\pi \text{ cm}^2} = \frac{2\pi}{6\pi}$$

$$A = \frac{1}{3}9\pi \text{ cm}^2 = 3\pi \text{ cm}^2 \approx 9.42 \text{ cm}^2$$

A **tangent line** intersects or touches a circle in exactly one point. If a radius is drawn to that point, the radius will be perpendicular to the tangent.

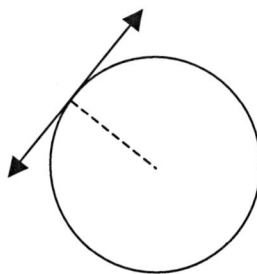

A **secant line** intersects a circle in two points and includes a **chord** which is a segment with endpoints on the circle. If a radius or diameter is perpendicular to a chord, the radius will cut the chord into two equal parts and vice-versa.

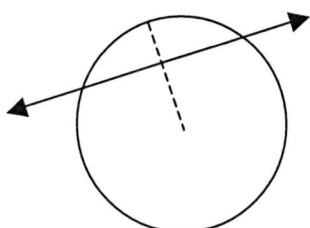

If **two chords** in the same circle have the same length, the two chords will have arcs that are the same length, and the two chords will be equidistant from the center of the circle. Distance from the center to a chord is measured by finding the length of a segment from the center perpendicular to the chord.

Example: $\overline{DB}$ is tangent to circle C at A. If $m\angle ADC = 40°$, find x.

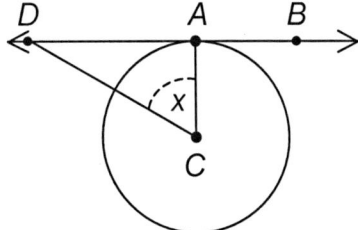

Since segment AC is perpendicular to the tangent line DB, angle DAC must be a right angle. Since $m\angle ADC = 40°$ and since the sum of the angles in a triangle must be 180º,

$90° + 40° + x = 180°$
$x = 50°$

An **inscribed angle** is an angle whose vertex is on the circumference circle. Such an angle could be formed by two chords, two diameters, two secants, or a secant and a tangent. An inscribed angle has one arc of the circle in its interior. The measure of the inscribed angle is one-half the measure of its intercepted arc. If two inscribed angles intercept the same arc, the two angles are congruent (i.e., their measures are equal). If an inscribed angle intercepts an entire semicircle, the angle is a right angle.

When two chords intersect inside a circle, two sets of vertical angles are formed in the interior if the circle. Each set of vertical angles intercepts two arcs that are across from each other. The measure of an angle formed by two chords in a circle is equal to one-half the sum of the arc intercepted by the angle and the arc intercepted by its vertical angle.

If an angle has its vertex outside of the circle and each side of the angle intersects the circle, then the angle contains two different arcs. The measure of the angle is equal to one-half the difference of the two arcs.

Example: Find x and y for the circle below.

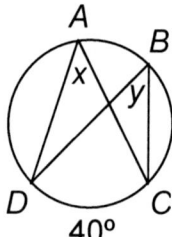

Since ∠DAC and ∠DBC are both inscribed angles, each has a measure equal to one-half of the measure of arc DC, or 20°. Thus, both x and y are equal to 20°.

Example: Find the measure of arc BC if the measure of arc DE is 30° and angle BAC is 20°.

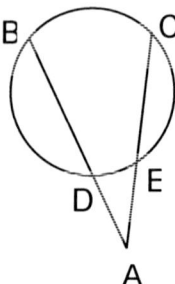

Use the following expression that relates the angle *BAC* to the lengths of the arcs *BC* and *DE*.

$$m\angle BAC = \frac{1}{2}(BC - DE)$$
$$2(20°) = 40° = BC - 30°$$
$$BC = 70°$$

If **two chords intersect inside a circle**, each chord is divided into two smaller segments. The product of the lengths of the two segments formed from one chord equals the product of the lengths of the two segments formed from the other chord.

If **two tangent segments intersect outside of a circle**, the two segments have the same length.

If **two secant segments intersect outside a circle**, a portion of each segment will lie inside the circle and a portion (called the exterior segment) will lie outside the circle. The product of the length of one secant segment and the length of its exterior segment equals the product of the length of the other secant segment and the length of its exterior segment.

If **a tangent segment and a secant segment intersect outside a circle**, the square of the length of the tangent segment equals the product of the length of the secant segment and its exterior segment.

Example: If $\overline{AB}$ and $\overline{CD}$ are chords and *CE* = 10, *ED* = *x*, *AE* = 5, and *EB* = 4, find *x*.

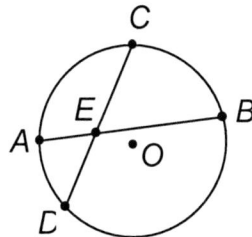

Because the chords intersect inside the circle, the product of the segment lengths for chord *CD* is equal to the product of the segment lengths for chord *AB*. Thus,

(*AE*)(*EB*) = (*CE*)(*ED*)
(5)(4) = (10)(*x*)
10*x* = 20
x = 2

The length of x (or segment ED) is 2.

Example: Find the lengths of chords AB and CB if $AB = x^2 + x - 2$ and $BC = x^2 - 3x + 5$.

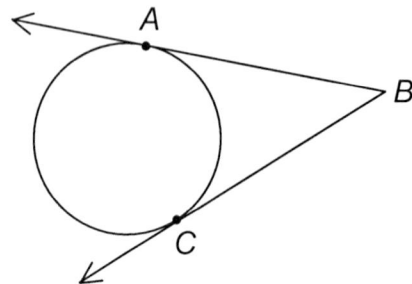

These chords are both tangent to the circle. Because intersecting tangents are equal in length, find x by setting the two quadratic expressions equal to one another:

$$x^2 + x - 2 = x^2 - 3x + 5$$
$$4x = 7$$
$$x = \frac{7}{4} = 1.75$$

Using this result, the lengths of the segments can be found by substitution.

$$AB = (1.75)^2 + (1.75) - 2 = 3.0625 + 1.75 - 2 = 2.8125$$
$$BC = (1.75)^2 - 3(1.75) + 5 = 3.0625 - 5.25 + 5 = 2.1825$$

The beginning teacher uses geometric patterns and properties (e.g., similarity, congruence) to make generalizations about two- and three-dimensional figures and shapes (e.g., relationships of sides, angles).

For a discussion of similarity and congruence as applied to geometric figures, see **Competency 012**.

The beginning teacher computes the perimeter, area, and volume of figures and shapes created by subdividing and combining other figures and shapes (e.g., arc length, area of sectors).

The attributes of geometric figures that are subdivided or combined are discussed lightly in **Competency 011**.

TEACHER CERTIFICATION STUDY GUIDE

The beginning teacher analyzes cross-sections and nets of three-dimensional shapes.

If a three-dimensional object is intersected by a plane, it forms a **cross section**. In the picture below, the striped portion is a cross section of a cube.

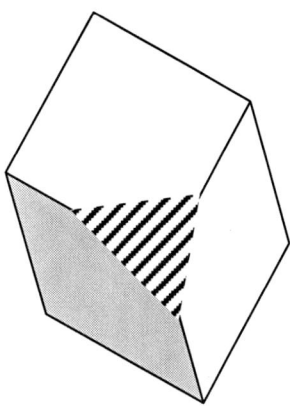

A conic section (e.g., a circle, ellipse, parabola, or hyperbola), for instance, is formed by the intersection of a cone with a plane.

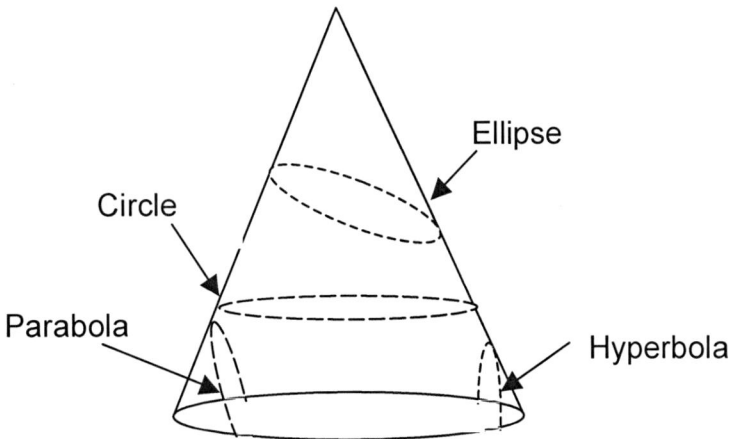

Circles and ellipses are both closed figures. The plane creating the circle cross-section in intersection with the cone is parallel to the base of the cone. Parabolas and hyperbolas are open figures. The plane creating the parabola cross-section in intersection with the cone is parallel to the slant side of the cone. For more on conic sections, see **Competency 014**.

A **net** is a two-dimensional figure that can be cut out and folded up to make a three-dimensional solid. Below are models of some regular solids with their corresponding face polygons and nets. Nets clearly show the shape and number of faces of a solid.

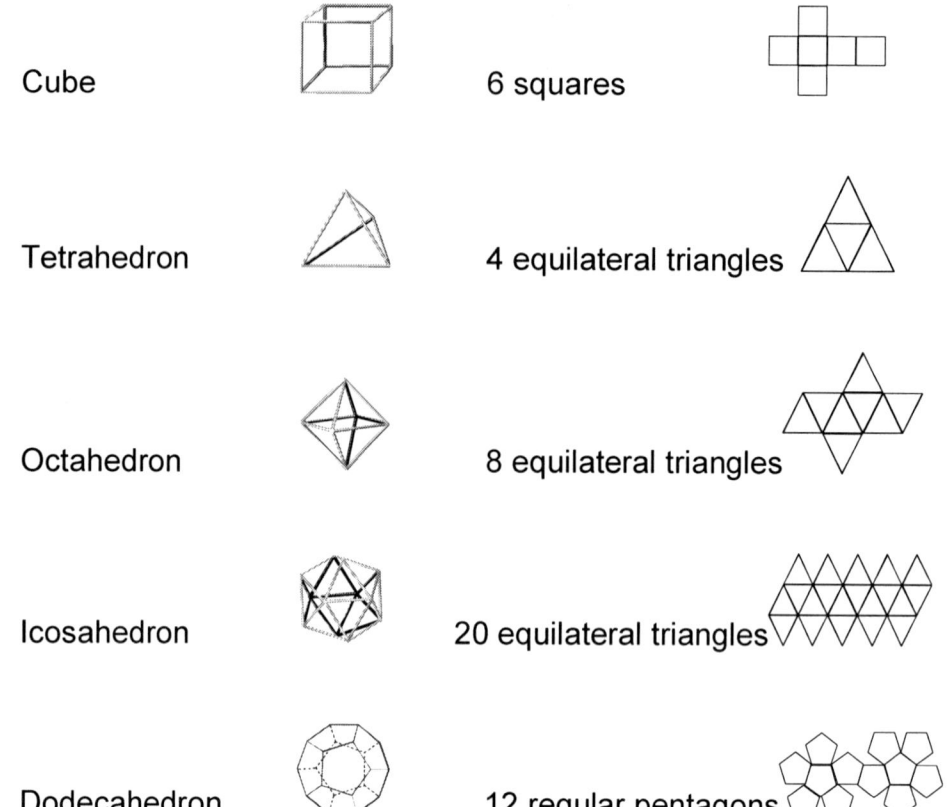

Cube	6 squares	
Tetrahedron	4 equilateral triangles	
Octahedron	8 equilateral triangles	
Icosahedron	20 equilateral triangles	
Dodecahedron	12 regular pentagons	

There can be more than one possible net for a particular three-dimensional figure. For instance, here are two more nets for a tetrahedron.

For a polyhedron, the numbers of vertices (*V*), faces (*F*), and edges (*E*) are related by **Euler's Formula**: $V + F = E + 2$.

Example: How many edges are in a pentagonal pyramid?

A pentagonal pyramid has six vertices and six faces. Using Euler's Formula, compute the number of edges:

$$V + F = E + 2$$
$$6 + 6 = E + 2$$
$$E = 10$$

Thus, the figure has 10 edges.

Example: Draw the net of a triangular prism and identify the polygons that make up the faces. How many vertices and edges do triangular prisms have?

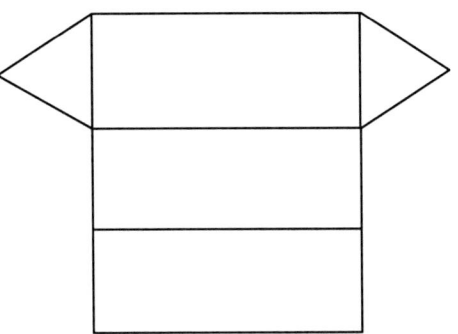

Two triangles and three rectangles are the faces of the figure. There are nine edges (three between the rectangle faces and three on each side where the triangle faces meet the rectangle faces). There are six vertices at the vertices of the two triangle faces.

The beginning teacher uses top, front, side, and corner views of three-dimensional shapes to create complete representations and solve problems.

When dealing with objects in real life, it is necessary to turn them in one or more directions to get a view of all sides. The same is true of geometric figures. To provide a complete and accurate representation, especially when using two-dimensional illustrations on paper or a computer screen, it is necessary to show several different views of the object. For instance, when representing a pyramid, it may be helpful to show several angles (an alternative, of course, is the use of transparency).

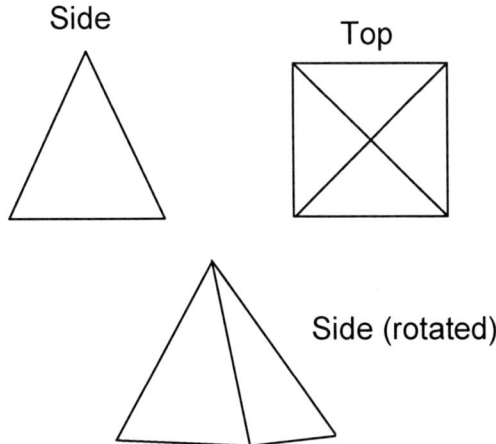

In certain computer programs that have geometry capabilities, shapes can be rotated or flipped easily with the click of a mouse button.

For more on three-dimensional objects, included solving problems associated with them, see **Competency 011**.

The beginning teacher applies properties of two- and three-dimensional shapes to solve problems across the curriculum and in everyday life.

Example problems involving two- and three-dimensional shapes are presented throughout this domain.

TEACHER CERTIFICATION STUDY GUIDE

Competency 014   **The teacher understands coordinate, transformational, and vector geometry and their connections.**

This section reviews various aspects of transformational, coordinate, and vector geometry. Different types of transformations and symmetries are discussed, followed by coordinate geometry (including the use of vectors and matrices for representing transformations) and conic sections.

The beginning teacher Identifies transformations (i.e., reflections, translations, glide-reflections, rotations, dilations) and explores their properties.

Transformational geometry is the study of manipulating objects through movement, rotation and scaling. The transformation of an object is called its *image.* If the original object was labeled with letters, such as *ABCD*, the image can be labeled with the same letters followed by a prime symbol: *A'B'C'D'*. Transformations can be characterized in different ways.

An **isometry** is a linear transformation that maintains the dimensions of a geometric figure. **Symmetry** is exact similarity between two parts or halves, as if one were a mirror image of the other. A **translation** is a transformation that "slides" an object a fixed distance in a given direction. The original object and its translation have the same shape and size, and they face in the same direction.

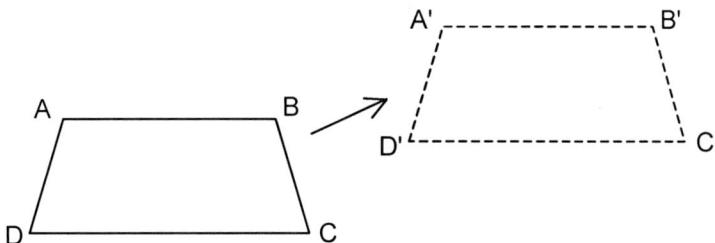

A **rotation** is a transformation that turns a figure about a fixed point, which is called the center of rotation. An object and its rotation are the same shape and size, but the figures may be oriented in different directions. Rotations can occur in either a clockwise or a counterclockwise direction.

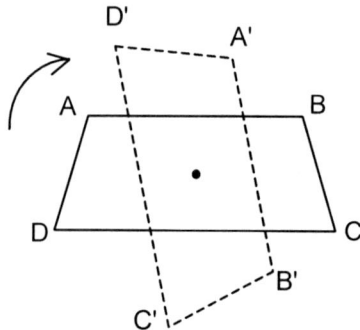

An object and its **reflection** have the same shape and size, but the figures face in opposite directions. The line (where a hypothetical mirror may be placed) is called the **line of reflection**. The distance from a point to the line of reflection is the same as the distance from the point's image to the line of reflection.

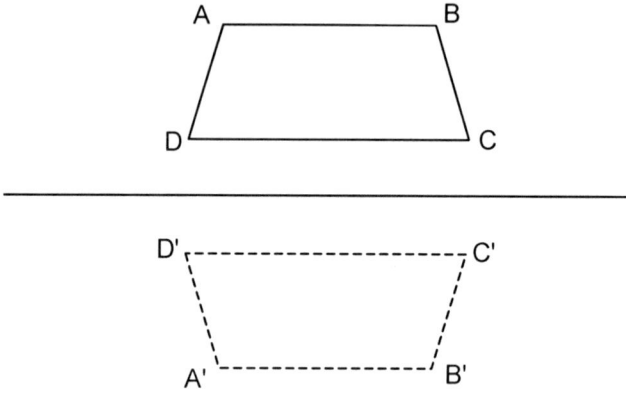

A **glide reflection** involves a combined translation along and a reflection across a single specified line. The characteristic that defines a glide reflection as opposed to a simple combination of an arbitrary translation and arbitrary reflection is that the direction of translation is parallel with the line of reflection. An example of a glide reflection is shown below.

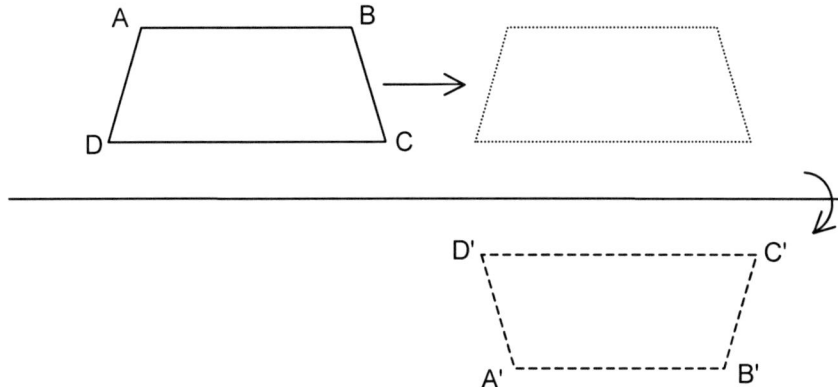

The examples of a translation, rotation, reflection, and glide reflection given above are for polygons, but the same principles apply to the simpler geometrical elements of points and lines. In fact, a transformation performed on a polygon can be viewed equivalently as the same transformation performed on the set of points (vertices) and lines (sides) that compose the polygon. Thus, to perform complicated transformations on a figure, it is helpful to perform the transformations on all the points (or vertices) of the figure, then reconnect the points with lines as appropriate.

Dilations involve an expansion of a figure and a translation of that figure (the translation may be for a distance zero). These two transformations are obtained by first defining a **center of dilation**, C, which is some point that acts like an origin for the dilation. The distance from C to each point in a figure is then altered by a **scale factor** s. If the magnitude of s is greater than zero, the size of the figure is increased; if the magnitude of s is less than zero, the size is decreased.

The expansion of a geometric figure is a result of the scale factor, s. For instance, if s = 2, the expanded figure will be twice the size of the original figure. (A dilation maintains the angles and relative proportions of a figure.) The translation of a geometric figure is a result of the location of the center of dilation, C. If C is located at the center of the figure, for instance, the figure is dilated without and translation of its center.

Example: Dilate the figure shown by a scale factor of 2 using the origin of the coordinate system as the center of dilation.

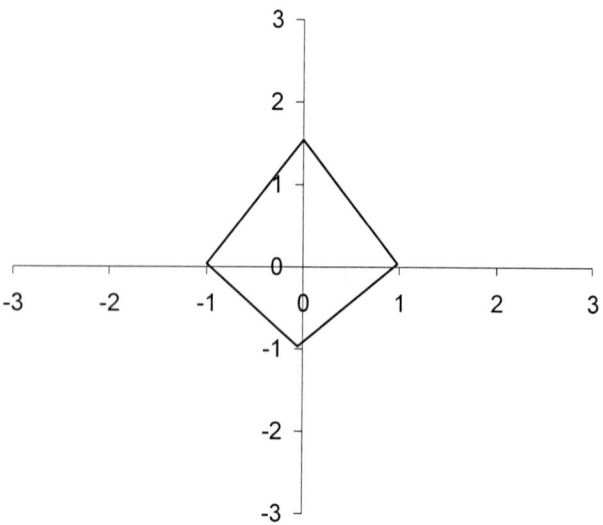

To perform this dilation, the distance between the origin and each point on the figure must be increased by a factor of 2. It is sufficient, however, to simply increase the distance of the vertices of the figure by a factor of 2 and then connect them to form the dilated figure.

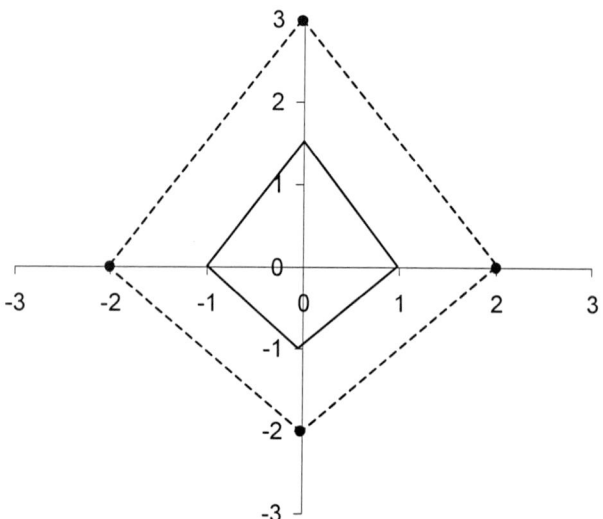

The resulting figure above (dashed line) is the dilation of the original figure.

The points on a figure are dilated by increasing or decreasing their respective distances from a center of dilation C. As a result, each point P on a figure is essentially translated along the line through P and C. To show that a dilation of this type preserves angles, consider some angle formed by two line segments, with a center of dilation at some arbitrary location. The dilation is for some scale factor s.

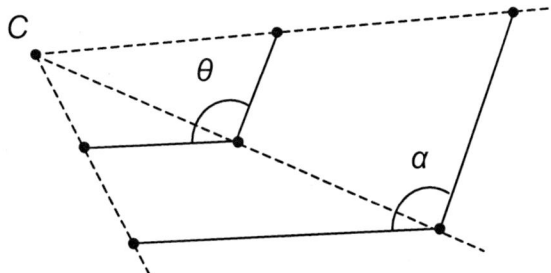

To show that the angles θ and α are equal, it is sufficient to show that the two pairs of overlapping triangles are similar. If they are similar, all the corresponding angles in the figure must be congruent.

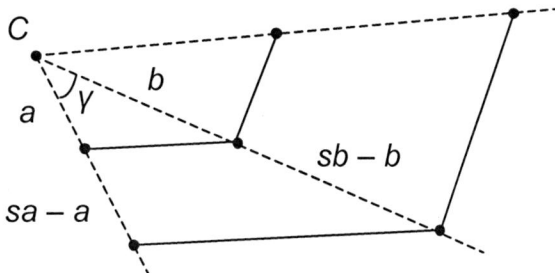

The smaller triangle in this case has sides of lengths a and b and an angle y between them. The larger triangle has sides of lengths sa (or $sa - a + a = sa$) and sb and an angle y between them. Thus, by SAS similarity, these two triangles are similar. Once this reasoning is applied to the other pair of overlapping triangles, it can be shown that angles θ and α are equal. Furthermore, due to the fact that these triangles have been shown to be similar, it is also true that line segments must scale by a factor s (this is necessary to maintain the similarity of the triangles above).

As a result of this reasoning, it can be shown that figures that are dilated using an arbitrary scale factor s and center of dilation C must maintain all angles through the dilation, and all line segments (or sides) of the figure must also scale by s. As a result, figures that are dilated are **similar** to the original figures.

Since dilations are transformations that maintain similarity of the figures being dilated, they can also be viewed as **changes of scale** about C. For instance, a dilation of a portion of a map would simply result in a change of the scale of the map.

The beginning teacher uses the properties of transformations and their compositions to solve problems.

Multiple transformations (or **compositions of transformations**) can be performed on a geometrical figure. The order of these transformations may or may not be important. For instance, multiple translations can be performed in any order, as can multiple rotations (around a single fixed point) or reflections (across a single fixed line). The order of the transformations becomes important when several types of transformations are performed or when the point of rotation or the line of reflection change among transformations. For example, consider a translation of a given distance upward and a clockwise rotation by 90° around a fixed point. Changing the order of these transformations changes the result.

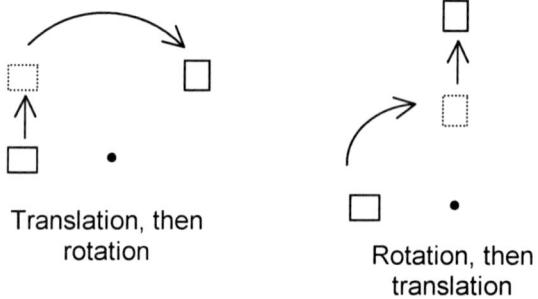

Translation, then rotation

Rotation, then translation

As shown, the final position of the box is different, depending on the order of the transformations. Thus, it is crucial that the proper order of transformations (whether determined by the details of the problem or some other consideration) be followed.

Example: Find the final location of a point at (1, 1) that undergoes the following transformations: rotate 90° counter-clockwise about the origin; translate distance 2 in the negative *y* direction; reflect about the *x*-axis.

First, draw a graph of the *x*- and *y*-axes and plot the point at (1, 1).

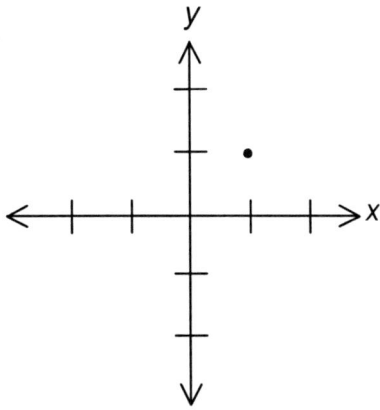

Next, perform the rotation. The center of rotation is the origin and is in the counter-clockwise direction. In this case, the even value of 90° makes the rotation simple to do by inspection. Next, perform a translation of distance 2 in the negative y direction (down). The results of these transformations are shown below.

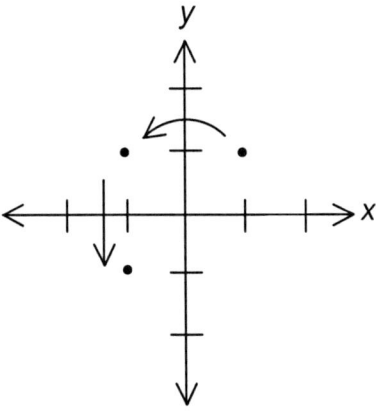

Finally, perform the reflection about the x-axis. The final result, shown below, is a point at (1, −1).

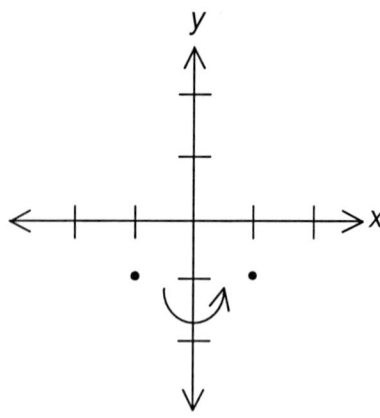

Using this approach, polygons can be transformed on a point-by-point basis.

For some problems, there is no need to work with coordinate axes. For instance, the problem may simply require transformations without respect to any absolute positioning.

Example: Rotate the following regular pentagon by 36° about its center and then reflect it about the horizontal line.

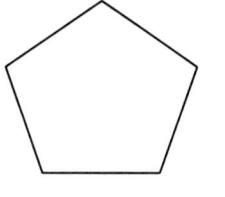

First, perform the rotation. In this case, the direction is not important because the pentagon is symmetric. As it turns out in this case, a rotation of 36° yields the same result as flipping the pentagon vertically (assuming the vertices of the pentagon are indistinguishable).

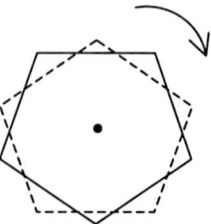

Finally, perform the reflection. Note that the result here is the same as a downward translation (assuming the vertices of the pentagon are indistinguishable).

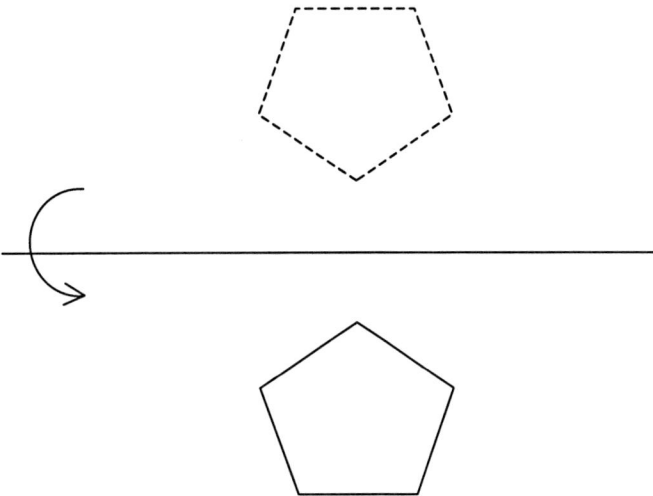

The beginning teacher uses transformations to explore and describe reflectional, rotational, and translational symmetry.

A figure has symmetry when there is an isometry that maps the figure onto itself. A figure has **rotational symmetry** if there is a rotation of 180 degrees or less that maps the figure onto itself. **Point symmetry** is where a plane figure can be mapped onto itself by a half-turn or a rotation of 180 degrees around some point. Thus, point symmetry is a specific type of rotational symmetry. An example of a figure with point symmetry is shown below.

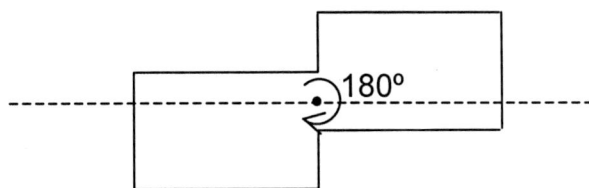

Reflectional or **line symmetry** is an isometry that maps the figure onto itself by reflection across a line. An alternative view is that if the figure is folded along a line of symmetry, the two halves will match perfectly. Examples of figures with line symmetry are shown below, with all the potential lines of symmetry marked as broken lines.

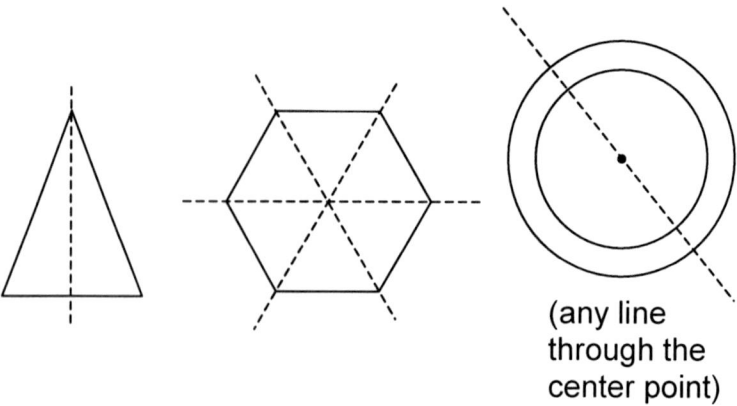

(any line through the center point)

Translational symmetry is where an image can be translated in a specific direction to produce the same image. Necessarily, this requires that the image be infinite in extent and repeating in nature. A **tessellation** is an image with translational symmetry. A tessellation or tiling, consists of a repeating pattern of figures, which completely cover an area. Below are a couple of example of art tessellations.

A regular tessellation is made by taking a pattern of polygons that are interlocked and can be extended infinitely. A portion of a tessellation made with hexagons is shown below. This image is made by taking congruent, regular polygons and using them to cover a plane in such a way that *there are no holes or overlaps*. A semi-regular tessellation is made with polygons arranged exactly the same way at every vertex point. Tessellations occur in frequently in nature. A bee's honeycomb is an example of a tessellation found in nature.

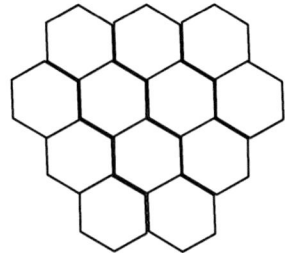

The beginning teacher applies transformations in the coordinate plane.

Examples of transformations (and compositions of transformations) in the coordinate plane are presented in the preceding skill sections.

The beginning teacher applies concepts and properties of slope, midpoint, parallelism, perpendicularity, and distance to explore properties of geometric figures and solve problems in the coordinate plane.

Coordinate geometry involves the application of algebraic methods to geometry. The locations of points in space are expressed in terms of coordinates on a Cartesian plane. The relationships between the coordinates of different points are expressed as equations.

Proofs using coordinate geometry techniques employ the following commonly used formulae and relationships:

1. **Midpoint formula**: The midpoint (x, y) of the line joining points (x_1, y_1) and (x_2, y_2) is given by

$$(x, y) = \left(\frac{x_1 + x_2}{2}, \frac{y_1 + y_2}{2} \right)$$

2. **Distance formula:** The distance between points (x_1, y_1) and (x_2, y_2) is given by

$$D = \sqrt{(x_2 - x_1)^2 + (y_2 - y_1)^2}$$

3. **Slope formula:** The slope m of a line passing through the points (x_1, y_1) and (x_2, y_2) is given by

$$m = \frac{y_2 - y_1}{x_2 - x_1}$$

4. **Equation of a line:** The equation of a line is given by $y = mx + b$, where m is the slope of the line and b is the y-intercept, i.e. the y-coordinate at which the line intersects the y-axis.

5. **Parallel and perpendicular lines:** Parallel lines have the same slope. The slope of a line perpendicular to a line with slope m is $-1/m$.

<u>Example:</u> Prove that quadrilateral ABCD with vertices A(–3,0), B(–1,0), C(0,3) and D(2,3) is in fact a parallelogram using coordinate geometry:

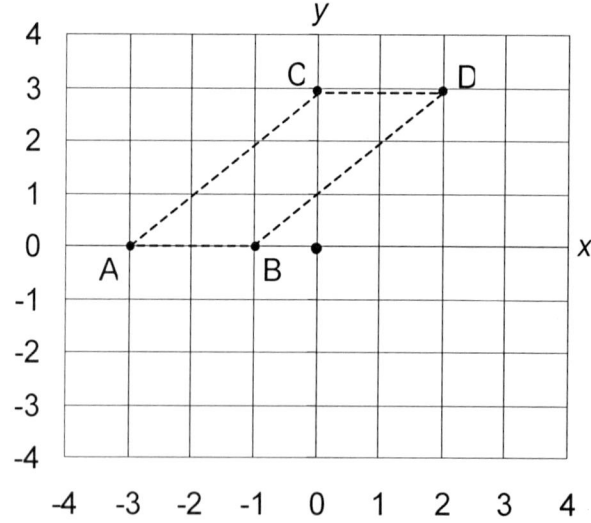

By definition, a parallelogram has diagonals that bisect each other. Using the midpoint formula, $(x, y) = \left(\dfrac{x_1 + x_2}{2}, \dfrac{y_1 + y_2}{2} \right)$, find the midpoints of $\overline{AD}$ and $\overline{BC}$.

The midpoint of $\overline{BC} = \left(\dfrac{-1+0}{2}, \dfrac{0+3}{2}\right) = \left(\dfrac{-1}{2}, \dfrac{3}{2}\right)$

The midpoint of $\overline{AD} = \left(\dfrac{-3+2}{2}, \dfrac{0+3}{2}\right) = \left(\dfrac{-1}{2}, \dfrac{3}{2}\right)$

Since the midpoints of the diagonals are the same, the diagonals bisect each other. Hence the polygon is a parallelogram.
In the above example the proof involved a specific geometrical figure with given coordinates. Coordinate geometry can also be used to prove more general results.

Example: Prove that the diagonals of a rhombus are perpendicular to each other.

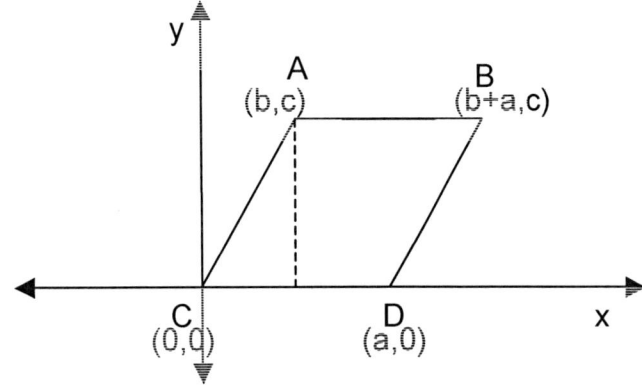

Draw a rhombus ABCD with side of length a such that the vertex C is at the origin and the side CD lies along the x-axis. The coordinates of the corners of the rhombus can then be written as shown above.

The slope m_1 of the diagonal AD is given by $m_1 = \dfrac{c}{b-a}$.

The slope m_2 of the diagonal BC is given by $m_2 = \dfrac{c}{b+a}$.

The product of the slopes is $m_1 \cdot m_2 = \dfrac{c}{b-a} \cdot \dfrac{c}{b+a} = \dfrac{c^2}{b^2-a^2}$.

The length of side AC = $\sqrt{b^2+c^2}$ = a (since each side of the rhombus is equal to a). Therefore,

$$b^2 + c^2 = a^2$$
$$\Rightarrow b^2 - a^2 = -c^2$$
$$\Rightarrow \frac{c^2}{b^2 - a^2} = -1$$

Thus the product of the slopes of the diagonals $m_1 \cdot m_2 = -1$. Hence the two diagonals are perpendicular to each other.

Example: Prove that the line joining the midpoints of two sides of a triangle is parallel to and half of the third side.

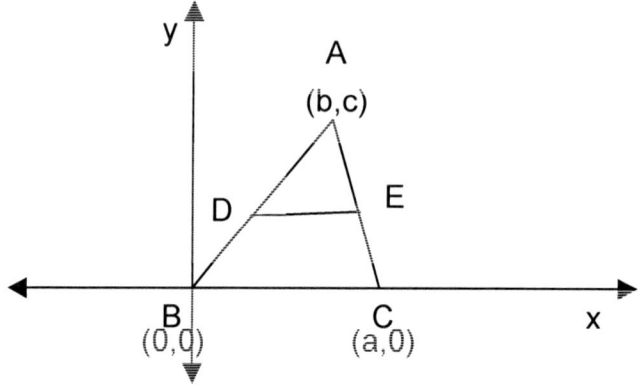

Draw triangle ABC on the coordinate plane in such a way that the vertex B coincides with the origin and the side BC lies along the x-axis. Let point C have coordinates (a,0) and A have coordinates (b,c). D is the midpoint of AB and E is the midpoint of AC.
We need to prove that DE is parallel to BC and is half the length of BC.

Using the midpoint formula,

$$\text{coordinates of D} = \left(\frac{b}{2}, \frac{c}{2}\right); \quad \text{coordinates of E} = \left(\frac{b+a}{2}, \frac{c}{2}\right)$$

The slope of the line DE is then given by $\dfrac{\frac{c}{2}-\frac{c}{2}}{\frac{b+a}{2}-\frac{b}{2}} = 0$, which is equal to the slope of the x-axis. Thus DE is parallel to BC.

The length of the line segment DE =

$$\sqrt{\left(\frac{b+a}{2}-\frac{b}{2}\right)^2+\left(\frac{c}{2}-\frac{c}{2}\right)^2}=\sqrt{\left(\frac{a}{2}\right)^2}=\frac{a}{2}$$

Thus the length of DE is half that of BC.

The beginning teacher explores the relationship between geometric and algebraic representations of vectors and uses this relationship to solve problems.

This skill section reviews the properties of vectors in both an algebraic and geometric sense. The application of vectors (along with matrices) to transformational geometry is discussed in the next skill section.

A vector is any quantity that has a **magnitude** (or length) and a **direction**. For instance, unlike temperature (which is just a scalar), velocity is a vector because it has a magnitude (speed) and a direction (the direction of travel). Vectors do not have specified locations, so they can be translated as long as their direction and magnitude are the same. A vector is often written in the same form as a point; for instance, a vector can be written as (x_1, y_1, z_1). In this case, the direction and magnitude of the vector are defined by a ray that starts at the origin and terminates at the point (x_1, y_1, z_1).

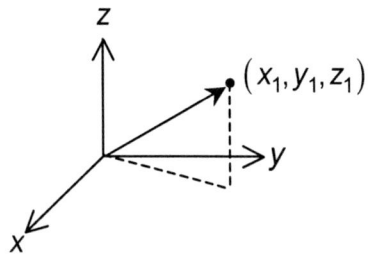

As noted before, however, the vector is not confined to the location shown above. The magnitude of a vector is simply the distance from the origin to the point (x_1, y_1, z_1). If $\vec{A} = (x_1, y_1, z_1)$, then the magnitude is written as $|\vec{A}|$.

$$|\vec{A}| = \sqrt{x_1^2 + y_1^2 + z_1^2}$$

Addition and Scalar Multiplication

Addition and subtraction of two vectors $\vec{A} = (x_1, y_1, z_1)$ and $\vec{B} = (x_2, y_2, z_2)$ can be performed by adding or subtracting corresponding components of the vectors.

$$\vec{A} + \vec{B} = (x_1 + x_2, y_1 + y_2, z_1 + z_2)$$
$$\vec{A} - \vec{B} = (x_1 - x_2, y_1 - y_2, z_1 - z_2)$$

Geometrically, addition involves placing the tail of $\vec{B}$ on the head of $\vec{A}$, as shown below. The result is a vector that starts from the tail of $\vec{A}$ and ends at the head of $\vec{B}$.

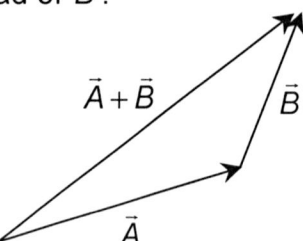

Subtraction of two vectors involves the same process, except that the direction of $\vec{B}$ must be reversed.

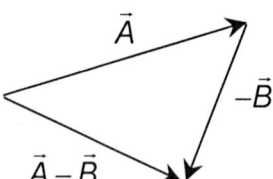

Multiplication of a vector by a scalar simply involves multiplying each component by the scalar.

$$c\vec{A} = (cx_1, cy_1, cz_1)$$

Geometrically, this operation extends the length of the vector by a factor c.

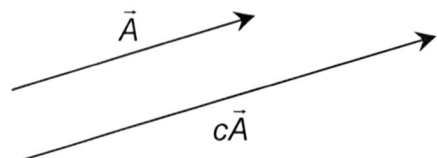

Note that the direction of a vector can be written as a unit vector $\vec{u}$ of length 1, such that

$$\vec{A} = \vec{u}|\vec{A}|$$

Vectors obey the laws of associativity, commutativity, identity and additive inverses:

$$\vec{A} + (\vec{B} + \vec{C}) = (\vec{A} + \vec{B}) + \vec{C}$$
$$\vec{A} + \vec{B} = \vec{B} + \vec{A}$$
$$\vec{A} + 0 = \vec{A}$$
$$\vec{A} + (-\vec{A}) = 0$$

As such, vectors have some of the same properties as real numbers.

Multiplication of Vectors

Vector multiplication takes two forms: the **dot product** (or scalar product) and the **cross product** (or vector product). The dot product is calculated by multiplying corresponding components of two vectors. The operator for this product is typically a small dot (·).

$$\vec{A} \cdot \vec{B} = x_1 x_2 + y_1 y_2 + z_1 z_2$$

Notice that the dot product yields a single scalar value. Also note that the magnitude of a vector can be written in terms of the dot product.

$$|\vec{A}| = \sqrt{\vec{A} \cdot \vec{A}}$$

Geometrically, the dot product is a **projection** of one vector onto another.

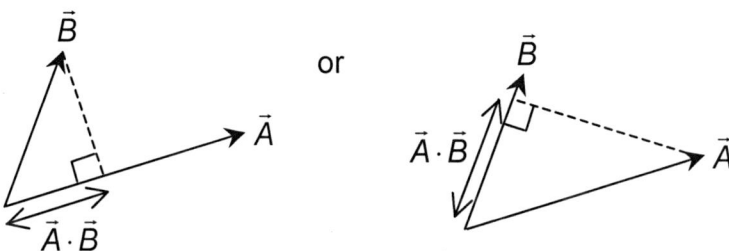

It can be shown that the dot product of two vectors is equivalent to the following:

$$\vec{A} \cdot \vec{B} = |\vec{A}||\vec{B}|\cos\theta$$

It is clear, both from the geometric and the algebraic definitions of the dot product, that if two vectors are perpendicular, then their dot product is zero.

The cross product of two vectors, typically symbolized by a × operator, yields a third vector. The cross product is defined as follows.

$$\vec{A} \times \vec{B} = (y_1 z_2 - y_2 z_1, z_1 x_2 - z_2 x_1, x_1 y_2 - x_2 y_1)$$

Geometrically, the cross product of $\vec{A}$ and $\vec{B}$ is a third vector that is perpendicular to both $\vec{A}$ and $\vec{B}$ (that is, to the plane formed by $\vec{A}$ and $\vec{B}$), with the direction defined by the so-called right-hand screw rule. If a right-hand screw is turned in the direction from $\vec{A}$ to $\vec{B}$, the direction in which the screw advances is the direction of the cross product.

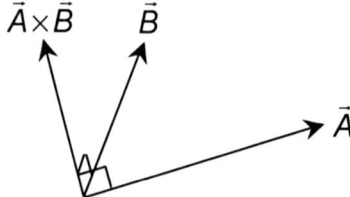

The magnitude of $\vec{A} \times \vec{B}$ is the area of the parallelogram defined by $\vec{A}$ and $\vec{B}$.

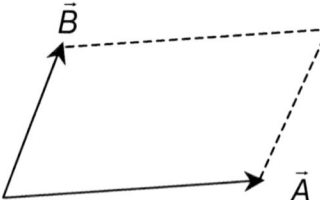

The magnitude of the cross product can also be written as follows.

$$|\vec{A} \times \vec{B}| = |\vec{A}||\vec{B}|\sin\theta$$

From both the algebraic and geometric definitions of the cross product, it is apparent that if two vectors are parallel, their cross product is zero.

Multiplication of vectors, either by the dot product or cross product, obeys the rule of distibutivity, where $*$ symbolizes either $\cdot$ or $\times$.

$$\vec{A} * (\vec{B} + \vec{C}) = \vec{A} * \vec{B} + \vec{A} * \vec{C}$$

Only the dot product obeys commutativity, however. There is a similar rule for cross products, though.

$$\vec{A} \cdot \vec{B} = \vec{B} \cdot \vec{A}$$
$$\vec{A} \times \vec{B} = -\vec{B} \times \vec{A}$$

Example: Find the cross product $\vec{a} \times \vec{b}$, where $\vec{a} = (-1, 4, 2)$ and $\vec{b} = (3, -1, 4)$.

Use the expression given to calculate the cross product.

$$\vec{a} \times \vec{b} = (4 \cdot 4 - (-1) \cdot 2, 2 \cdot 3 - 4 \cdot (-1), (-1) \cdot (-1) - 3 \cdot 4)$$
$$\vec{a} \times \vec{b} = (16 + 2, 6 + 4, 1 - 12) = (18, 10, -11)$$

Example: Find the following for $\vec{a} = (1, 2, 3)$, $\vec{b} = (2, 3, 1)$ and $c = 2$: $c(\vec{a} \cdot \vec{b})$.

First, find the dot product, then multiply. Note that the result must be a scalar.

$$c(\vec{a} \cdot \vec{b}) = 2[1(2) + 2(3) + 3(1)] = 2[11] = 22$$

The beginning teacher relates geometry and algebra by representing transformations as matrices and uses this relationship to solve problems.

The different types of transformations of geometric figures in the plane, such as translations, rotations, reflections and dilations, can be represented in the coordinate plane or space (especially when the transformations are complicated or when multiple sequential transformations are to be performed) using vectors and matrices. The basic properties of vectors are reviewed in the previous skill section. See **Competency 003** and **Competency 006** for more on matrices.

Points in a geometric figure, such as vertices, can be treated as a vector with two elements—(x_1, y_1), for instance—that specify the location of the point in some coordinate system. Consider the example below.

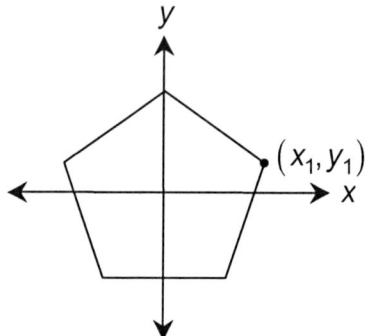

Given that each point on the figure is represented by an ordered pair, as is the case for one of the vertices of the pentagon in the example graph above, a 2 × 2 matrix can be used to perform the transformation. For any given transformation, a particular 2 × 2 matrix $\overline{T}$ can be determined that transforms all the points in the correct manner.

Consider a **reflection** about the x-axis. This transformation results in the following:

$$(x, y) \rightarrow (x, -y)$$

The result of this transformation is shown for the pentagon in the diagram above.

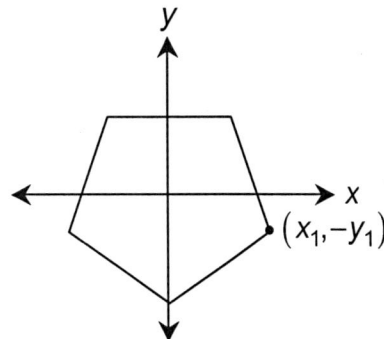

A matrix $\bar{T}$ can be constructed that performs this transformation.

$$\bar{T} = \begin{pmatrix} 1 & 0 \\ 0 & -1 \end{pmatrix}$$

Thus:

$$\bar{T}\begin{pmatrix} x \\ y \end{pmatrix} = \begin{pmatrix} 1 & 0 \\ 0 & -1 \end{pmatrix}\begin{pmatrix} x \\ y \end{pmatrix} = \begin{pmatrix} x \\ -y \end{pmatrix}$$

A similar transformation matrix can be constructed for reflections about the y-axis or about an arbitrary line (although this latter case is significantly more difficult).

For a **translation**, it is sufficient to simply construct a vector that is added to each point (x, y) in the figure. This vector is composed of a length for the translation in the x direction and a length for translation in the y direction. For instance, to translate a figure a distance a in the positive x direction and a distance b in the negative y direction, use a vector $(a, -b)$. The result is shown below for the pentagon.

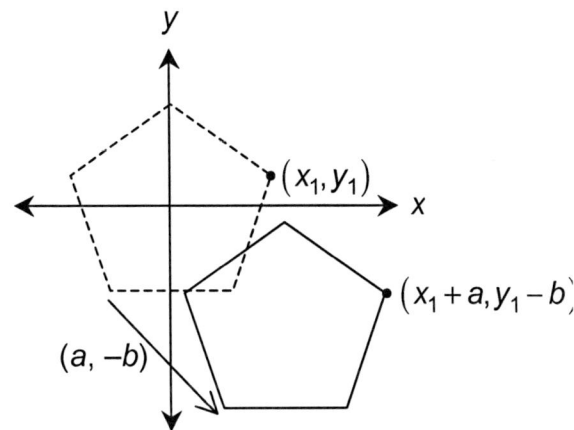

Algebraically, the transformation for the transformed figure in terms of the points (x, y) on the original figure is the following for a general translation (c, d):

$$\begin{pmatrix} x \\ y \end{pmatrix} + \begin{pmatrix} c \\ d \end{pmatrix}$$

Rotations are slightly more complicated transformations. Consider a rotation around the origin of a point specified as (x_1, y_1).

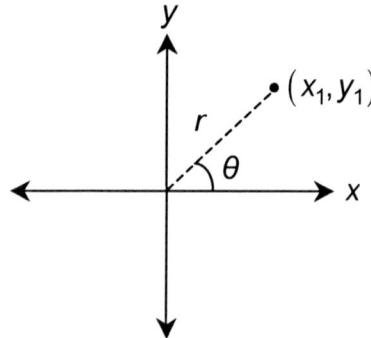

Using polar coordinates, the point can likewise be represented as a distance r from the origin and an angle θ from the x-axis.

$$r = \sqrt{x_1^2 + y_1^2}$$

$$\theta = \arctan \frac{y_1}{x_1}$$

A rotation around the origin simply involves, in this case, changing θ but holding r constant. Consider a rotation α in the counterclockwise direction.

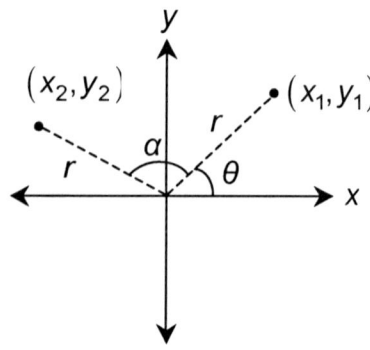

The coordinates of the new point (x_2, y_2) are then the following:

$$x_2 = r\cos(\alpha + \theta)$$
$$y_2 = r\sin(\alpha + \theta)$$

Use the sum formulas for trigonometric functions (see **Competency 009**) to expand and then simplify these expressions.

$$x_2 = r(\cos\alpha\cos\theta - \sin\alpha\sin\theta) = (r\cos\theta)\cos\alpha - (r\sin\theta)\sin\alpha$$
$$y_2 = r(\sin\alpha\cos\theta + \cos\alpha\sin\theta) = (r\cos\theta)\sin\alpha + (r\sin\theta)\cos\alpha$$

But $r\cos\theta$ is simply x_1, and $r\sin\theta$ is simply y_1.

$$x_2 = x_1\cos\alpha - y_1\sin\alpha$$
$$y_2 = x_1\sin\alpha + y_1\cos\alpha$$

Clearly, these two equations can be written in matrix form. Thus, a rotation of point (x_1, y_1) about the origin by angle α can be expressed as follows.

$$\begin{pmatrix} x_2 \\ y_2 \end{pmatrix} = \begin{pmatrix} \cos\alpha & -\sin\alpha \\ \sin\alpha & \cos\alpha \end{pmatrix} \begin{pmatrix} x_1 \\ y_1 \end{pmatrix}$$

This result can be tested using simple cases. For instance, consider a point (1, 0) rotated by π radians.

$$\begin{pmatrix} x_2 \\ y_2 \end{pmatrix} = \begin{pmatrix} \cos\pi & -\sin\pi \\ \sin\pi & \cos\pi \end{pmatrix} \begin{pmatrix} 1 \\ 0 \end{pmatrix} = \begin{pmatrix} -1 & 0 \\ 0 & -1 \end{pmatrix} \begin{pmatrix} 1 \\ 0 \end{pmatrix} = \begin{pmatrix} -1 \\ 0 \end{pmatrix}$$

This result makes intuitive sense. To rotate a figure, simply rotate a set of representative points (such as the vertices), then reconnect them after the rotation. In cases where a point of rotation is chosen that is not the origin, a change of coordinates to make the origin and the point of rotation coincide may simplify the problem and eliminate the need to handle complicated transformation matrices.

Dilations involve a change in the distance r from some point (such as the origin), rather than a change in the angle θ. As with rotations, it is sometimes convenient to perform a change of coordinates so that the center of dilation and the origin of the coordinate system coincide. In such a case, a dilation simply involves multiplying both coordinates of each point in the figure by the dilation factor. Consider the pentagon used above with a dilation factor of 2. For each point (x, y) on the pentagon, the corresponding point (x', y') on the dilated pentagon is simply the following:

$$\begin{pmatrix} x' \\ y' \end{pmatrix} = d \begin{pmatrix} x \\ y \end{pmatrix}$$

where, in this case, the dilation factor d is 2. The result of the dilation is shown below for the pentagon.

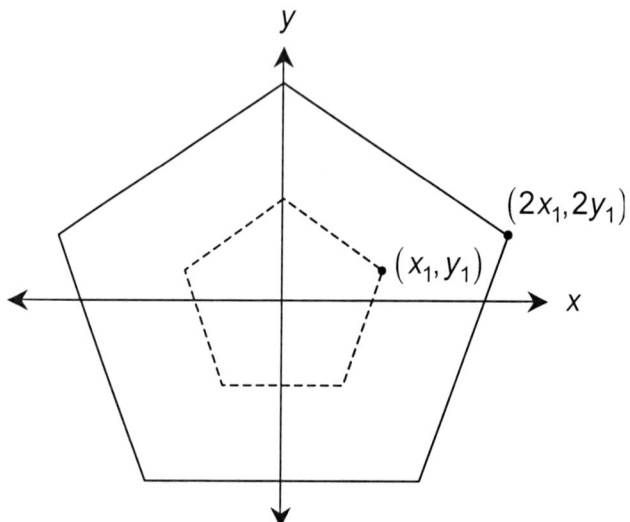

Compound transformations (or **composition of transformations**) can be made by simply concatenating several transformation operators (whether a multiplicative matrix, a multiplicative constant or an additive vector).

Example: Find the formula for a transformation involving first a rotation of α counterclockwise around the origin, then dilation by a factor of 4, then translation in the positive y direction by 2 and then clockwise rotation by α.

To solve this problem, consider a point (x, y) on the plane. To rotate counterclockwise by α, use the rotation matrix.

$$\begin{pmatrix} x' \\ y' \end{pmatrix} = \begin{pmatrix} \cos\alpha & -\sin\alpha \\ \sin\alpha & \cos\alpha \end{pmatrix} \begin{pmatrix} x \\ y \end{pmatrix}$$

The dilation simply involves multiplication by a factor of 4.

$$\begin{pmatrix} x'' \\ y'' \end{pmatrix} = 4 \begin{pmatrix} x' \\ y' \end{pmatrix}$$

The translation can be represented as follows.

$$\begin{pmatrix} x''' \\ y''' \end{pmatrix} = \begin{pmatrix} x'' \\ y'' \end{pmatrix} + \begin{pmatrix} 0 \\ 2 \end{pmatrix}$$

The final result requires use of the rotation matrix for $-\alpha$.

$$\begin{pmatrix} \bar{x} \\ \bar{y} \end{pmatrix} = \begin{pmatrix} \cos\alpha & \sin\alpha \\ -\sin\alpha & \cos\alpha \end{pmatrix} \begin{pmatrix} x''' \\ y''' \end{pmatrix}$$

Rewriting the equation in full yields

$$\begin{pmatrix} \bar{x} \\ \bar{y} \end{pmatrix} = \begin{pmatrix} \cos\alpha & \sin\alpha \\ -\sin\alpha & \cos\alpha \end{pmatrix} \left\{ 4 \begin{pmatrix} \cos\alpha & -\sin\alpha \\ \sin\alpha & \cos\alpha \end{pmatrix} \begin{pmatrix} x \\ y \end{pmatrix} + \begin{pmatrix} 0 \\ 2 \end{pmatrix} \right\}$$

Using this formula, any point or set of points can be transformed in the manner specified by the question.

The beginning teacher uses coordinate geometry to derive and explore the equations, properties, and applications of conic sections (i.e., lines, circles, hyperbolas, ellipses, parabolas).

Conic sections are aptly named as the various cross sections of a cone. Even a line and a point may in some sense be considered conic sections, although these are usually not included explicitly as such (they are sometimes called **degenerate conics**, because they include the vertex of the cone). Geometrically, a conic section is the intersection of an infinite cone with a plane.

An **ellipse** is a conic section that is formed as shown below. The standard algebraic expression for an ellipse is also included, along with some important parameters associated with the ellipse.

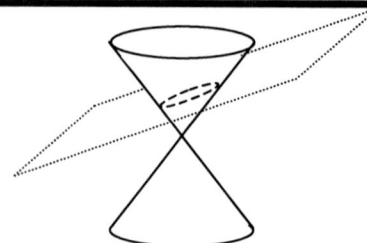

Equation: $\dfrac{(x-h)^2}{a^2} + \dfrac{(y-k)^2}{b^2} = 1$

Center: (h, k)

Foci: $(h \pm c, k)$, where $c = \sqrt{|a^2 - b^2|}$

Major axis: Maximum(a, b)
Minor axis: Minimum(a, b)

Example graph:

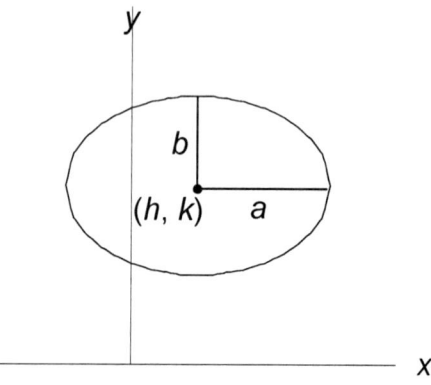

The **foci** of an ellipse are located along the major axis such that the sum of the distances from each foci to any given point on the ellipse is always constant, as shown below.

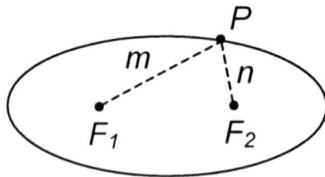

In the above diagram, $m + n$ is a constant, regardless of the position of point P on the ellipse. The equation for an ellipse can actually be derived from this condition using the distance formula. Consider an ellipse where point P is located distance b from the center (that is, P is located directly "above" the center).

Since the foci are each located a distance c from the center and since $m = n$ for this particular case, then

$$m = n = \sqrt{b^2 + c^2}$$

Also note that $m + n$ must be the following (when P is located at a distance a from the center):

$$m + n = (a - c) + (a + c) = 2a$$

Equating these results leads to the expression below:

$$m + n = 2a = 2\sqrt{b^2 + c^2}$$
$$a^2 = b^2 + c^2$$

Using this result, the formula for the ellipse can be derived as follows. Use the distance formula to express m and n for an arbitrary point $P = (x, y)$ on the ellipse.

$$m + n = 2a = \sqrt{[x - (h - c)]^2 + (y - k)^2} + \sqrt{[x - (h + c)]^2 + (y - k)^2}$$

Define $(y - k)$ as y' and define $(x - h)$ as x'. Simplify the terms in brackets as follows.

$$[x - (h - c)]^2 = x^2 - 2hx + 2cx + h^2 - 2hc + c^2$$
$$= (x - h)^2 + 2c(x - h) + c^2 = x'^2 + 2cx' + c^2$$
$$[x - (h + c)]^2 = x^2 - 2hx - 2cx + h^2 + 2hc + c^2$$
$$= (x - h)^2 - 2c(x - h) + c^2 = x'^2 - 2cx' + c^2$$

Rearrange the original expression and square both sides:

$$2a - \sqrt{x'^2 + 2cx' + c^2 + y'^2} = \sqrt{x'^2 - 2cx' + c^2 + y'^2}$$
$$4a^2 - 4a\sqrt{x'^2 + 2cx' + c^2 + y'^2} + x'^2 + 2cx' + c^2 + y'^2 =$$
$$= x'^2 - 2cx' + c^2 + y'^2$$

Simplify as follows.

$$4a^2 - 4a\sqrt{x'^2 + 2cx' + c^2 + y'^2} + 4cx' = 0$$

$$\sqrt{x'^2 + 2cx' + c^2 + y'^2} = a + \frac{c}{a}x'$$

$$x'^2 + 2cx' + c^2 + y'^2 = a^2 + 2cx' + \frac{c^2}{a^2}x'^2$$

$$\left(\frac{a^2 - c^2}{a^2}\right)x'^2 + y'^2 = a^2 - c^2$$

Note from above that $a^2 - c^2$ is b^2. Thus,

$$\left(\frac{b^2}{a^2}\right)x'^2 + y'^2 = b^2$$

$$\frac{x'^2}{a^2} + \frac{y'^2}{b^2} = 1$$

$$\frac{(x-h)^2}{a^2} + \frac{(y-k)^2}{b^2} = 1$$

This is the equation for an ellipse.

Note that a **circle**, which is another conic section, is simply an ellipse for which the major and minor axes are of equal lengths (that is, $a = b$). The foci then coincide with the center of the ellipse, forming a circle. The equation can then be written in terms of radius r.

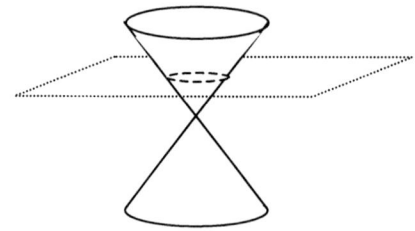

$$(x-h)^2 + (y-k)^2 = r^2$$

This expression is also the same as the distance formula for all points that are a distance r from a center point (h, k).

A **parabola** is a conic section that involves the cross section shown below. A parabola is defined by a set of points that are equidistant from a fixed line (the **directrix**) and a non-collinear fixed point (the **focus**). The cross section and important parameters are shown below.

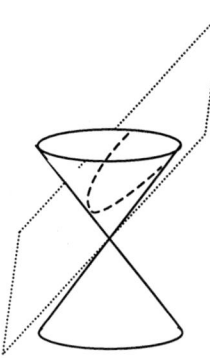

Equation: $(x-h)^2 = 4p(y-k)$

Vertex: (h, k)

Directrix: $y = k - p$

Focus: $(h, k + p)$

Major axis: Maximum(a, b)
Minor axis: Minimum(a, b)

Example graph:

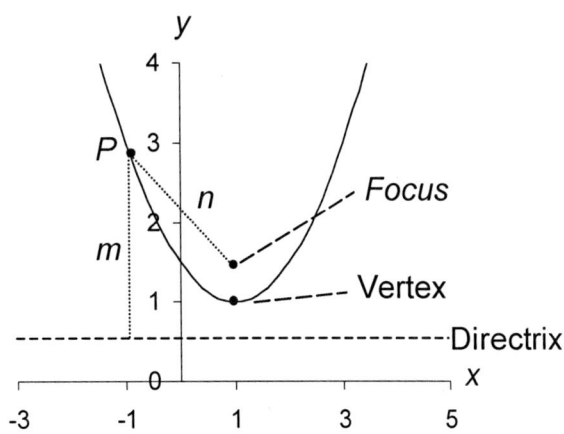

Note that the equation form above is for a parabola that is concave up. The form for concave down (or for horizontal concavity) can be deduced from this expression. Also note that $m = n$ for any point P on the parabola.

The equation of a parabola can be derived from a given focus and directrix along with the fact that $m = n$. Thus, for any point $P = (x, y)$ on the parabola (assuming a parabola that is concave up), the equation $m = n$ can be expressed using the distance formula.

$$m = y - (k - p) = n = \sqrt{(x-h)^2 + [y-(k+p)]^2}$$

By manipulating this expression as follows, the formula for a parabola can be derived.

$$\{y - (k-p)\}^2 = (x-h)^2 + [y-(k+p)]^2$$
$$y^2 - 2(k-p)y + (k-p)^2 = (x-h)^2 + y^2 - 2(k+p)y + (k+p)^2$$
$$(-2ky + 2py) + (k^2 - 2kp + p^2) =$$
$$= (x-h)^2 - (2ky + 2py) + (k^2 + 2kp + p^2)$$
$$2py - 2kp = (x-h)^2 - 2py + 2kp$$
$$4py - 4kp = (x-h)^2$$
$$(x-h)^2 = 4p(y-k)$$

A **hyperbola** is a conic section as shown below. The hyperbola has two foci and two separate curves, and is similar in some ways to an ellipse. The defining characteristic of a hyperbola is that, for any point P on the hyperbola, the difference between the distances from P to each focus is a constant. The important characteristics of a hyperbola are shown below.

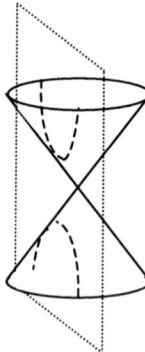

Equation: $\dfrac{(x-h)^2}{a^2} - \dfrac{(y-k)^2}{b^2} = 1$

Center: (h, k)

Foci: $(h \pm c, k)$, where $c = \sqrt{|a^2 + b^2|}$

Asymptotes: $y = k + \dfrac{b}{a}(x - h)$, $y = k - \dfrac{b}{a}(x - h)$

Example graph:

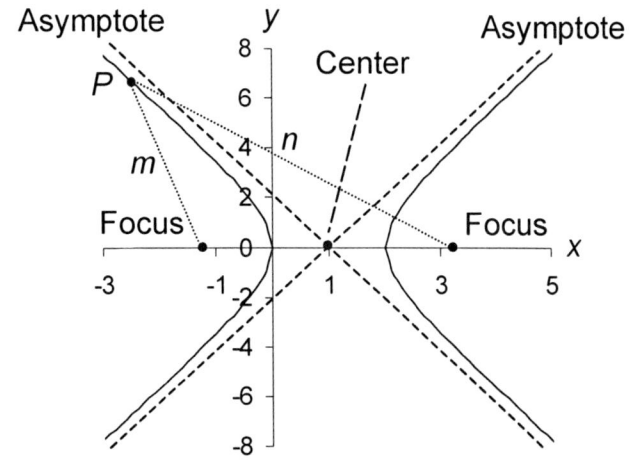

Note that $m - n$ is a constant for any point P on the hyperbola. The asymptotes are also illustrated above. The hyperbola can also be oriented in an up-down manner by multiplying the left side of the equation by -1. The corresponding expressions for the foci and asymptotes must also be changed accordingly.

Example: Find the equation of an ellipse with foci located at (0, 2) and (0, 0) and a minor axis of 1.

First, note that this ellipse is oriented along the y-axis. The expression for the ellipse is then the following, where $b > a$ and $a = 1$.

$$\dfrac{(x-h)^2}{a^2} + \dfrac{(y-k)^2}{b^2} = 1$$

The center of the ellipse is located halfway between the foci; by inspection, the center (h, k) is then (0, 1). (Generally, the midpoint formula can be used to find the center when the foci are known.) The length of the major axis can be found using the formula

$$c = \sqrt{|a^2 - b^2|}$$

where c can be deduced from the fact that the foci are located at $(h \pm c, k)$.

$$c = 1 = \sqrt{|a^2 - b^2|}$$
$$c^2 = 1 = |a^2 - b^2| = b^2 - a^2$$
$$b^2 - (1)^2 = 1$$
$$b^2 = 2$$
$$b = \sqrt{2}$$

Finally, construct the equation for the ellipse in terms of x and y.

$$x^2 + \frac{(y-1)^2}{2} = 1$$

The plot of this ellipse is shown below.

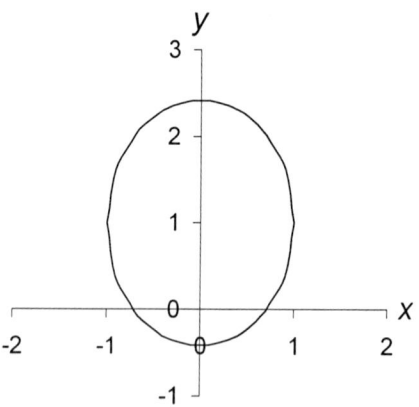

TEACHER CERTIFICATION STUDY GUIDE

DOMAIN IV.	PROBABILITY AND STATISTICS

Competency 015 The teacher understands how to use appropriate graphical and numerical techniques to explore data, characterize patterns, and describe departures from patterns.

This competency reviews basic statistical concepts for analyzing and appropriately describing data distributions. Measurement scales are discussed, followed by methods of displaying data, quantitative measures of central tendency and dispersion, linear transformations of data, and other considerations focusing on the use and application of statistics.

The beginning teacher selects and uses an appropriate measurement scale (i.e., nominal, ordinal, interval, ratio) to answer research questions and analyze data.

The four main types of **measurement scales** used in statistical analysis are nominal, ordinal, interval, and ratio. The type of variable measured and the research questions asked determine the appropriate measurement scale. The different measurement scales have distinctive qualities and attributes.

The **nominal measurement scale** is the most basic measurement scale. When measuring using the nominal scale, responses are simply classified into categories. Examples of variables measured on the nominal scale are gender, religion, ethnicity, and marital status. The essential attribute of the nominal scale is that the classifications have no numerical or comparative value. For example, when classifying people by marital status, there is no sense in which "single" is greater or less than "married." The only measure of central tendency applicable to the nominal scale is mode, and the only applicable arithmetic operation is counting.

The **ordinal measurement scale** is more descriptive than the nominal scale in that it allows comparison between categories. Examples of variables measured on the ordinal scale are movie ratings, consumer satisfaction surveys, and any rank or order. Although categories of responses can be compared on the ordinal scale (e.g., "highly satisfied" indicates a higher level of satisfaction than "somewhat satisfied"), there are no clear differences between the categories. In other words, one cannot presume that the difference between two particular categories is the same as the distance between two other categories.

Even if the responses are in numeric form (e.g., 1 = good, 2 = fair, and 3 = poor), there is no specific information regarding the intervals separating the groups. Ordinal scales allow greater-than or less-than comparisons, and the applicable measures of central tendency are range and median.

The next measurement scale is the **interval scale**. Interval scales are numeric scales where intervals have a fixed, uniform value throughout the scale. An example of an interval scale is the Fahrenheit temperature scale. On the Fahrenheit scale, the difference between 40 degrees and 50 degrees is the same as the difference between 70 degrees and 80 degrees. The major limitation of interval scales is that there is no fixed zero point. For example, even though the Fahrenheit scale has a value of zero degrees, this assignment is arbitrary because the measurement does not represent the absence of heat. Because interval scales lack a true zero point, ratio comparison of values has no meaning. Thus, the arithmetic operations of addition and subtraction are applicable to interval scales, but multiplication and division are not. The measures of central tendency applicable to interval scales are mode, median, and arithmetic mean.

The final (and most informative) measurement scale is the **ratio scale**. The ratio scale is essentially an interval scale with a true zero point. Examples of ratio scales are measurement of length (meters, inches, etc.), monetary systems, and the Kelvin temperature scale. The zero value of each of these scales represents the absence of length, money, and heat, respectively. The presence of a true zero point allows proportional comparisons. For example, one can say that someone with one dollar has twice as much money as someone with fifty cents. Because ratios have meaning on ratio scales, we can apply the arithmetic operations of multiplication and division to the data sets.

The beginning teacher organizes, displays, and interprets data in a variety of formats (e.g., tables, frequency distributions, scatter plots, stem-and-leaf plots, box-and-whisker plots, histograms, pie charts).

There are many graphical ways in which to represent data, such as line plots, line graphs, scatter plots, stem and leaf plots, histograms, bar graphs, pie charts, and pictographs.

A **line plot** organizes data in numerical order along a number line. An x is placed above the number line for each occurrence of the corresponding number. Line plots allow viewers to see at a glance a range of data and where typical and atypical data falls. These plots are generally used to summarize relatively small sets of data.

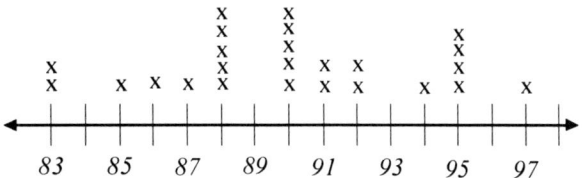

A **line graph** compares two variables, and each variable is plotted along an axis. A line graph highlights trends by drawing connecting lines between data points. This representation is particularly appropriate for data that varies continuously. Line graphs are sometimes referred to as **frequency polygons**.

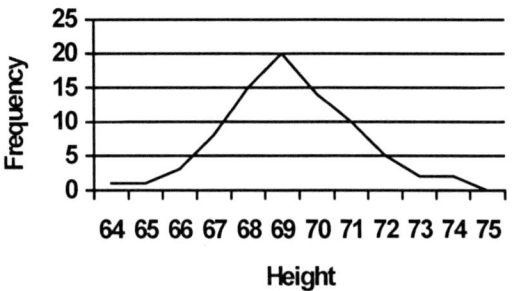

Bar graphs are similar to histograms. However, bar graphs are often used to convey information about categorical data where the horizontal scale represents a non-numeric attributes such as cities or years. Another difference is that the bars in bar graphs rarely touch. Bar graphs are also useful in comparing data about two or more similar groups of items.

**Production for ACME
June-September 2006
(in thousands)**

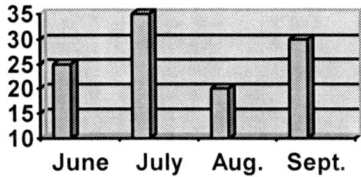

A **pie chart**, also known as a **circle graph**, is used to represent relative amounts of a whole.

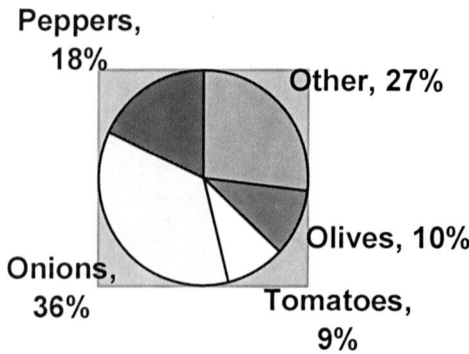

Scatter plots compare two characteristics of the same group of things or people and usually consist of a large body of data. They show how much one variable is affected by another. The relationship between the two variables is their **correlation**. The closer the data points come to making a straight line when plotted, the closer the correlation.

Stem and leaf plots are visually similar to line plots. The **stems** are the digits in the greatest place value of the data values, and the **leaves** are the digits in the next greatest place values. Stem and leaf plots are best suited for small sets of data and are especially useful for comparing two sets of data. The following is an example using test scores:

4	9
5	4 9
6	1 2 3 4 6 7 8 8
7	0 3 4 6 6 6 7 7 7 8 8 8 8
8	3 5 5 7 8
9	0 0 3 4 5
10	0 0

A **box-and-whisker plot** displays five statistics: a minimum and maximum score (neither of which should be considered outliers) and three quartiles. The box is composed of the first quartile, the median (or second quartile) and the third quartile, as shown in the example below. The (non-outlier) minimum and maximum values are shown at the end of the "whiskers" attached to the box.

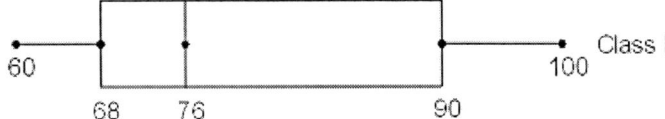

The above box-and-whisker example summarizes the scores on a mathematics test for the students in Class I. It indicates that the lowest score is 60 and the highest score is 100. Twenty-five percent of the class scored 68 or lower (the first quartile), 50% scored 76 or lower (the second quartile), and 25% scored 90 or higher (the third quartile).

Histograms are used to summarize information from large sets of data that can be naturally grouped into intervals. The vertical axis indicates **frequency** (the number of times any particular data value occurs), and the horizontal axis indicates data values or ranges of data values. The number of data values in any interval is the **frequency of the interval**.

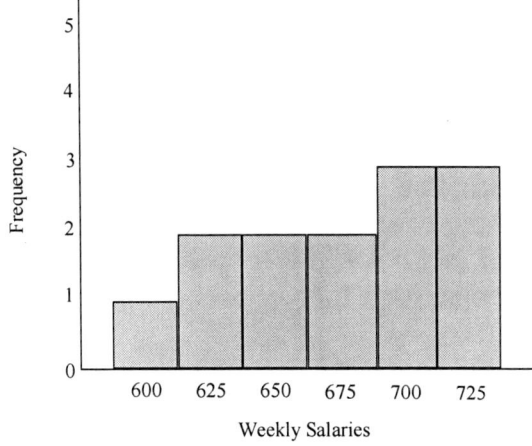

A **pictograph** uses small figures or icons to represent data. Pictographs are used to summarize relative amounts, trends, and data sets, andhey are useful in comparing quantities.

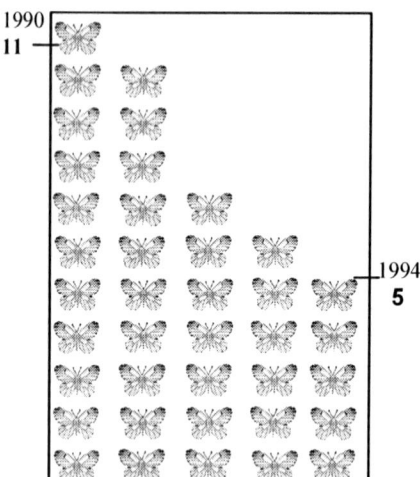

The data in this graph is not accurate. It is for illustration purposes only.

The beginning teacher understands measures of central tendency (i.e., mean, median, mode) and dispersion (i.e., range, interquartile range, variance, standard deviation).

Measures of Central Tendency

The mean, median and mode are **measures of central tendency** (i.e., the average or typical value) in a data set. They can be defined both for discrete and continuous data sets. A **discrete variable** is one that can only take on certain specific values. For instance, the number of students in a class can only be a whole number (e.g., 15 or 16, but not 15.5). A **continuous variable**, such as the weight of an object, can take on a continuous range of values.

For discrete data, the **mean** is the average of the data items, or the value obtained by adding all the data values and dividing by the total number of data items. For a data set of n items with data values $x_1, x_2, x_3, \ldots, x_n$, the mean is given by

$$\overline{X} = \frac{x_1 + x_2 + x_3 + \ldots + x_n}{n}$$

The **median** is found by putting the data in order from smallest to largest and selecting the item in the middle (or the average of the two values in the middle if the number of data items is even). The **mode** is the most frequently occurring datum. There can be more than one mode in a data set.

Example: Find the mean, median, and mode of the test scores listed below:

```
85    77    65
92    90    54
88    85    70
75    80    69
85    88    60
72    74    95
```

Mean: sum of all scores ÷ number of scores = 78
Median: put numbers in order from smallest to largest. Pick the middle number.

54, 60, 65, 69, 70, 72, 74, 75, 77, 80, 85, 85, 85, 88, 88, 90, 92, 95

Both values are in the middle.

Therefore, median is average of two numbers in the middle, or 78.5. The mode, or most frequent number is 85.

Discrete data is typically displayed in a table as shown in the example above. If the data set is large, it may be expressed in compact form as a **frequency distribution**. The number of occurrences of each data point is the **frequency** of that value. The **relative frequency** is defined as the frequency divided by the total number of data points. Since the sum of the frequencies equals the number of data points, the relative frequencies add up to 1. The relative frequency of a data point, therefore, represents the probability of occurrence of that value. Thus, a distribution consisting of relative frequencies is known as a **probability distribution**.

For data expressed as a frequency distribution, the mean is given by

$$\bar{x} = \frac{\sum x_i f_i}{\sum f_i} = \sum x_i f'_i$$

where x_i represents a data value, f_i the corresponding frequency and f'_i the corresponding relative frequency.

The **cumulative frequency** of a data point is the sum of the frequencies from the beginning up to that point. The median of a frequency distribution is the point at which the cumulative frequency reaches half the value of the total number of data points.

The mode is the point at which the frequency distribution reaches a maximum. There can be more than one mode in which case the distribution is **multimodal**.

Example: The frequency distribution below shows the summary of some test results where people scored points ranging from 0 to 45 in increments of 5. One person scored 5 points, 4 people scored 10 points and so on. Find the mean, median and mode of the data set.

Points	Frequency	Cumulative Frequency	Relative Frequency
5	1	1	0.009
10	4	5	0.035
15	12	17	0.105
20	22	39	0.193
25	30	69	0.263
30	25	94	0.219
35	13	107	0.114
40	6	113	0.053
45	1	114	0.009

The mean score is the following:
(5x1+10x4+15x12+20x22+25x30+30x25+35x13+40x6+45x1)/114
= (5+40+180+440+750+750+455+240+45)/114 = 25.5.

The median score (the point at which the cumulative frequency reaches or surpasses the value 57) is 25.

The mode (or value with the highest frequency) is 25.

The frequency distribution from the above example is displayed below as a histogram.

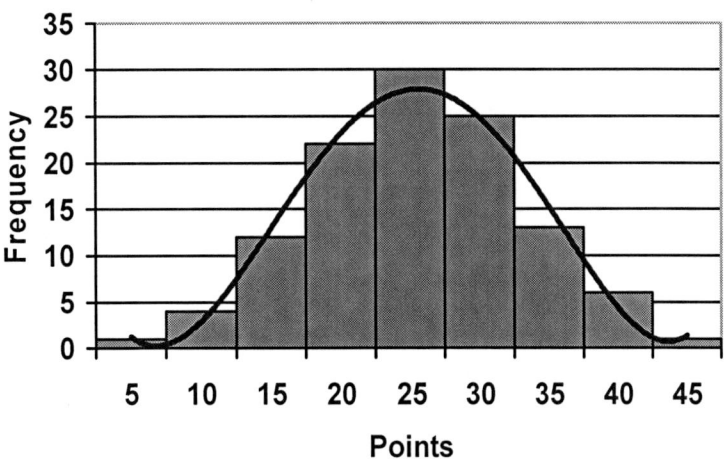

The histogram shows the reason why the mean, median and mode of this distribution are practically identical. This is due to the distribution being symmetric with one peak exactly in the middle. A trend line has been added to the histogram. Notice that this approximates the most common **continuous distribution**, a **normal or bell curve** for which the mean, median and mode are identical.

A frequency distribution may also be created by subdividing the range of the data into sub-ranges. In this case, the count in each subdivision or **bin** is the frequency. Discrete as well as continuous data may be represented in this way.

A large data set of continuous data is often represented using a **probability distribution** expressed as a **probability density function.** The integral of the probability density function over a certain range gives the probability of a data point being in that range of values. The integral of the probability density function over the whole range of values is equal to 1.

The **mean** value for a distribution of a variable x represented by a probability density function f(x) is given by

$$\int_{-\infty}^{+\infty} xf(x)dx$$

(Compare this with its discrete counterpart $\bar{x} = \sum x_i f_i'$).

The **median** is the upper bound for which the integral of the probability density function is equal to 0.5; i.e., if $\int_{-\infty}^{a} f(x)dx = 0.5$, then a is the median of the distribution.

The **mode** is the maximum value or values of the probability density function within the range of the function.

As mentioned before, the mean and median are very close together for symmetric distributions. **If the distribution is skewed to the right, the mean is greater than the median. If the distribution is skewed to the left, the mean is smaller than the median.** Distributions can also be described in terms of a center (a distribution has a clear center if it is symmetric about some vertical line), spread (a distribution may be relatively flat, or it may be sharply peaked), and shape (a distribution can have a single peak, multiple peaks, or a variety of other shapes).

Example: Find the mean, median and mode for the distribution given by the probability density function

$$f(x) = \begin{cases} 4x(1-x^2) & 0 \le x \le 1 \\ 0 & \text{otherwise} \end{cases}$$

$$\text{Mean} = \int_0^1 4x^2(1-x^2)dx = \frac{4x^3}{3}\Big|_0^1 - \frac{4x^5}{5}\Big|_0^1 = \frac{4}{3} - \frac{4}{5} = \frac{20-12}{15} = \frac{8}{15} = 0.53$$

If $x = a$ is the median, then

$$\int_0^a 4x(1-x^2)dx = 0.5$$

$$\Rightarrow \frac{4x^2}{2}\Big|_0^a - \frac{4x^4}{4}\Big|_0^a = 0.5$$

$$\Rightarrow 2a^2 - a^4 = 0.5$$

$$\Rightarrow 2a^4 - 4a^2 + 1 = 0$$

Solving for a yields

$$a^2 = \frac{4 \pm \sqrt{16-8}}{4} = 1 \pm \frac{2\sqrt{2}}{4} = 1 - \frac{\sqrt{2}}{2}$$ (to keep x within the range 0 to 1)

$$a = \sqrt{1 - \frac{1}{\sqrt{2}}} = 0.54$$

The mode is obtained by taking the derivative of the probability density function and setting it to zero as shown below. (Notice that the second derivative is negative at $x = 0.58$, and, hence, this is clearly a maximum.)

$$\frac{d}{dx}(4x - 4x^3) = 4 - 12x^2 = 0$$
$$\Rightarrow 12x^2 = 4$$
$$\Rightarrow x^2 = \frac{1}{3}$$
$$\Rightarrow x = \frac{1}{\sqrt{3}} = 0.58$$

Measures of Dispersion

The following discussion reviews measures of data dispersion, such as standard deviation, variance, skewness, and interquartile range. These characteristics are considered for both discrete and continuous distributions.

Statistics for Discrete Distributions

Range is a measure of variability that is calculated by subtracting the smallest value from the largest value in a set of discrete data.

The **variance** and **standard deviation** are measures of the "spread" (or **dispersion**) of data around the mean. It is noteworthy that descriptive statistics involving such parameters as variance and standard deviation can be applied to a set of data that spans the entire population (parameters, typically represented using Greek symbols) or to a set of data that only constitutes a portion of the population (sample statistics, typically represented by Latin letters).

The mean of a set of data, whether for a population (μ) or for a sample ($\bar{x}$), uses the formula discussed above, and it can be represented as either a set of individual data or as a set of data with associated frequencies. The variance and standard deviation for the population differ slightly from those of a sample. The population variance (σ^2) and the population standard deviation (σ) are as follows.

$$\sigma^2 = \frac{1}{n}\sum(x_i - \mu)^2$$
$$\sigma = \sqrt{\sigma^2}$$

Another statistic is the so-called skewness (here represented by the symbol γ), which is calculated according to the following formula.

$$\gamma = \frac{1}{n}\sum\left(\frac{x_i - \mu}{\sigma}\right)^3$$

Skewness quantifies the asymmetry of a distribution. A normal distribution, for instance, is symmetric about its mean and would therefore have a skewness of zero.

Example: Calculate and describe the skewness of the following ordered data set (consider this set a population rather than a sample): {2, 5, 8, 14, 7, 4, 3, 2, 1, 0}.

To describe qualitatively the skewness, prepare a plot of the data. In this case, a histogram is one possibility:

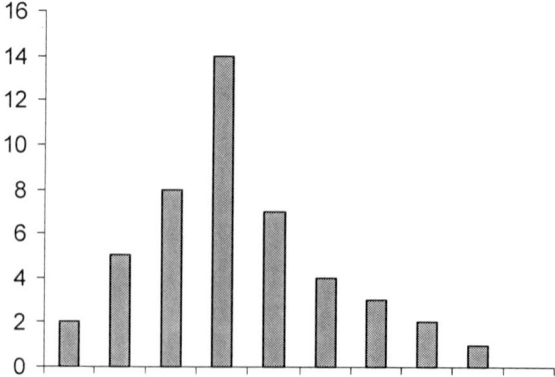

Qualitatively, this data set is skewed to the left, since the peak is off center and the general trend is asymmetric. To calculate the skewness quantitatively, use the skewness formula above. First, calculate the mean, µ, and the standard deviation, σ. Since the data set is simply described as "ordered," simply assume that each datum receives an equal weight.

$$\mu = \frac{2+5+8+14+7+4+3+2+1+0}{10} = 4.6$$

$$\sigma^2 = \frac{1}{n}\sum(x_i - \mu)^2 = \frac{1}{10}\{(2-4.6)^2 + (5-4.6)^2 + \ldots + (0-4.6)^2\}$$

$$\sigma^2 \approx 15.64$$

$$\sigma = \sqrt{\sigma^2} \approx \sqrt{15.64} \approx 3.95$$

Using this information, the skewness, γ, can now be calculated.

$$\gamma = \frac{1}{n}\sum\left(\frac{x_i - \mu}{\sigma}\right)^3 = \frac{1}{10}\left\{\left(\frac{2-4.6}{3.95}\right)^3 + \ldots + \left(\frac{0-4.6}{3.95}\right)^3\right\}$$

$$\gamma \approx 1.13$$

Thus, the data distribution has a skewness of about 1.13 to the left (as shown on the histogram).

For a sample, the data does not include the entire population. As a result, it should be expected that the sample data might not be perfectly representative of the population. To account for this shortcoming in the sample variance (s^2) and standard deviation (s), the sum of the squared differences between the data and the mean is divided by ($n - 1$) instead of just n. This increases the variance and standard deviation slightly, which in turn increases slightly the data spread to account for the possibility that the sample may not accurately represent the population.

$$s^2 = \frac{1}{n-1}\sum(x_i - \bar{x})^2$$

$$s = \sqrt{s^2}$$

Example: Calculate the range, variance and standard deviation for the following data set: {3, 3, 5, 7, 8, 8, 8, 10, 12, 21}.

The range is simply the largest data value minus the smallest. In this case, the range is 21 – 3 = 18.

To calculate the variance and standard deviation, first calculate the mean. If it is not stated whether a data set constitutes a population or sample, assume it is a population. (In this case, if the data was labeled as "ages of the 10 people in a room," this would be a population. If the data was labeled "ages of males at a crowded circus event," the data would be a sample.)

$$\mu = \frac{3+3+5+7+8+8+8+10+12+21}{10} = 8.5$$

Use this mean to calculate the variance.

$$\sigma^2 = \frac{1}{10}\sum(x_i - 8.5)^2$$

$$\sigma^2 = \frac{1}{10}\{(3-8.5)^2 + (3-8.5)^2 + (5-8.5)^2 + \ldots + (21-8.5)^2\}$$

$$\sigma^2 = \frac{246.5}{10} = 24.65$$

The standard deviation is

$$\sigma = \sqrt{\sigma^2} = \sqrt{24.65} \approx 4.96$$

The **interquartile range** is a measure of dispersion that uses **quartiles**, which divide the data into four segments. To find the quartile of a particular datum, first determine the median of the data set (which is labeled Q2), then find the median of the upper half (labeled Q3) and the median of the lower half (labeled Q1) of the data set. There is some confusion in determining the upper and lower quartile, and statisticians do not agree on the appropriate method to use. Tukey's method for finding the quartile values is to find the median of the data set, then find the median of the upper and lower halves of the data set. If there is an odd number of values in the data set, include the median value in both halves when finding the quartile values. For example, consider the following data set:

{1, 4, 9, 16, 25, 36, 49, 64, 81}

First, find the median value, which is 25. This is the value Q2. Since there is an odd number of values in the data set (nine), include the median in both halves. To find the quartile values, find the medians of the two sets

{1, 4, 9, 16, 25} and {25, 36, 49, 64, 81}

Since each of these subsets has an odd number of elements (five), use the middle value. Thus, the lower quartile value (Q1) is 9 and the upper quartile value (Q3) is 49.

Another method to find quartile values (if the total data set has an odd number of values) excludes the median from both halves when finding the quartile values. Using this approach on the data set above, exclude the median (25) from each half. To find the quartile values, find the medians of

{1, 4, 9, 16} and {36, 49, 64, 81}

Since each of these data sets has an even number of elements (four), average the middle two values. Thus the lower quartile value (Q1) is (4+9)/2 = 6.5 and the upper quartile value (Q3) is (49+64)/2 = 56.5. The middle quartile value (Q2) remains 25.

Other methods for calculating quartiles also exist, but these two methods are the most straightforward. To calculate the interquartile range (R_{IQ}), simply subtract Q1 from Q3. Thus,

$$R_{IQ} = Q1 - Q3$$

Statistics for Continuous Distributions

The range for a continuous data distribution is the same as that for a discrete distribution: the largest value minus the smallest value. Calculation of the mean, variance and standard deviation are similar, but slightly different. Since a continuous distribution does not permit a simple summation, integrals must be used. The mean μ of a distribution function $f(x)$ is expressed below (and is discussed previously in this section).

$$\mu = \int_{-\infty}^{\infty} x f(x) dx$$

The variance σ^2 over also has an integral form, and has a form similar to that of a discrete distribution.

$$\sigma^2 = \int_{-\infty}^{\infty} (x - \mu)^2 f(x) dx$$

The standard deviation σ is simply

$$\sigma = \sqrt{\sigma^2}$$

The skewness of a continuous distribution is given below.

$$\gamma = \int_{-\infty}^{\infty} \left(\frac{x-\mu}{\sigma}\right)^3 f(x)\,dx$$

The interquartile range can also be found for continuous distributions. Quartiles are typically applied to discrete data distributions, but application to continuous distributions is also possible. In such a case, quartiles would be calculated by dividing the area under the curve of the distribution into four even (or approximately even) segments. The boundaries of these segments are the quartile values. Thus, formula for interquartile range (I_{QR}) in the case of continuous distributions is the same as that for discrete distributions.

$$I_{QR} = Q1 - Q3$$

Example: Calculate the standard deviation of a data distribution function $f(x)$ where

$$f(x) = \begin{cases} 0 & x < 1 \\ -2x^2 + 2 & -1 \leq x \leq 1 \\ 0 & x > 1 \end{cases}$$

First calculate the mean of the function. Since the function is zero except between 1 and –1, the integral can likewise be evaluated from –1 to 1. (For further discussion of integrals, see **Competency 010**.)

$$\mu = \int_{-1}^{1} \left(-2x^2 + 2\right) x \, dx$$

$$\mu = -2 \int_{-1}^{1} \left(x^3 - x\right) dx$$

$$\mu = -2 \left[\frac{x^4}{4} - \frac{x^2}{2}\right]_{x=-1}^{x=1}$$

$$\mu = -2 \left\{\left[\frac{(1)^4}{4} - \frac{(1)^2}{2}\right] - \left[\frac{(-1)^4}{4} - \frac{(-1)^2}{2}\right]\right\} = 0$$

The mean can also be seen clearly by the fact that the graph of the function $f(x)$ is symmetric about the y-axis, indicating that its center (or mean) is at $x = 0$. Next, calculate the variance of f.

$$\sigma^2 = \int_{-1}^{1} (x-0)^2 \left(-2x^2 + 2\right) dx = -2 \int_{-1}^{1} x^2 \left(x^2 - 1\right) dx$$

$$\sigma^2 = -2 \int_{-1}^{1} \left(x^4 - x^2\right) dx$$

$$\sigma^2 = -2 \left[\frac{x^5}{5} - \frac{x^3}{3}\right]_{x=-1}^{x=1} = -2 \left\{\left[\frac{(1)^5}{5} - \frac{(1)^3}{3}\right] - \left[\frac{(-1)^5}{5} - \frac{(-1)^3}{3}\right]\right\}$$

$$\sigma^2 = -2 \left\{\frac{1}{5} - \frac{1}{3} - \left(-\frac{1}{5}\right) + \left(-\frac{1}{3}\right)\right\} = -2 \left(\frac{2}{5} - \frac{2}{3}\right)$$

$$\sigma^2 = \frac{8}{15} \approx 0.533$$

The standard deviation is

$$\sigma = \sqrt{\sigma^2} = \sqrt{\frac{8}{15}} \approx 0.730$$

The beginning teacher applies concepts of center, spread, shape, and skewness to describe a data distribution.

Application of statistical concepts such as center, spread, shape, and skewness are made in example problems and discussions throughout this skill section.

The beginning teacher applies linear transformations (i.e., translating, stretching, shrinking) to convert data and describes the effect of linear transformations on measures of central tendency and dispersion.

Statisticians use linear transformations of data sets to facilitate comparisons between data sets and to equalize aberrant differences at very large and very small values. For example, to compare scores from different intelligence tests, it is necessary to transform raw test scores (i.e., number of correct answers) into scaled scores with uniform means and standard deviations. Linear transformations take the form

$$X' = a + bX_0.$$

where X' is the transformed value, X_0 is the original value, a is the additive component and b is the multiplicative component.

The additive component of the transformation, a, shifts the original distribution to the right if a is positive, or to the left if a is negative. The additive component increases or decreases the mean of the distribution by a units, but it does not affect the standard deviation. The following is an example of an additive transformation of a data set and a generic graphical representation, where $a = 20$ and $b = 1$.

	X_0	X' $\longrightarrow$	
	12	32	
	14	34	
	21	41	Difference of a units
	23	43	
	27	47	
mean	19.4	39.4	
standard deviation	6.27	6.27	

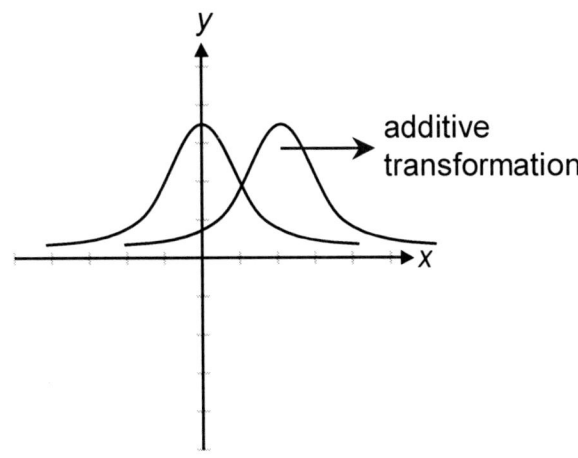

The multiplicative component of a linear transformation, b, horizontally stretches the distribution if b is greater than one, and it shrinks the distribution if b is between zero and one. The multiplicative component increases or decreases the mean and standard deviation of a data set by a factor of b. The following is an example of a multiplicative transformation of a data set and a generic graphical representation for $a = 0$ and $b = 3$.

	X_0	X'	
	12	36	
	14	42	
	21	63	
	23	69	Increase by factor of b
	27	81	
mean	19.4	58.2	
standard deviation	6.27	18.8	

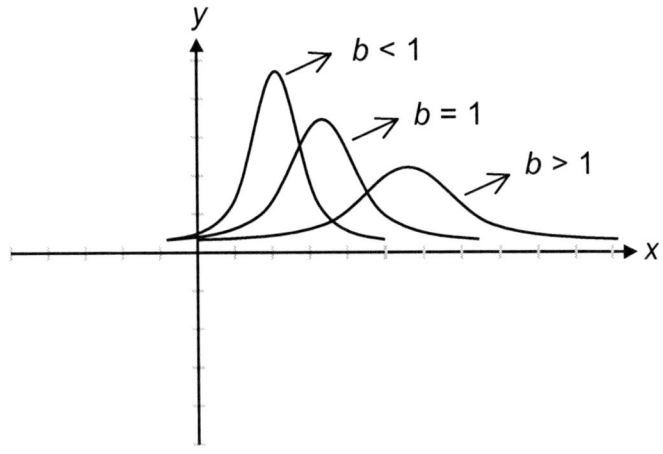

TEACHER CERTIFICATION STUDY GUIDE

The beginning teacher analyzes connections among concepts of center and spread, data clusters and gaps, data outliers, and measures of central tendency and dispersion.

The connections among statistical concepts such as measures of central tendency and dispersion are considered throughout the discussion in this competency and are illustrated in the various example problems.

The beginning teacher supports arguments, makes predictions, and draws conclusions using summary statistics and graphs to analyze and interpret one-variable data.

Statistics are cited almost invariably, in one form or another, in studies and publications that seek to prove a point about people, animals, food, drugs, or any number of other subjects. Using summary statistics and graphs appropriately to support conclusions and predictions is a crucial skill to this end.

Summary statistics and graphs for one-variable data include the types of information (mean, median, mode, variance, etc.) and methods of display (the various types of data plots and diagrams) discussed above. Appropriate application of these statistics and methods of display requires a knowledge of the particular area to which they are being applied as well as how well each aspect of the statistics represents the information. For example, consider the following frequency data distribution presented in histogram form. The frequencies (vertical axis) are shown for several discrete values (horizontal axis).

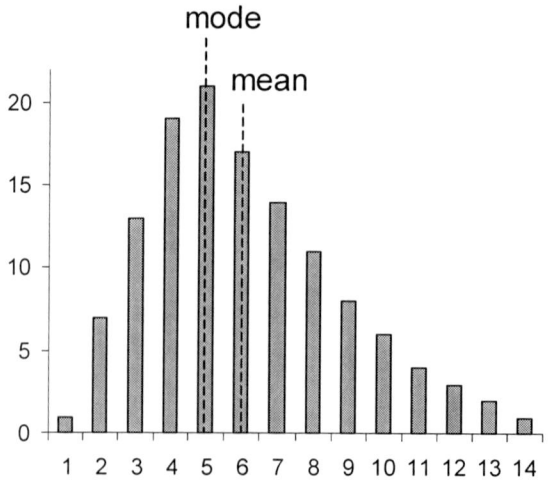

The (weighted) mean and mode of the data are labeled in the plot. Note that the use of the mean for describing the data distribution is somewhat descriptive, it does not seem to be the best way to describe the data since it is off center. In some situations, this seemingly minor difference could drastically affect the conclusions or predictions made based on the data. Thus, means (or averages) can be deceptive or can lead to incorrect conclusions when they are used as summary statistics. If an appropriate graph or other method of displaying the data is included, however, the use of the mean as a measure of central tendency could be a little less deceptive or otherwise misleading.

As can be seen from the preceding example, then, selecting the appropriate set of numbers and visual displays for a given set of data is necessary to accurately summarize a set of data. Such accurate summary statistics are necessary to making proper predictions and to drawing warranted conclusions. The same considerations applied to measures of central tendency in the example also apply to data spread and distribution shape or skewness.

In addition, arguments based on hypothesis testing require appropriate selection of test statistics. For more on hypothesis testing and some potential statistical tests that can be performed, see **Competency 017**.

Competency 016 **The teacher understands concepts and applications of probability.**

Many phenomena in social or physical sciences can be modeled using probability theory. The results of an experiment or set of experiments, whether in the scientific realm, political realm or a range of other areas, can be analyzed using statistics and associated methods of data analysis. This review covers first the elements of probability theory, including sample spaces, combinations and permutations, and then covers the applications of these concepts to random experiments. The fundamental counting principles are also discussed, along with probability distributions and simulations of probability.

The beginning teacher uses the concepts and principles of probability to describe the outcomes of simple and compound events.

The **probability** of an outcome, given a random experiment (a structured, repeatable experiment where the outcome cannot be predicted—or, alternatively, where the outcome is dependent on "chance"), is the relative frequency of the outcome. The relative frequency of an outcome is the number of times an experiment yields that outcome for a very large (ideally, infinite) number of trials. For instance, if a "fair" coin is tossed a very large number of times, then the relative frequency of a "heads-up" outcome is 0.5, or 50% (that is, one out of every two trials, on average, should be heads up). The probability is this relative frequency.

In probability theory, the **sample space** is a list of all possible outcomes of an experiment. For example, the sample space of tossing two coins is the set {HH, HT, TT, TH}, where H is heads and T is tails, and the sample space of rolling a six-sided die is the set {1, 2, 3, 4, 5, 6}. When conducting experiments with a large number of possible outcomes, it is important to determine the size of the sample space. The size of the sample space can be determined by using the fundamental counting principles and the rules of combinations and permutations.

A **random variable** is a function that corresponds to the outcome of some experiment or event, which is in turn dependent on "chance." For instance, the result of a tossed coin is a random variable: the outcome is either heads or tails, and each outcome has an associated probability. A **discrete variable** is one that can only take on certain specific values. For instance, the number of students in a class can only be a whole number (e.g., 15 or 16, but not 15.5).

A **continuous variable**, such as the weight of an object, can take on a continuous range of values.

The probabilities for the possible values of a random variable constitute the **probability distribution** for that random variable. Probability distributions can be discrete, as with the case of the tossing of a coin (there are only two possible distinct outcomes), or they can be continuous, as with, for instance, the outside temperature at a given time of day. In this latter case, the probability is represented as a continuous function over a range of possible temperatures, and finite probabilities can only be measured in terms of ranges of temperatures rather than specific temperatures. This is to say that, for a continuous distribution, it is not meaningful to say "the probability that the outcome is x"; instead, only "the probability that the outcome is between x and Δx" is meaningful. (Note that if each potential outcome in a continuous distribution has a non-zero probability, then the sum of all the probabilities would be greater than one, since there are an infinite number of potential outcomes.) Specific probability distributions are presented later in this discussion.

Example: Find the sample space and construct a probability distribution for a six-sided die (with numbers 1 through 6) where the even numbers are twice as likely as the odd numbers to come up on a given roll (assume the even numbers are equally likely and the odd numbers are equally likely).

The sample space is simply the set of all possible outcomes that can arise in a given trial. For this die, the sample space is {1, 2, 3, 4, 5, 6}. To construct the associated probability distribution, note first that the sum of the probabilities must equal 1. Let the probability of rolling an odd number (1, 3, or 5) be x; the probability of rolling an even number (2, 4, or 6) is then $2x$.

$$1 = p(1) + p(2) + p(3) + p(4) + p(5) + p(6) = 3x + 6x = 9x$$
$$x = \frac{1}{9}$$

The probability distribution can be shown as a histogram below.

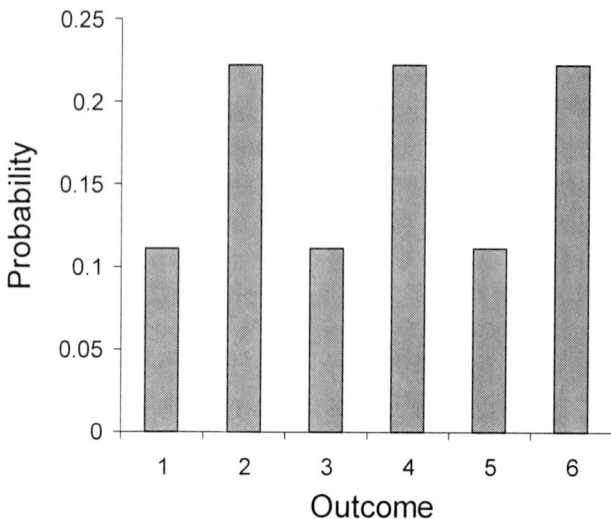

The sum of the probabilities for all the possible outcomes of a discrete distribution (or the integral of the continuous distribution over all possible values) must be equal to unity. The **expected value** of a probability distribution is the same as the **mean value** of a probability distribution. (See below for more discussion of mean values.) The expected value is thus a measure of the central tendency or average value for a random variable with a given probability distribution.

A **Bernoulli trial** is an experiment whose outcome is random and can be either of two possible outcomes, which are called "success" or "failure." Tossing a coin would be an example of a Bernoulli trial. The probability of success is represented by p, with the probability of failure being $q = 1 - p$. Bernoulli trials can be applied to any real-life situation in which there are only two possible outcomes. For example, concerning the birth of a child, the only two possible outcomes for the sex of the child are male or female.

Probability can also be expressed in terms of **odds**. Odds are defined as the ratio of the number of favorable outcomes to the number of unfavorable outcomes. The sum of the favorable outcomes and the unfavorable outcomes should always equal the total possible outcomes.

For example, given a bag of 12 red marbles and 7 green marbles, compute the odds of randomly selecting a red marble.

$$\text{Odds of red} = \frac{12}{7}$$

$$\text{Odds of not getting red} = \frac{7}{12}$$

In the case of flipping a coin, it is equally likely that a head or a tail will be tossed. The odds of tossing a head are 1:1. This is called **even odds**.

The beginning teacher solves a variety of probability problems using combinations and permutations.

A **permutation** is the number of possible arrangements of items, without repetition, where order of selection is important.

A **combination** is the number of possible arrangements, without repetition, where order of selection is not important.

Example: If any two numbers are selected from the set {1, 2, 3, 4}, list the possible permutations and combinations.

Combinations	Permutations
12, 13, 14, 23, 24, 34	12, 21, 13, 31, 14, 41, 23, 32, 24, 42, 34, 43,
six ways	twelve ways

Note that the list of permutations includes 12 and 21 as separate possibilities since the order of selection is important. In the case of combinations, however, the order of selection is not important and, therefore, 12 is the same combination as 21. Hence, 21 is not listed separately as a possibility.

The number of permutations and combinations may also be found by using the formulae given below.

The number of possible permutations in selecting r objects from a set of n is given by

$$_nP_r = \frac{n!}{(n-r)!}$$

The notation $_nP_r$ is read "the number of permutations of n objects taken r at a time."

In our example, two objects are being selected from a set of four.

$$_4P_2 = \frac{4!}{(4-2)!}$$ Substitute known values.

$$_4P_2 = 12$$

The number of possible combinations in selecting r objects from a set of n is given by

$$_nC_r = \frac{n!}{(n-r)!r!}$$ The number of combinations when r objects are selected from n objects.

In our example,

$$_4C_2 = \frac{4!}{(4-2)!2!}$$ Substitute known values.

$$_4C_2 = 6$$

It can be shown that $_nP_n$, the number of ways n objects can be arranged in a row, is equal to $n!$. We can think of the problem as n positions being filled one at a time. The first position can be filled in n ways using any one of the n objects. Since one of the objects has already been used, the second position can be filled only in $n-1$ ways. Similarly, the third position can be filled in $n-2$ ways and so on. Hence, the total number of possible arrangements of n objects in a row is given by

$$_nP_n = n(n-1)(n-2)........1 = n!$$

Example: Five books are placed in a row on a bookshelf. In how many different ways can they be arranged?

The number of possible ways in which 5 books can be arranged in a row is $5! = 1 \times 2 \times 3 \times 4 \times 5 = 120$.

The formula given above for $_nP_r$, **the number of possible permutations of r objects selected from n objects** can also be proven in a similar manner. If r positions are filled by selecting from n objects, the first position can be filled in n ways, the second position can be filled in $n-1$ ways and so on (as shown before). The rth position can be filled in $n-(r-1) = n-r+1$ ways.

Hence,

$$_nP_r = n(n-1)(n-2)\ldots(n-r+1) = \frac{n!}{(n-r)!}$$

The formula for the **number of possible combinations of *r* objects selected from *n*,** $_nC_r$, may be derived by using the above two formulae. For the same set of *r* objects, the number of permutations is *r*!. All of these permutations, however, correspond to the same combination. Hence,

$$_nC_r = \frac{_nP_r}{r!} = \frac{n!}{(n-r)!r!}$$

The number of permutations of n objects in a ring is given by (*n* – 1)!. This can be demonstrated by considering the fact that the number of permutations of *n* objects in a row is *n*!. When the objects are placed in a ring, moving every object one place to its left will result in the same arrangement. Moving each object two places to its left will also result in the same arrangement. We can continue this kind of movement up to *n* places to get the same arrangement. Thus the count *n*! is *n* times too many when the objects are arranged in a ring. Hence, the number of permutations of *n* objects in a ring is given by $\frac{n!}{n} = (n-1)!$.

Example: There are 20 people at a meeting. Five of them are selected to lead a discussion. How many different combinations of five people can be selected from the group? If the five people are seated in a row, how many different seating permutations are possible? If the five people are seated around a circular table, how many possible permutations are there?

The number of possible combinations of 5 people selected from the group of 20 is

$$_{20}C_5 = \frac{20!}{15!5!} = \frac{16 \times 17 \times 18 \times 19 \times 20}{1 \times 2 \times 3 \times 4 \times 5} = \frac{1860480}{120} = 15504$$

The number of possible permutations of the five seated in a row is

$$_{20}P_5 = \frac{20!}{15!} = 16 \times 17 \times 18 \times 19 \times 20 = 1860480$$

The number of possible permutations of the five seated in a circle is

$$\frac{_{20}P_5}{5} = \frac{20!}{5 \times 15!} = \frac{16 \times 17 \times 18 \times 19 \times 20}{5} = 372096$$

If the set of n objects contains some objects that are exactly alike, the number of permutations will again be different than $n!$. For instance, if n_1 of the n objects are exactly alike, then switching those objects among themselves will result in the same arrangement. Since we already know that n_1 objects can be arranged in $n_1!$ ways, $n!$ must be reduced by a factor of $n_1!$ to get the correct number of permutations. Thus, the number of permutations of n objects of which n_1 are exactly alike is given by $\frac{n!}{n_1!}$. Generalizing this, we can say that **the number of different permutations of n objects of which n_1 are alike, n_2 are alike,… n_j are alike, is**

$$\frac{n!}{n_1! n_2! \ldots n_j!} \text{ where } n_1 + n_2 \ldots + n_j = n$$

Example: A box contains 3 red, 2 blue and 5 green marbles. If all the marbles are taken out of the box and arranged in a row, how many different permutations are possible?

The number of possible permutations is

$$\frac{10!}{3!2!5!} = \frac{6 \times 7 \times 8 \times 9 \times 10}{6 \times 2} = 2520$$

Dependent and Independent Events

Dependent events occur when the probability of the second event depends on the outcome of the first event. For example, consider the two events: the home team wins the semifinal round (event A) and the home team wins the final round (event B). The probability of event B is contingent on the probability of event A. If the home team fails to win the semifinal round, it has a zero probability of winning in the final round. On the other hand, if the home team wins the semifinal round, then it may have a finite probability of winning in the final round. Symbolically, the probability of event B given event A is written $P(B|A)$. The conditional probability can be calculated according to the following definition (The symbol $\cap$ means "and," $\cup$ means "or" and $P(x)$ means "the probability of x").

$$P(B|A) = \frac{P(A \cap B)}{P(A)}$$

Consider a pair of dice: one red and one green. First the red die is rolled, followed by the green die. It is apparent that these events do not depend on each other, since the outcome of the roll of the green die is not affected by the outcome of the roll of the red die. The total probability of the two independent events can be found by multiplying the separate probabilities.

$$P(A \cap B) = P(A)P(B)$$
$$P(A \cap B) = \left(\frac{1}{6}\right)\left(\frac{1}{6}\right) = \frac{1}{36}$$

In many instances, however, events are not independent. Suppose a jar contains 12 red marbles and 8 blue marbles. If a marble is selected at random and then replaced, the probability of picking a certain color is the same in the second trial as it is in the first trial. If the marble is *not* replaced, then the probability of picking a certain color is *not* the same in the second trial, because the total number of marbles is decreased by one. This is an illustration of conditional probability. If R_n signifies selection of a red marble on the *n*th trial and B_n signifies selection of a blue marble on the *n*th trial, then the probability of selecting a red marble in two trials *with replacement* is

$$P(R_1 \cap R_2) = P(R_1)P(R_2) = \left(\frac{12}{20}\right)\left(\frac{12}{20}\right) = \frac{144}{400} = 0.36$$

The probability of selecting a red marble in two trials *without replacement* is

$$P(R_1 \cap R_2) = P(R_1)P(R|R_1) = \left(\frac{12}{20}\right)\left(\frac{11}{19}\right) = \frac{132}{360} \approx 0.367$$

Example: A car has a 75% probability of traveling 20,000 miles without breaking down. It has a 50% probability of traveling 10,000 additional miles without breaking down if it first makes it to 20,000 miles without breaking down. What is the probability that the car reaches 30,000 miles without breaking down?

Let event A be that the car reaches 20,000 miles without breaking down.

$$P(A) = 0.75$$

Event B is that the car travels an additional 10,000 miles without breaking down (assuming it didn't break down for the first 20,000 miles). Since event B is contingent on event A, write the probability as follows:

$$P(B|A) = 0.50$$

Use the conditional probability formula to find the probability that the car travels 30,000 miles $(A \cap B)$ without breaking down.

$$P(B|A) = \frac{P(A \cap B)}{P(A)}$$

$$0.50 = \frac{P(A \cap B)}{0.75}$$

$$P(A \cap B) = (0.50)(0.75) = 0.375$$

Thus, the car has a 37.5% probability of traveling 30,000 consecutive miles without breaking down.

The beginning teacher calculates probabilities using the axioms of probability and related theorems and concepts such as the addition rule, multiplication rule, conditional probability, and independence.

The following discussion of finite probability uses the symbols $\cap$ to mean "and," $\cup$ to mean "or" and $P(x)$ to mean "the probability of x." Also, $N(x)$ means "the number of ways that x can occur."

The Addition Principle of Counting states that if A and B are arbitrary events, then

$$N(A \cup B) = N(A) + N(B) - N(A \cap B)$$

Furthermore, if A and B are **mutually exclusive events** (that is, A and B cannot *both* take place—usually implying that they cannot both take place simultaneously), then

$$N(A \cup B) = N(A) + N(B)$$

Correspondingly, the probabilities associated with arbitrary events are

$$P(A \cup B) = P(A) + P(B) - P(A \cap B)$$

For mutually exclusive events,

$$P(A \cup B) = P(A) + P(B)$$

Example: In how many ways can you select a black card or a Jack from an ordinary deck of playing cards?

Let B denote selection of a black card and let J denote selection of a jack. Then, since half the cards (26) are black and four are jacks,

$$N(B) = 26$$
$$N(J) = 26$$

Also, since a card can be both black and a jack (the jack of spades and the jack of clubs),

$$N(B \cap J) = 2$$

Thus, the solution is

$$N(B \cup J) = N(B) + N(J) - N(B \cap J) = 26 + 4 - 2 = 28$$

Example: A travel agency offers 40 possible trips: 14 to Asia, 16 to Europe and 10 to South America. In how many ways can you select a trip to Asia or Europe through this agency?

Let A denote selection of a trip to Asia and let E denote selection of a trip to Europe. Since these are mutually exclusive events, then

$$N(A \cup E) = N(A) + N(E) = 14 + 16 = 30$$

Therefore, there are 30 ways you can select a trip to Asia or Europe.

The Multiplication Principle of Counting states that if A and B are arbitrary events, then the number of ways that A and B can occur in a two-stage experiment is given by

$$N(A \cap B) = N(A)N(B|A)$$

where $N(B|A)$ is the number of ways B can occur given that A has already occurred. This expression is also known as the **joint probability** of events A and B. If A and B are **independent** events (events for which the probability of one event is not dependent on the outcome of another event), then

$$N(A \cap B) = N(A)N(B)$$

Also, the probabilities associated with arbitrary events are

$$P(A \cap B) = P(A)P(B|A)$$

For independent events,

$$P(A \cap B) = P(A)P(B)$$

Example: How many ways from an ordinary deck of 52 cards can two jacks be drawn in succession if the first card is not replaced into the deck before the second card is drawn (that is, without replacement)?

This is a two-stage experiment. Let A be selection of a jack in the first draw and let B be selection of a jack in the second draw. It is clear that

$$N(A) = 4$$

If the first card drawn is a jack, however, then there are only three remaining jacks remaining for the second draw. Thus, drawing two cards without replacement means the events A and B are dependent, and

$$N(B|A) = 3$$

The solution is then

$$N(A \cap B) = N(A)N(B|A) = (4)(3) = 12$$

Example: How many six-letter code "words" can be formed if repetition of letters is not allowed?

Since these are code words, a word does not have to be in the dictionary; for example, *abcdef* could be a code word. Since the experiment requires choosing each letter without replacing the letters from previous selections, the experiment has six stages.

Repetition is not allowed; thus, there are 26 choices for the first letter, 25 for the second, 24 for the third, 23 for the fourth, 22 for the fifth and 21 for the sixth. Therefore, if A is the selection of a six-letter code word without repetition, then

$$N(A) = (26)(25)(24)(23)(22)(21) = 165,765,600$$

There are over 165 million ways to choose a six-letter code word with six unique letters.

Finite Probability

Using the fundamental counting principles described above, finite probability problems can be solved. Generally, finding the probability of a particular event or set of events involves dividing the number of ways the particular event can take place by the total number of possible outcomes for the experiment. Thus, by appropriately counting these possible outcomes using the above rules, probabilities can be determined.

Example: Determine the probability of rolling three even numbers on three successive rolls of a six-sided die.

This is a three-stage experiment. First, determine the total number of possible outcomes for three rolls of a die. For each roll,

$$N(\text{roll}) = 6$$

There are three possible even rolls for a die: 2, 4 and 6.

$$N(\text{even}) = 3$$

The probability of rolling an even number on any particular roll is then

$$P(\text{even}) = \frac{N(\text{even})}{N(\text{roll})} = \frac{3}{6} = \frac{1}{2}$$

For three successive rolls, use the multiplication rule for mutually exclusive events.

$$P(3 \text{ even rolls}) = P(\text{even})^3 = \left(\frac{1}{2}\right)^3 = \frac{1}{8} = 0.125$$

Thus, the probability of rolling three successive even numbers using a six-sided die is 0.125.

The beginning teacher understands expected value, variance, and standard deviation of probability distributions (e.g., binomial, geometric, uniform, normal).

The following discussion presents three common probability distributions and examples that illustrate their application. These distributions are the normal distribution, the binomial distribution and the exponential distribution. Some fundamental parameters of probability distributions are the **expected value**, variance, and standard deviation. The expected value $E(X)$, given a random variable X and an associated probability distribution $f(x)$ is the following for continuous and discrete distributions, respectively:

$$E(X) = \int_{-\infty}^{\infty} x f(x) \, dx$$

$$E(X) = \sum_i x_i f(x_i)$$

The expected (or expectation) value for a random variable X is also sometimes written as <X>. The expected value can be applied to random variables such as the mean µ (written as E(µ) or <µ>), the variance (written as E(X − µ) or <X − µ>), or any other parameter. The variance and standard deviation of a probability distribution are defined in the same manner as those in **Competency 015** for statistics.

The Normal Distribution

A **normal distribution** is the distribution associated with most sets of real-world data. It is frequently called a **bell curve**. A normal distribution has a **continuous random variable** X with mean μ and variance σ^2. The normal distribution has the following form.

$$f(x) = \frac{1}{\sigma\sqrt{2\pi}} e^{-\frac{1}{2}\left(\frac{x-\mu}{\sigma}\right)^2}$$

The total area under the normal curve is one. Thus,

$$\int_{-\infty}^{\infty} f(x)\,dx = 1$$

Since the area under the curve of this function is one, the distribution can be used to determine probabilities through integration. If a continuous random variable x follows the normal distribution, then the probability that x has a value between a and b is

$$P(a < X \leq b) = \int_a^b f(x)\,dx = F(b) - F(a)$$

Since this integral is difficult to evaluate analytically, tables of values are often used. Often, however, the tables use the integral

$$\frac{1}{\sqrt{2\pi}} \int_a^b e^{-\frac{t^2}{2}}\,dt = F(b) - F(a)$$

To use this form, simply convert x values to t values using

$$t = \frac{x_i - \mu}{\sigma}$$

where x_i is a particular value for the random variable X. This formula is often called the **z-score**.

Example: Albert's Bagel Shop's morning customer load follows a normal distribution, with **mean** (average) 50 and **standard deviation** 10. Determine the probability that the number of customers on a particular morning will be less than 42.

First, convert to a form that allows use of normal distribution tables:

$$t = \frac{x - \mu}{\sigma} = \frac{42 - 50}{10} = -0.8$$

Next, use a table to find the probability corresponding to the z-score. The actual integral in this case is

$$P(X < 42) = \frac{1}{\sqrt{2\pi}} \int_{-\infty}^{-0.8} e^{-\frac{t^2}{2}} dt$$

The table gives a value for $x = 0.8$ of 0.7881. To find the value for $x < -0.8$, subtract this result from one.

$$P(X < 42) = 1 - 0.7881 = 0.2119$$

This means that there is about a 21.2% chance that there will be fewer than 42 customers in a given morning.

Example: The scores on Mr. Rogers' statistics exam follow a normal distribution with mean 85 and standard deviation 5. A student is wondering what is the probability that she will score between a 90 and a 95 on her exam.

To compute $P(90 < x < 95)$, first compute the z-scores for each raw score.

$$z_{90} = \frac{90 - 85}{5} = 1$$
$$z_{95} = \frac{95 - 85}{5} = 2$$

Use the tables to find $P(1 < z < 2)$. To do this, subtract as follows.

$$P(1 < z < 2) = P(z < 2) - P(z < 1)$$

The table yields

$$P(1 < z < 2) = 0.9772 - 0.8413 = 0.1359$$

It can then be concluded that there is a 13.6% chance that the student will score between a 90 and a 95 on her exam.

The Binomial Distribution

The **binomial distribution** is a probability distribution for discrete random variables and is expressed as follows.

$$f(x) = \binom{n}{x} p^x q^{n-x}$$

where a sequence of *n* trials of an experiment are performed and where *p* is the probability of "success" and *q* is the probability of "failure." The value *x* is the number of times the experiment yields a successful outcome. Notice that this probability function is the product of p^x (the probability of successful outcomes in *x* trials) and q^{n-x} (the probability of unsuccessful outcomes in the remainder of the trials). The factor $\binom{n}{x}$ indicates that the *x* successful trials can be chosen $\binom{n}{x}$ ways (combinations) from the *n* total trials. (In other words, the successful trials may occur at different points in the sequence.)

Example: A loaded coin has a probability 0.6 of landing heads up. What is the probability of getting three heads in four successive tosses?

Use the binomial distribution. In this case, *p* is the probability of the coin landing heads up, and *q* = 1 − *p* is the probability of the coin landing tails up. Also, the number of "successful" trials (heads up) is 3.

Then,

$$f(3) = \binom{4}{3}(0.6)^3(1-0.6)^{4-3}$$

$$f(3) = \frac{4!}{3!(4-3)!}(0.6)^3(0.4)^1$$

$$f(3) = \frac{24}{6(1)}(0.216)(0.4) = 0.3456$$

Thus, there is a 34.56% chance that the loaded coin will land heads up three out of four times.

The Exponential Distribution

The exponential distribution is for continuous random variables and has the following form.

$$f(x) = \lambda e^{-\lambda x}$$

Here, $x \geq 0$. The parameter λ is called the **rate parameter**. For instance, the exponential distribution is often applied to failure rates. If a certain device has a failure rate λ failures per hour, then the probability that a device has failed at time T hours is

$$P(T) = \lambda \int_0^T e^{-\lambda t} dt = -\lambda \frac{1}{\lambda} e^{-\lambda t}\Big|_0^T = -\left[e^{-\lambda T} - e^0\right] = 1 - e^{-\lambda T}$$

Example: Testing has revealed that a newly designed widget has a failure rate of 1 per 5,000 hours of use. What is the probability that a particular part will be operational after a year of continual use?

Use the formula given above for the exponential distribution.

$$P(1 \text{ year}) = 1 - e^{-\lambda(1 \text{ year})}$$

Write λ in terms of failures per year.

$$\lambda = \frac{1 \text{ failure}}{5,000 \text{ hours}}\left(\frac{24 \text{ hours}}{1 \text{ day}}\right)\left(\frac{365 \text{ days}}{1 \text{ year}}\right) = 1.752 \frac{\text{failures}}{\text{year}}$$

Then

$$P(1\text{ year}) = 1 - e^{-1.752(1)} = 1 - 0.173 = 0.827$$

Thus, there is an 82.7% probability that the device will be operational after one year of continual use.

The beginning teacher solves a variety of probability problems using ratios of areas of geometric regions.

Probabilities can be calculated by finding the ratios of areas of geometric regions. In the context of probability distributions, probabilities correspond to areas under the probability density function. For instance, the total area under the probability density curve should be unity (since the probability of some outcome for an experiment must be one); the ratio of the area under the curve for some particular value or range of values for the random variable to the total area (unity) is the probability of that value or range of values.

Likewise, probabilities can be represented using the **Venn diagram**, which represents events or sets of events as shapes that depict the relationships of these events by overlapping (or not overlapping). For example, let the rectangle below represent all the possible outcomes of the random selection of a card from a standard deck. Let oval A be the all the outcomes for which a spade is chosen, and let oval B be all the outcomes for which a jack is chosen. Since there is one choice that falls within both of these categories (the jack of spades), the ovals overlap.

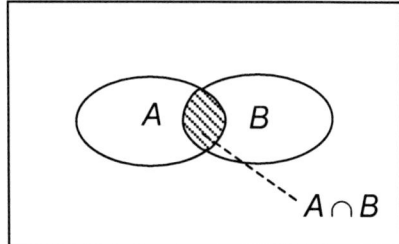

If the shapes correspond to areas that are to scale with their probabilities, then a Venn diagram can be used to calculate probabilities using ratios of these areas. Consider, for instance, the flip of a fair coin. The diagram for this case is shown below. (Although this may not strictly be considered a Venn diagram, depending on the definition of such, it does relay the same idea.)

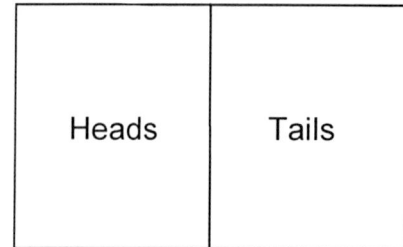

Notice that the total area A is divided evenly between "heads" (A/2) and "tails" (A/2). Thus, the probability of heads (or tails) is

$$\frac{A/2}{A} = \frac{1}{2}$$

Another Venn diagram is shown below for a six-sided die.

1	4
2	5
3	6

Again, the possibility of a particular outcome or range of outcomes can be found by using ratios of the associated areas. In both the cases above, there are no possible outcomes beyond those shown, so the Venn diagram does not show any area outside these outcomes.

The beginning teacher understands how to explore concepts of probability through sampling, experiments, and simulations, and generates and uses probability models to represent situations.

Simulations of random events or variables can be helpful in making informal inferences about theoretical probability distributions. Although simulations can involve use of physical situations that bear some similarity to the situation of interest, often times simulations involve computer modeling.

One of the crucial aspects of modeling probability using a computer program is the need for a random number that can be used to "randomize" the aspect of the program that corresponds to the event or variable. Although there is no function on a computer that can provide a truly random number, most programming languages have some function designed to produce a **pseudorandom number**. A pseudorandom number is not truly random, but it is sufficiently unpredictable that it can be used as a random number in many contexts.

Pseudorandom numbers can serve as the basis for simulation of rolling a die, flipping a coin, selecting an object from a collection of different objects, and a range of other situations. If, for instance, the pseudorandom number generator produces a number between zero and 1, simply divide up that range in accordance with the probabilities of each particular outcome. (For instance, assign 0 to 0.5 as heads and 0.5 to 1 as tails for the flip of a fair coin.) By performing a number of simulated trials and tallying the results, empirical probability distributions can be created.

Ideally, as the number of trials goes to infinity, the empirical probability distribution should approach the theoretical distribution. As a result, by performing a sufficiently large number of trials (this number must be at least somewhat justified for the particular situation) should allow informal inferences based on that data. Such inferences, however, must take into account the limitations of the computer, such as the inability to perform an infinite number of trials in finite time and the numerical inaccuracies that are an inherent part of computer programming.

The use of probability models to represent various situations is illustrated through the various example problems throughout this competency.

The beginning teacher determines probabilities by constructing sample spaces to model situations.

The concept and use of sample spaces is discussed in the first skill section of this competency.

TEACHER CERTIFICATION STUDY GUIDE

The beginning teacher applies concepts and properties of discrete and continuous random variables to model and solve a variety of problems involving probability and probability distributions (e.g., binomial, geometric, uniform, normal).

Discrete and continuous random variables are discussed throughout this competency, and numerous example problems are solved. The following example problems provide some further reinforcement of the concepts presented above.

Example: If the height of a certain population follows a normal distribution and has a mean of 5'6" and a standard deviation of 4", what is the probability that a randomly selected individual is taller than 6'?

Use the formula for the normal probability density distribution, which is given below, and a mean value µ = 66" and a standard deviation value σ = 4". Define X as the continuous random variable associated with height.

$$f(x) = \frac{1}{\sigma\sqrt{2\pi}} e^{-\frac{1}{2}\left(\frac{x-\mu}{\sigma}\right)^2} = \frac{1}{4\sqrt{2\pi}} e^{-\frac{1}{2}\left(\frac{x-66}{4}\right)^2}$$

The probability that a person is taller than 6' (or 72") is expressed as follows:

$$P(X \geq 72") = \int_{72"}^{\infty} f(x)\,dx$$

Using the previously defined probability density function and a table of values to evaluate the integral yields the resulting probability, where t has been defined as $\dfrac{x-66}{4}$ (thus, $dt = dx/4$):

$$P(X \geq 72") = \int_{72"}^{\infty} \frac{1}{4\sqrt{2\pi}} e^{-\frac{1}{2}\left(\frac{x-66}{4}\right)^2} dx = \frac{1}{4\sqrt{2\pi}} 4 \int_{72"}^{\infty} e^{-\frac{1}{2}t^2} dt$$

The probability integral is expressed in two slightly different forms since many tables list the integral from $-\infty$ to certain values of t.

$$P(X \geq 72") = \frac{1}{\sqrt{2\pi}} \int_{1.5}^{\infty} e^{-\frac{1}{2}(t)^2} dt = 1 - \frac{1}{\sqrt{2\pi}} \int_{-\infty}^{1.5} e^{-\frac{1}{2}(t)^2} dt$$

$$P(X \geq 72") \approx 1 - 0.9332 = 0.0668$$

Thus, there is only a 6.68% probability that a person selected at random from the population will be taller than 6'.

Example: A six-sided die is loaded to roll an even number 60% of the time. If each even number has an equal likelihood of being rolled and each odd has an equal likelihood of being rolled, what is the probability that exactly three out of four consecutive rolls will be greater than 3?

This problem involves a discrete random variable: the roll of a die. First, use the information presented in the problem statement to find the probability that a single roll yields a number greater than 3. Note that the evens are equally likely, and the odds are equally likely. If the probability of an even roll is 0.6, then the probability of numbers 2, 4, and 6 is 0.2 each. For the odd numbers, where the total probability is 1 – 0.6 = 0.4, the probability of rolling numbers 1, 3, and 5 is 4/30 each.

The total probability that a number greater than 3 (4, 5, or 6) is rolled is then the sum of the individual probabilities of these numbers.

$$P(4, 5, \text{ or } 6) = \frac{2}{10} + \frac{4}{30} + \frac{2}{10} = \frac{16}{30} = \frac{8}{15}$$

Thus, the probability of a "successful" roll (a number greater than 3) is 8/15, and the probability of a "failed" roll (a number 3 or less) is 1 – 8/15 = 7/15. Next, apply the binomial distribution probability function for three successful trials out of four.

$$P(3 \text{ out of } 4) = \binom{4}{3}\left(\frac{8}{15}\right)^3\left(\frac{7}{15}\right)^1$$

$$P(3 \text{ out of } 4) = 4\left(\frac{8}{15}\right)^3\left(\frac{7}{15}\right)^1 = \frac{4 \cdot 8^3 \cdot 7}{15^4} = \frac{14,336}{50,625} \approx 0.283$$

Thus, the probability for three out of four rolls greater than 3 is about 0.283.

Competency 017 The teacher understands the relationships among probability theory, sampling, and statistical inference, and how statistical inference is used in making and evaluating predictions.

This section reviews the application of statistics in surveys and sampling, bivariate data analysis, nonlinear data analysis, hypothesis testing, and other areas. Since a number of skills in this competency overlap with those of previous competencies, certain other skill sections are cross-referenced and should be consulted.

The beginning teacher analyzes and interprets statistical information (e.g., the results of polls and surveys) and recognizes misleading as well as valid uses of statistics.

In cases where the number of events or individuals is too large to collect data on each one, scientists collect information from only a small percentage. This is known as **sampling** or **surveying**. If sampling is done correctly, it should give the investigator nearly the same information he would have obtained by testing the entire population. The survey must be carefully designed, considering both the sampling technique and the size of the sample.

There are a variety of sampling techniques, both **random** and **non-random**. Random sampling is also known as **probability sampling** since the methods of probability theory can be used to ascertain the odds that the sample is representative of the whole population. Statistical methods may be used to determine how large a sample is necessary to give an investigator a specified level of certainty (95% is a typical confidence interval). Conversely, if an investigator has a sample of certain size, those same statistical methods can be used to determine how confident one can be that the sample accurately reflects the whole population.

A truly **random** sample must choose events or individuals without regard to time, place or result. **Simple random sampling** is ideal for populations that are relatively homogeneous with respect to the data being collected.

In some cases an accurate representation of distinct sub-populations requires **stratified random sampling** or **quota sampling**. For instance, if men and women are likely to respond very differently to a particular survey, the total sample population can be separated into these two subgroups and then a random group of respondents selected from each subgroup. This kind of sampling not only provides balanced representation of different subgroups, it also allows comparison of data between subgroups. Stratified sampling is sometimes **proportional**; i.e., the number of samples selected from each subgroup reflects the fraction of the whole population represented by the subgroup.

Sometimes compromises must be made to save time, money or effort. For instance, when conducting a phone survey, calls are typically only made in a certain geographical area and at a certain time of day. This is an example of **cluster random sampling**. There are three stages to cluster or area sampling: the target population is divided into many regional clusters (groups), a few clusters are randomly selected for study and a few subjects are randomly chosen from within a cluster

Systematic random sampling involves the collection of a sample at defined intervals (for instance, every tenth part to come off a manufacturing line). Here, it is assumed that the population is ordered randomly and there is no hidden pattern that may compromise the randomness of the sampling.

Non-random sampling is also known as **non-probability sampling**. **Convenience sampling** is the method of choosing items arbitrarily and in an unstructured manner from the frame. **Purposive sampling** targets a particular section of the population. **Snowball sampling** (e.g., having survey participants recommend others) and **expert sampling** are other types of non-random sampling. Obviously, non-random samples are far less representative of the whole population than random ones. They may, however, be the only methods available or may meet the needs of a particular study.

Statistics can be used both to inform and to mislead; thus, it is necessary to be able to discern between appropriate and inappropriate use of statistics. For instance, an improperly chosen measure of central tendency can mislead. Consider a case where a population is divided almost exclusively into extremely poor and extremely rich. In some cases, the mean income (or other measure of wealth) might lead a reader to think that a significant number of people are in the middle class, even if no one qualifies for this categorization, simply because the average income happens to fall between rich and poor. Likewise, extremely broad distributions of particular variables can make measures of central tendency misleading as well.

In addition to the mathematical considerations associated with appropriate use of statistics, the linguistic aspect must also be considered. The numbers used in a statistical statement are interpreted in light of the language used with them. Thus, to say "Half of population X in this area contracts disease Y, therefore everyone should get tested" might not include the fact that the only ones who contract the particular disease are those who (for instance) work at a certain chemical plant—thus, only those who work at the plant would need testing for the disease. Thus, it is necessary that the use of statistics includes all relevant information.

Another common fallacy of statistics is mistaking association for causation.

The beginning teacher applies knowledge of designing, conducting, analyzing, and interpreting statistical experiments to investigate real-world problems.

The concepts of the preceding skill section provide insight into the design, conducting, analysis, and interpretation of statistical experiments.

The beginning teacher understands random samples and sample statistics (e.g., the relationship between sample size and confidence intervals, biased or unbiased estimators).

Random sampling is the process of studying an aspect of a population by selecting and gathering data from a segment of that population and making inferences and generalizations based on the results. Two main types of random sampling are simple and stratified. With simple random sampling, each member of the population has an equal chance of selection to the sample group. With stratified random sampling, each member of the population has a known but unequal chance of selection to the sample group, as the study selects a random sample from each population demographic. In general, stratified random sampling is more accurate because it provides a more representative sample group. Sample statistics are important generalizations about the entire sample such as mean, median, mode, range, and sampling error (standard deviation). Various factors affect the accuracy of sample statistics and the generalizations made from them about the larger population.

Sample size is one important factor in the accuracy and reliability of sample statistics. As sample size increases, sampling error (standard deviation) decreases. Sampling error is the main determinant of the size of the confidence interval. Confidence intervals decrease in size as sample size increases. A confidence interval gives an estimated range of values, which is likely to include a particular population parameter. The confidence level associated with a confidence interval is the probability that the interval contains the population parameter. For example, a poll reports 60% of a sample group prefers candidate A with a margin of error of ±3% and a confidence level of 95%. In this poll, there is a 95% chance that the preference for candidate A in the whole population is between 57% and 63%.

The ultimate goal of sampling is to make generalizations about a population based on the characteristics of a random sample. Estimators are sample statistics used to make such generalizations. For example, the mean value of a sample is the estimator of the population mean. Unbiased estimators, on average, accurately predict the corresponding population characteristic. Biased estimators, on the other hand, do not exactly mirror the corresponding population characteristic. Although most estimators contain some level of bias, limiting bias to achieve accurate projections is the goal of statisticians.

The beginning teacher makes inferences about a population using binomial, normal, and geometric distributions.

Some fundamental parameters of probability distributions are the **expected value**, variance, and standard deviation. The expected value $E(X)$, given a random variable X and an associated probability distribution $f(x)$ is the following for continuous and discrete distributions, respectively:

$$E(X) = \int_{-\infty}^{\infty} x f(x) dx$$

$$E(X) = \sum_i x_i f(x_i)$$

The expected (or expectation) value for a random variable X is also sometimes written as $<X>$. The expected value can be applied to random variables such as the mean μ (written as $E(X)$ or $<X>$), the variance (written as $E((X - \mu)^2)$ or $<(X - \mu)^2>$), or any other parameter.

The variance and standard deviation of a probability distribution are defined in the same manner as those in **Competency 015** for statistics.

Additional concepts and applications of various probability distributions are discussed at length in **Competency 016**. Example problems involving calculating population parameters are also provided therein.

The beginning teacher describes and analyzes bivariate data using various techniques (e.g., scatterplots, regression lines, outliers, residual analysis, correlation coefficients).

Bivariate data involves information that corresponds to two different variables. Methods of displaying bivariate data are discussed above (such as the scatterplot).

Regression for bivariate data allows description of data and offers the ability to interpolate (and, in some cases, extrapolate) additional data. It is often helpful to use regression to construct a more general trend or distribution based on sample data. To select an appropriate model for the regression, a representative set of data must be examined. It is often helpful, in this case, to plot the data and review it visually on a graph. In this manner, it is relatively simple to select a general class of functions (linear, quadratic, exponential, etc.) that might be used to model the data. There are two basic aspects of regression: selection of an appropriate curve that best fits the data and quantification of the "goodness of fit" of that curve. For instance, if a line can be constructed that passes through every data point of a distribution, then that line is a perfect fit to the data (and, obviously, linear regression is an appropriate choice for the model). If the distribution of data points seems to bear no particular resemblance to the line, then linear regression is probably not a wise choice, and a quantification of the goodness of fit should reflect this fact.

An important consideration prior to performing regression is the presence of **outliers**. An outlier is a piece of data that does not seem to fit with the general trend of the balance of the data. Outliers can result from particular cases that run contrary to a general trend (for instance, the presence of an extremely tall person or extremely short person in a crowd of otherwise average people), or they can result from an error in measurement. Often, it is appropriate to remove outliers from data when performing regression, although the choice to do so requires careful consideration and (sometimes) statistical testing.

The Method of Least Squares

Given a set of data, a curve approximation can be fitted to the data by using the **method of least squares**. The best-fit curve, defined by the function $f(x)$, is assumed to approximate a set of data with coordinates (x_i, y_i) by minimizing the sum of squared differences between the curve and the data. Mathematically, the sum of these squared differences (errors) can be written as follows for a data set with n points.

$$S = \sum_{i=1}^{n} \left[f(x_i) - y_i \right]^2$$

Thus, the best-fit curve approximation to a set of data (x_i, y_i) is $f(x)$ such that S is minimized.

Shown below is a set of data and a linear function that approximates it. The vertical distances between the data points and the line are the errors that are squared and summed to find S.

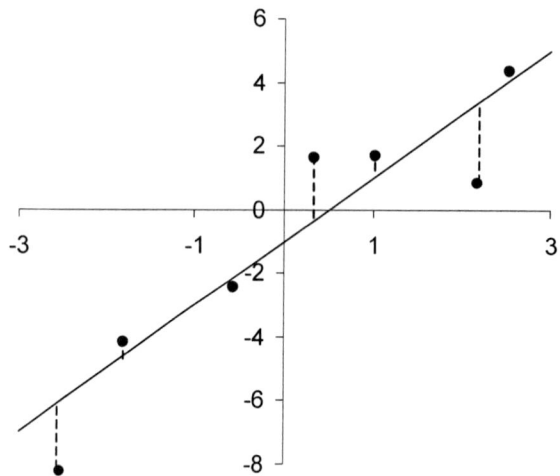

Linear least squares regression

If the curve f(x) that is used to approximate a set of data by minimizing the sum of squared errors (or **residuals**), S, then f(x) is called a **least squares regression line**. The process of determining f(x) is called **linear least squares regression**. In this case, f(x) has the form $f(x) = ax + b$.

Given a set of data $\{(x_1, y_1), (x_2, y_2), (x_3, y_3), \ldots, (x_n, y_n)\}$, the sum S for linear regression is the following.

$$S = \sum_{i=1}^{n}[ax_i + b - y_i]^2$$

To find f(x), it is necessary to find a and b. This can be done by minimizing S. Since S is a function of both a and b, S must be minimized through the use of partial derivatives. (A partial derivative is exactly the same as a full derivative, except that all variables other than the one being differentiated are treated as constants. Partial derivatives often use the symbol ∂ in place of d.) Therefore, find the partial derivative with respect to a and the partial derivative with respect to b.

$$\frac{\partial S}{\partial a} = \frac{\partial}{\partial a}\sum_{i=1}^{n}[ax_i + b - y_i]^2 \qquad \frac{\partial S}{\partial b} = \frac{\partial}{\partial b}\sum_{i=1}^{n}[ax_i + b - y_i]^2$$

$$\frac{\partial S}{\partial a} = \sum_{i=1}^{n} 2x_i[ax_i + b - y_i] \qquad \frac{\partial S}{\partial b} = \sum_{i=1}^{n} 2[ax_i + b - y_i]$$

Set these equal to zero. This yields a system of equations that can be solved to find *a* and *b*. Although the algebra is somewhat involved, it is not conceptually difficult. The results are given below.

$$a = \frac{n\sum_{i=1}^{n} x_i y_i - \sum_{i=1}^{n} x_i \sum_{i=1}^{n} y_i}{n\sum_{i=1}^{n} x_i^2 - \left[\sum_{i=1}^{n} x_i\right]^2}$$

Note that the average *x* value for the data (which is the sum of all *x* values divided by *n*) and the average *y* value for the data (which is the sum of all *y* values divided by *n*) can be used to simplify the expression. The average *x* value is defined as $\bar{x}$ and the average *y* value is defined as $\bar{y}$.

$$a = \frac{\sum_{i=1}^{n} x_i y_i - n\bar{x}\bar{y}}{\sum_{i=1}^{n} x_i^2 - n\bar{x}^2}$$

Since the expression for *b* is complicated, it suffices to the above expression for *b* in terms of *a*.

$$b = \frac{1}{n}\left(\sum_{i=1}^{n} y_i - a\sum_{i=1}^{n} x_i\right)$$
$$b = \bar{y} - a\bar{x}$$

Thus, given a set of data, the linear least squares regression line can be found by calculating *a* and *b* as shown above.

The **correlation coefficient**, r, can be used as a measure of the quality of f(x) as a fit to the data set. The value of r ranges from zero (for a poor fit) to one (for a good fit). The correlation coefficient formula is given below.

$$r^2 = \frac{\left[\sum_{i=1}^{n} x_i y_i - \frac{1}{n}\sum_{i=1}^{n} x_i \sum_{i=1}^{n} y_i\right]^2}{\left[\sum_{i=1}^{n} x_i^2 - \frac{1}{n}\left(\sum_{i=1}^{n} x_i\right)^2\right]\left[\sum_{i=1}^{n} y_i^2 - \frac{1}{n}\left(\sum_{i=1}^{n} y_i\right)^2\right]}$$

$$r^2 = \frac{\left(\sum_{i=1}^{n} x_i y_i - n\overline{xy}\right)^2}{\left(\sum_{i=1}^{n} x_i^2 - n\overline{x}^2\right)\left(\sum_{i=1}^{n} y_i^2 - n\overline{y}^2\right)}$$

Example: A company has collected data comparing the age of its employees to their respective income (in thousands of dollars). Find the line that best fits the data (using a least squares approach). Also calculate the correlation coefficient for the fit. The data is given below in the form of (age, income).

$$\{(35,42),(27,23),(54,43),(58,64),(39,51),(31,40)\}$$

The data are plotted in the graph below.

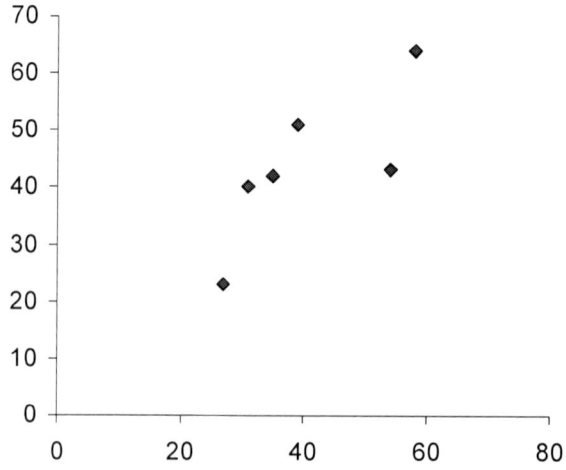

Note that there are six pieces of data. It is helpful to first calculate the following sums:

$$\sum_{i=1}^{6} x_i = 35 + 27 + 54 + 58 + 39 + 31 = 244$$

$$\sum_{i=1}^{6} y_i = 42 + 23 + 43 + 64 + 51 + 40 = 263$$

$$\sum_{i=1}^{6} x_i y_i = 35(42) + 27(23) + 54(43) + 58(64) + 39(51) + 31(40)$$
$$= 11354$$

$$\sum_{i=1}^{6} x_i^2 = 35^2 + 27^2 + 54^2 + 58^2 + 39^2 + 31^2 = 10716$$

$$\sum_{i=1}^{6} y_i^2 = 42^2 + 23^2 + 43^2 + 64^2 + 51^2 + 40^2 = 12439$$

Based on these values, the average x and y values are given below.

$$\bar{x} = \frac{244}{6} \approx 40.67$$

$$\bar{y} = \frac{263}{6} \approx 43.83$$

To find the equation of the least squares regression line, calculate the values of a and b.

$$a = \frac{\sum_{i=1}^{n} x_i y_i - n\bar{x}\bar{y}}{\sum_{i=1}^{n} x_i^2 - n\bar{x}^2} = \frac{11354 - 6(40.67)(43.83)}{10716 - 6(40.67)^2} \approx 0.832$$

$$b = \bar{y} - a\bar{x} = 43.83 - 0.832(40.67) = 9.993$$

Thus, the equation of the least squares regression line is

$$f(x) = 0.832x + 9.993$$

This result can be displayed on the data graph to ensure that there are no egregious errors in the result.

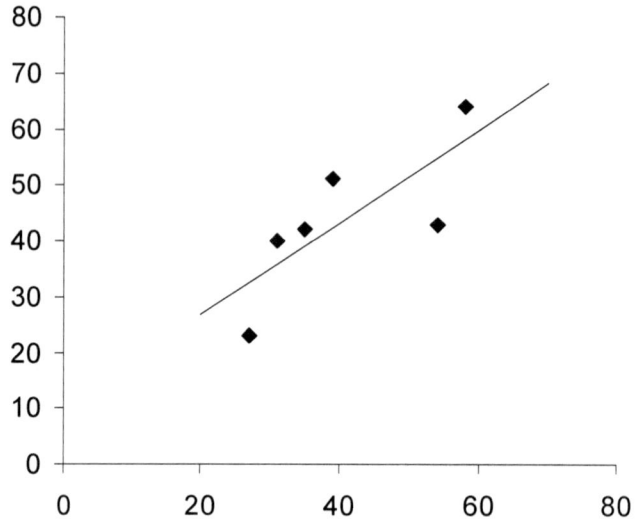

The regression line in the graph above appears to do a good job of approximating the trend of the data. To quantify how well the line fits the data, calculate the correlation coefficient using the formula given above.

$$r^2 = \frac{(11354 - 6(40.67)(43.83))^2}{(10716 - 6(40.67)^2)(12439 - 6(43.83)^2)}$$

$$r^2 = \frac{(658.603)^2}{(791.707)(912.587)} = 0.600$$

$$r = 0.775$$

Thus, the fit to the data is reasonably good.

The beginning teacher understands how to transform nonlinear data into linear form in order to apply linear regression techniques to develop exponential, logarithmic, and power regression models.

Linear regression is discussed in detail above. In some cases, however, data does not demonstrate a linear trend, so linear regression is not possible (directly). Nevertheless, if the data trend is exponential, logarithmic, or power-based, the data can be transformed to a linear form so that linear regression can be used.

To transform non-linear data to a linear form, it is necessary to use the inverse operation for the particular data trend. After applying linear regression to the results, the line can then be converted back to the nonlinear form. Consider the following general nonlinear forms.

$$y = ae^{bx} \quad \text{Exponential}$$
$$y = a\ln(bx) \quad \text{Logarithmic}$$
$$y = ax^b \quad \text{Power}$$

Using either an exponential or logarithmic operator, these nonlinear forms can be converted to linear.

$$\ln y = \ln a + bx \quad \text{Linear (from exponential)}$$
$$e^{y/a} = bx \quad \text{Linear (from logarithmic)}$$
$$\left(\frac{y}{a}\right)^{\frac{1}{b}} = x \quad \text{Linear (from power)}$$

If the left-hand side of each of the above equations is renamed y', then these forms are clearly linear. Nevertheless, despite the ease of transforming data to a linear form, these transformations can affect the regression. Thus, in some cases nonlinear regression is preferable to a transformation to linear data followed by linear regression.

If a linear transformation is an acceptable route, however, the regression line can then be transformed back to the nonlinear form, yielding a data fit that should follow the observed trend.

The beginning teacher uses the law of large numbers and the central limit theorem in the process of statistical inference.

The law of large numbers and the central limit theorem are two fundamental concepts in statistics. The **law of large numbers** states that the larger the sample size of a random process (or the more times a random variable is measured), the closer the sample mean will come to the population mean (or, in other words, the sample mean will approach the expected value). For example, the average weight of 40 apples out of a population of 100 will more closely approximate the population average weight than will a sample of 5 apples. Thus, the process of statistical inference should take note of the importance of a sufficiently large sample size when making conclusions about a general random variable or process. Although time or monetary considerations can limit sample size, greater confidence is achieved through a larger sample.

The **central limit theorem** expands on the law of large numbers by stating that as the number of samples increases, the distribution of sample means (averages) approaches a normal distribution. This holds true regardless of the distribution of the population. Thus, as the number of samples taken increases, the sample mean becomes closer to the population mean. This property of statistics allows for the analysis of the properties of populations with unknown distributions.

In conclusion, the law of large numbers and central limit theorem show the importance of large sample size and large number of samples to the process of statistical inference. As sample size and the number of samples taken increase, the accuracy of conclusions about the population drawn from the sample data increases.

The beginning teacher estimates parameters (e.g., population mean and variance) using point estimators (e.g., sample mean and variance).

Point estimators are estimates of population parameters and are based on sample statistics. The relationship of population parameters and sample statistics is discussed in **Competency 015** in connection with measures of central tendency and dispersion.

The beginning teacher understands principles of hypotheses testing.

Statistical hypothesis testing is a method of determining, to within a certain confidence level, whether a hypothesis can be accepted according to a certain set of data.

The first step of hypothesis testing is to formulate the so-called **null hypothesis**, which is assumed to be true and is to be accepted unless sufficient evidence warrants its rejection. Thus, the null hypothesis is often a simple or readily accepted statement, and it is typically labeled H_0. The opposite of the null hypothesis is the so-called alternate hypothesis, which is typically labeled H_1 or H_a. If the null hypothesis is rejected, the alternate hypothesis is then accepted.

The next step involves computing some test statistic using the associated sample data. Comparison of this data with a critical value for the test statistic (which is a threshold value for a given confidence) allows determination of whether to accept or reject the null hypothesis.

Common test statistics include the t-test, the z-test, and the χ^2 (chi-square) "goodness of fit" test.

The (Student's) **t-test** is the most commonly used method to evaluate the difference in means between two groups. The t-test assesses whether the means of two groups are statistically different from each other. The formula for the t-test is a ratio: the numerator of the ratio is the difference between the two means or averages, and the denominator is a measure of the variability or dispersion of the scores.

$$\frac{\text{difference between group means}}{\text{variability of groups}} = \frac{\overline{X}_T - \overline{X}_C}{SE(\overline{X}_T - \overline{X}_C)} = \text{t-value}$$

In this example T refers to a treatment group and C refers to a control group. To compute the numerator of the formula, find the difference between the means. The denominator is called the **standard error of the difference**. To compute this quantity, take the variance for each group and divide it by the population of that group. Add these two values and then take the square root of the sum.

$$SE(\overline{X}_T - \overline{X}_C) = \sqrt{\frac{\text{var}_T}{n_T} + \frac{\text{var}_C}{n_C}}$$

The final formula for the t-test is

$$t = \frac{\overline{X}_T - \overline{X}_C}{\sqrt{\dfrac{\text{var}_T}{n_T} + \dfrac{\text{var}_C}{n_C}}}$$

Once the t-value is computed, a Student's t distribution table is needed to compare the t-test statistic with the threshold value for a given confidence level. This confidence level is often expressed as a risk, or alpha level, which is usually 0.05. Another important value is the number of degrees of freedom (*df*) for the test. In the t-test, *df* is the sum of the populations in both groups less two. Using the t-value, the alpha level, and *df*, it is possible to look up the t-value in a standard table to determine whether the t-value is large enough to be significant. If it is, one can conclude (to a confidence defined by the alpha level) that the difference between the means for the two groups is statistically significant.

Example: The national average household income, which is based on a random sample of 500 households, is $43,000 with a standard deviation of $12,000. In a particular region, a 25-household sample indicates that the average household income is $39,000 with a standard deviation of $15,000. Determine if the discrepancy between average incomes is statistically significant.

Define the null and alternate hypotheses as follows:

H_0: There is no significant difference between the average incomes.
H_a: There is a significant difference between the average incomes.

Use the Student's t-test to calculate a test statistic. Assume an alpha level of 0.05. Also, convert standard deviations to variances by squaring.

$$t = \frac{\overline{X}_T - \overline{X}_C}{\sqrt{\dfrac{\text{var}_T}{n_T} + \dfrac{\text{var}_C}{n_C}}} = \frac{\$39,000 - \$43,000}{\sqrt{\dfrac{(\$15,000)^2}{25} + \dfrac{(\$12,000)^2}{500}}}$$

$$t = \left| -\frac{4,000}{\sqrt{9,000,000 + 288,000}} \right| \approx 1.31$$

The number of degrees of freedom in this case is 500 + 25 − 2 = 523. In many cases, the table of values for the Student's t distribution will not list this high a number, so ∞ is usually close enough. The critical t-value for an alpha value of 0.05 is 1.65. Since 1.31 < 1.65, there is not sufficient reason to reject the null hypothesis. Thus, it can be concluded to a confidence of 95% that there is no statistically significant difference between the regional average income and the national average income.

Chi-square tests are used to determine the acceptability of a null hypothesis. They are useful for testing the "goodness-of-fit" between an observed distribution and an expected distribution or for testing the independence of two variables.

The chi-square value is calculated using the observed and expected frequencies of a distribution. Given that O_i is the observed frequency for the ith value of a data set and E_i is the expected frequency, then the chi-square (χ^2) value is the following.

$$\chi^2 = \sum_{i=1}^{N} \frac{(O_i - E_i)^2}{E_i}$$

Here, N is the total number of values in the observed data set. Notice that NO_i and NE_i are the total number of observed and expected samples, respectively, of the ith value of the data set. Notice also that χ^2 is a measure of the deviation of the observed values from the expected values. If the observed values are precisely the same as the expected values, then χ^2 is zero. A high χ^2 means that there is large variation.

Another important parameter in the chi-square test is the **degrees of freedom**, n. The number of degrees of freedom is the total number of possible values or value classes, less one. Thus, if a problem involves rolling a six-sided die, the number of degrees of freedom is the six possible outcomes minus one, or five.

Problems may either involve finding or using the significance parameter α, which is the probability that χ^2 will exceed a certain amount c.

$$\alpha = P(\chi^2 > c) = 1 - P(\chi^2 \leq c)$$

Again, if the hypothesis is true, α is the probability that χ^2 will exceed c. A common value for α is 0.05, which is intended to ensure that a hypothesis is rejected only if it can reasonably be assumed that the lack of a fit or dependence results from intention rather than chance.

The critical values, c, corresponding to various significance levels, α, and degrees of freedom, n, can be found in χ^2 tables. A sample portion of the χ^2 table is shown below. Note that the probability values, P, correspond to $1 - \alpha$. (The values in the table have limited numerical accuracy, and cases where greater accuracy is needed may require direct calculation of the critical value.)

$$P(\chi^2 \leq c)$$

	0.99	0.95	0.90	0.75	0.50	0.25	0.10	0.05
1	6.63	3.84	2.71	1.32	0.46	0.10	0.02	0.00
2	9.21	5.99	4.61	2.77	1.39	0.58	0.21	0.10
3	11.3	7.81	6.25	4.11	2.37	1.21	0.58	0.35
4	13.3	9.49	7.78	5.39	3.36	1.92	1.06	0.71
5	15.1	11.1	9.24	6.63	4.35	2.67	1.61	1.15

(rows labeled by n)

Example: A coin is flipped 200 times and lands heads up 92 times and tails up 108 times. Determine if the coin is a fair coin.

The "fairness" of the coin can be determined using a chi-square test. First, establish a null hypothesis. In this example, the null hypothesis is that the coin should be equally likely to land heads up or tails up for a given flip. This null hypothesis allows us to state expected frequencies. For 200 tosses we would expect 100 heads and 100 tails.

Next, prepare a table.

	Heads	Tails	Total
Observed	92	108	200
Expected	100	100	200
Total	192	208	400

The observed values are the data gathered. The expected values are the frequencies expected, based on the null hypothesis. Calculate χ^2 as follows, noting that there are two sample values: heads up and tails up.

$$\chi^2 = \sum_{i=1}^{2} \frac{(O_i - E_i)^2}{E_i} = \frac{(92-100)^2}{100} + \frac{(108-100)^2}{100}$$

$$\chi^2 = \frac{64}{100} + \frac{64}{100} = \frac{128}{100} = 1.28$$

Next, determine the degrees of freedom (n) by subtracting one from the number of sample values. In this example, there are two possible sample values (heads and tails), so $n = 1$.

An appropriate significance level (α) must also be chosen. For instance, assume that there should only be a 5% chance that the χ^2 value is greater than the critical value c. Next, consult the table of critical values of the chi-squared distribution.

The chi-squared value corresponding to $n = 1$ is 1.28. In the table above, the corresponding critical value c is 3.84. Since 1.28 < 3.84, the hypothesis should not be rejected.

Notice, however, that if the number of heads was 114 and the number of tails 86, then the χ^2 value would be 3.92, which exceeds the critical value of $c = 3.84$. In this case, for the chosen significance level, there would be evidence that the coin was loaded, and the hypothesis (the coin is fair) would be rejected.

DOMAIN V. **MATHEMATICAL PROCESSES AND PERSPECTIVES**

Competency 018 **The teacher understands mathematical reasoning and problem solving.**

This competency discusses the application of different aspects of reasoning to mathematics. Concepts of proof, deduction and induction, correct mathematical inference, formal and informal reasoning, problem-solving strategies, and validity of mathematical models are presented.

The beginning teacher understands the nature of proof, including indirect proof, in mathematics.

A **proof** is an argument that demonstrates the truth (or falsity) of a proposition. Mathematical proofs begin with certain axioms or known propositions and, by some line of reasoning, deduce a particular conclusion. (For more on deduction and induction, see the next skill section.)

Because not all concepts in mathematics can be proven or otherwise defined (if this were the case, then either circular definitions/proofs would be required—but these are not informative—or an infinite regression of definitions/proofs would be required—but these are impossible), mathematical proofs necessarily start from certain unproven or undefined concepts. In geometry, for instance, the concept of a point is undefined.

Mathematical proof thus attempts to reason in a consistent and orderly way from known premises (as long as they are either defined to be true or are proven to be true) to non-trivial conclusions. This process may involve positively demonstrating a proposition through **direct proof**, which involves showing that the proposition is true, or it may involve **indirect proof**, which involves showing that the negation (opposite) of the proposition is false. Both approaches are valid, but one or the other may be simpler. Because some propositions can be shown to be false through a single counter example, indirect proof is sometimes the simplest method of proof. Indirect proof may also involve using the opposite of the proposition being proved to demonstrate that a contradiction is reached. In such a case, assuming that all other premises are true, then the opposite of the proposition being proved must be false. Consider the following example.

Example: Prove that there are no even prime numbers other than the number 2.

A direct proof of this proposition may be possible, but an indirect proof is much simpler. Assume that the opposite is true: there is an even prime number other than 2. Let this number be called *x* (an integer). Since *x* is even, then the following must be true, where *y* is an integer.

$$\frac{x}{2} = y$$

But if *x* is evenly divisible by 2, it cannot be a prime number. Thus, a contradiction is reached with the originally assumed proposition that *x* is prime. This then proves that there are no even prime numbers other than 2.

The beginning teacher uses inductive reasoning to make conjectures and uses deductive methods to evaluate the validity of conjectures.

Two forms of reasoning are inductive and deductive. **Inductive reasoning** involves making inference from specific facts to general principles; **deductive reasoning** involves making inference from general principles to specific facts. As such, inductive reasoning is generally weaker than deductive reasoning. (Inductive reasoning— or induction—should not be confused with mathematical induction, which is not an example of inductive reasoning, strictly speaking.)

Inductive Reasoning

Inductive reasoning generally involves finding a representative set of examples that support the general application of a broader principle. In a common context, an example of inductive reasoning would be inferring from the fact that only black crows have ever been spotted to the general statement that all crows are black. This inference has a foundation in numerous observations, and it thereby gains significant weight. Nevertheless, it is feasible that somewhere a white (or other colored) crow does exist but simply hasn't yet been spotted. Thus, inductive inferences can never acquire 100% certainty, regardless of the amount of information in support of them.

Regardless of the uncertainty associated with induction, inductive inferences can be helpful for building a theory or for making a conjecture about some aspect of life, mathematics, or any other area. The physical sciences are a particular example where induction is commonly used to develop theories about the universe. Although these theories may be founded on a large body of empirical and mathematical evidence, a single counterexample could topple their status. Thus, again, inductive reasoning can be helpful, but it is much weaker than deductive reasoning.

Inductive reasoning, because it is weaker than deduction, is also less rigorous in its application of specific rules for the process of arriving at conclusions. For instance, there is no rule concerning how much evidence constitutes a sufficient reason to inductively accept a particular hypothesis. (Thus, there is no minimum number of sightings of black crows that is required prior to making an inference that all crows are black.) The particular area in which inductive reasoning is applied and the amount of potential evidence that could reasonably be gathered are factors that help determine what constitutes an acceptable inductive inference.

In a mathematical context, induction can serve to make conjectures for which a proof (or a proof of the contrary) can then be sought. For instance, Fermat's Last Theorem states that there are no integer solutions x, y, and z to the expression $x^n + y^n = z^n$ for $n > 2$. Although this theorem was suspected to be true (largely by induction from numerous test cases) for hundreds of years, only recently was a deductive proof discovered. Thus, induction can serve as a less rigorous method of making tentative conclusions pending a formal proof.

Deductive Reasoning

Deductive reasoning is a method of reasoning that is stronger and more rigorous than inductive reasoning. Deductive arguments reason from a set of premises to a conclusion and are classified as invalid, valid, and sound. An **invalid argument** is one in which the conclusion does not necessarily follow from the premises. A **valid argument** is one in which the conclusion necessarily follows from the premises. A **sound argument** is a valid argument for which all the premises are true. Thus, the following argument is valid but not sound:

>Premise 1: All dogs are black.
>Premise 2: Rover is a dog.
>Conclusion: Rover is black.

Were premises 1 and 2 both true, the conclusion would necessarily be true as well. Premise 1 is false, however, so the argument is valid but not sound. On the other hand, the following argument is both valid *and* sound.

>Premise 1: All integers are real numbers.
>Premise 2: 1 is an integer.
>Conclusion: 1 is a real number.

Both premises 1 and 2 are true, and the conclusion follows from the premises. Specific examples of deductive logical steps that can be taken in developing or evaluating an argument include modus ponens ("if A, then B" and "A is true" necessarily implies "B is true") and modus tollens ("if A, then B" and "B is false" necessarily implies "A is false").

Because deductive reasoning is more rigorous and the rules clearer, the process of arriving at an acceptable conclusion from a given set of premises (or the process of evaluating a deductive argument) is likewise clearer. Demonstrating the truth of the premises, however, may still be a complicated process. The premises may even require inductive reasoning to demonstrate their truth (at least tentatively). Thus, whether a deductive argument is sound can still be a matter that rests on the strength of a particular instance of inductive reasoning.

The beginning teacher applies correct mathematical reasoning to derive valid conclusions from a set of premises.

Given a set of premises, deductive reasoning can be used to reach valid conclusions. The key to correct reasoning in this regard is an understanding of the fundamental concepts of the particular area of interest. In number theory, for instance, it is necessary to have a grasp of the properties of real numbers if some conclusion about the set of real numbers (or a subset thereof) is being sought.

In addition to specific knowledge about the particular area about which conclusions are being sought, general knowledge of the process of correct mathematical reasoning is required. The use of deduction in the process of mathematical reasoning is discussed in the previous skill section.

Derivation of valid conclusions first requires consistent application of known rules and principles. Use of unproven or controversial approaches can result in questionable results (unless that approach is also justified). Consistency, even with seemingly strange premises, can lead to interesting conclusions. For instance, although Euclidean geometry requires that the sum of the interior angles of any triangle is 180°, spherical geometry allows other sums (a triangle in spherical geometry could even have three right angles). Thus, even though spherical geometry dispenses with certain results that are seldom questioned, consistency in this context leads to interesting and useful results.

Deriving a valid conclusion requires justification for each step in the reasoning process. Although such justifications need not always be explicitly expressed, making note of them at least mentally helps to avoid reasoning errors.

The beginning teacher uses formal and informal reasoning to justify mathematical ideas.

Formal reasoning, which includes deductive reasoning, follows a structured and orderly approach according to various rules of inference. **Informal reasoning**, which includes inductive reasoning, is less structured and tends not to be as rigorous as formal reasoning. Both of these types of reasoning, however, can be used to justify mathematical ideas.

Formal reasoning is applied for rigorous proofs and deriving conclusions in a way that provides certainty of the results. Informal reasoning is applied to situations where it is necessary to lend evidence to a conjecture or to build a strong (but not necessarily conclusive) case for some conclusion. Because it is less rigorous, informal reasoning tends to provide less certainty, but informal reasoning can be very helpful for finding potential solutions or possible avenues of approach for otherwise intractable problems.

The other skill sections in this competency discuss various particular aspects and methods of formal and informal reasoning and their application to mathematics.

The beginning teacher understands the problem-solving process (i.e., recognizing that a mathematical problem can be solved in a variety of ways, selecting an appropriate strategy, evaluating the reasonableness of a solution).

The process of problem solving in mathematics is similar to that of other areas. One of the first steps is to identify what is known about the problem. Each problem for which a solution can be found should provide enough information to form a starting point from which a valid sequence of reasoning leads to the desired conclusion: a solution to the problem. Between identification of known information and identification of a solution to the problem is a somewhat gray area that, depending on the problem, could potentially involve myriad different approaches. Two potential approaches that do not involve a "direct" solution method are discussed below.

The **guess-and-check** strategy calls for making an initial guess of the solution, checking the answer, and using the outcome of this check to inform the next guess. With each successive guess, one should get closer to the correct answer. Constructing a table from the guesses can help organize the data.

Example: There are 100 coins in a jar: 10 are dimes, and the rest are pennies and nickels. If there are twice as many pennies as nickels, how many pennies and nickels are in the jar?

Based on the given information, there are 90 total nickels and pennies in the jar (100 coins – 10 dimes = 90 nickels and pennies). Also, there are twice as many pennies as nickels. Using this information, guess results that fulfill the criteria and then adjust the guess in accordance with the result. Continue this iterative process until the correct answer is found: 60 pennies and 30 nickels. The table below illustrates this process.

Number of Pennies	Number of Nickels	Total Number of Pennies and Nickels
40	20	60
80	40	120
70	35	105
60	30	90

Another non-direct approach to problem solving is **working backwards**. If the result of a problem is known (for example, in problems that involve proving a particular result), it is sometimes helpful to begin from the conclusion and attempt to work backwards to a particular known starting point. A slight variation of this approach involves both working backwards and working forwards until a common point is reached somewhere in the middle. The following example from trigonometry illustrates this process.

Example: Prove that $\sin^2\theta = \frac{1}{2} - \frac{1}{2}\cos 2\theta$.

If the method for proving this result is not clear, one approach is to work backwards and forwards simultaneously. The following two-column approach organizes the process. Judging from the form of the result, it is apparent that the Pythagorean identity is a potential starting point.

$\sin^2\theta + \cos^2\theta = 1$	$\sin^2\theta = \frac{1}{2} - \frac{1}{2}\cos 2\theta$
$\sin^2\theta = 1 - \cos^2\theta$	$\sin^2\theta = \frac{1}{2} - \frac{1}{2}(2\cos^2\theta - 1)$
	$\sin^2\theta = \frac{1}{2} - \cos^2\theta + \frac{1}{2}$
	$\sin^2\theta = 1 - \cos^2\theta$

Thus, a proof is apparent based on the combination of the reasoning in these two columns.

Selection of an appropriate problem-solving strategy depends largely on the type of problem being solved and the particular area of mathematics with which the problem deals. For instance, problems that involve proving a specific result often require different approaches than do problems that involve finding a numerical result.

When solving any problem, it is helpful to evaluate the **reasonableness** of the solution. Often, errors in the solution lead to final results that do not make any sense in the context of the problem. Thus, checking the reasonableness of the solution can be a fast way to help determine if an error was made at some point in the process. Although such checks help to raise confidence in a solution, they do not necessarily guarantee that a solution is correct.

For instance, an error can result in a relatively small deviation in a numerical result; although the answer may still seem reasonable, it could still be incorrect. Thus, the reasonableness of a solution is a necessary but not sufficient check of its correctness.

Two characteristics of a numerical answer that can be quickly evaluated are sign and magnitude. If a problem calls for determining the length of a side of some geometric figure, for example, then a negative number should indicate an error at some point in the solution. Similarly, a result that is magnitudes larger that would seem appropriate to the other aspects of the problem (such as the lengths of other measurements) could also indicate an error.

Additionally, the problem may provide information that limits the answer to a certain range. For example, if a problem asks for the average speed of an automobile over some distance and range of speeds, it is clear that the average speed should not exceed the maximum speed, nor should it be less than the minimum speed. Again, although this type of evaluation does not necessarily help to judge answers that fall within this range, it does help rule out results containing particularly egregious errors. On the other hand, if a speed distribution is shown that is weighted heavily toward faster speeds than slower speeds, it would then be reasonable to assume that the correct solution should be at the higher end of the speed range of the car. Similar types of qualitative evaluation or rough estimation for judging the reasonableness of a solution can be applied to other problems as well.

The beginning teacher evaluates how well a mathematical model represents a real-world situation.

Mathematical models of real-world situations are just that: models. As a result, any model is likely to contain some deviation from observation, since it is seldom (if ever) that a model is able to take into account all the relevant variables. For instance, an attempt to model the probability of a so-called fair coin is not always as simple as it may seem. Imperfections or weight imbalances in the coin can cause deviations in the probability distribution, as can the method of flipping the coin and other factors such as air currents and even electric or magnetic fields (whose effect on the coin depend on the materials contained in the coin). On the other hand, such factors may have such a small effect on the model for a particular situation that they are negligible. It is thus often necessary to determine which variables or factors must be considered and which can be ignored.

Perhaps the surest way to evaluate how well a model represents a real-world situation is to compare the description provided by the model (whether predictions or other information) and compare it to observations of the phenomenon being modeled. Although this is an important first step in validating a model, it is not the only step. The model may correctly describe certain phenomena, but it may also be an inordinately complicated model. (For instance, epicycles may to some extent correctly predict the motion of planets, but much simpler models do the same task.)

In addition to correct results and simplicity, the accuracy of a model can be measured to some extent by its applicability beyond the situation of immediate interest. For instance, Newtonian mechanics in physics is sufficient to explain a wide variety of common and observable phenomena, but it is apparently inadequate for some extreme cases, and it furthermore does not apparently describe the behavior of matter at the atomic and subatomic levels.

Another consideration is of a philosophical nature: does the mathematics feasibly correspond to the observable reality that it describes. For instance, imaginary numbers are a crucial ingredient in a variety of models of phenomena in science and engineering; nevertheless, the correspondence of the imaginary number i to reality is unclear. Whether or not a particular model actually has any meaningful correspondence with reality, regardless of whether it produces correct results, is almost always a matter for debate.

Thus, evaluating a mathematical model involves multiple levels of consideration, some of which may be considered more or less relevant depending on the purpose of the model. (For example, if the model is simply intended to provide a fast prediction of some physical quantity, the philosophical considerations may be unimportant.) An appropriate evaluation requires weighing a number of factors and giving them the suitable influence in the evaluation.

Competency 019   **The teacher understands mathematical connections both within and outside of mathematics and how to communicate mathematical ideas and concepts.**

This section discusses various representations of mathematical ideas and concepts as well as the application of mathematics to other fields. In addition, the proper use of mathematical terminology is briefly reviewed. Many of the themes presented in this section are illustrated throughout the guide, rather than being discussed here at length.

The beginning teacher recognizes and uses multiple representations of a mathematical concept (e.g., a point and its coordinates, the area of a circle as a quadratic function of the radius, probability as the ratio of two areas, area of a plane region as a definite integral).

The recognition and use of multiple representations of mathematical concepts can be useful skills on a number of levels. From a pedagogical perspective, the use of several different representations of a concept can help provide perspective that illuminates the concept, and it can be a way to present the information to different students who may each have slightly different conceptual strengths or approaches to learning mathematics. From a practical perspective, solving problems can often be facilitated by using slightly different representations of the information presented in the problem statement.

Although there are innumerable instances of mathematical concepts that can be expressed using a number of different representations, the following examples suffice to illustrate this idea. Throughout this guide, additional examples can be found, especially as tools and concepts from multiple areas of mathematics are applied to certain problems.

The concept of a point, for instance, is interchangeable with a set of coordinates. In certain cases where a coordinate system is helpful in solving a problem or explicating a concept, the use of coordinates to represent a point is the best representation. In other cases, the use of a system of coordinates may be an unneeded factor that would simply complicate the problem.

In addition, a plane region can be viewed as a geometric concept or as a definite integral of a function in a coordinate plane. In either case, the same concept is being presented, but one or the other of the representations might be better in the context of a particular problem or task.

The beginning teacher understands how mathematics is used to model and solve problems in other disciplines (e.g., art, music, science, social science, business).

Artists, musicians, scientists, social scientists, and those in business use mathematical modeling to solve problems in their disciplines. These disciplines rely on the tools and symbols of mathematics to model natural events and manipulate data. Mathematics is a key aspect of visual art.

Artists use the geometric properties of shapes, ratios, and proportions in creating paintings and sculptures. For example, mathematics is essential to the concept of perspective. Artists must determine the appropriate lengths and heights of objects to portray three-dimensional distance in two dimensions.

Mathematics is also an important part of music. Many musical terms have mathematical connections. For example, the musical octave contains twelve notes and spans a factor of two in frequency. In other words, the frequency—the speed of vibration that determines tone and sound quality—doubles from the first note in an octave to the last. Thus, starting from any note we can determine the frequency of any other note using the following formula.

$$\text{Freq} = \text{note} \times 2^{N/12}$$

Here, N is the number of notes from the starting point and note is the frequency of the starting note. Mathematical understanding of frequency plays an important role in the tuning of musical instruments.

In addition to the visual and auditory arts, mathematics is an integral part of most scientific disciplines. The uses of mathematics in science are almost endless, and the following are but a few examples of how scientists use mathematics. Physical scientists use vectors, functions, derivatives, and integrals to describe and model the movement of objects. Biologists and ecologists use mathematics to model ecosystems and study DNA. Also, chemists use mathematics to study the interaction of molecules and to determine proper amounts and proportions of reactants.

Many social science disciplines use mathematics to model and solve problems. Economists, for example, use functions, graphs, and matrices to model the activities of producers, consumers, and firms. Political scientists use mathematics to model the behavior and opinions of the electorate. In addition, sociologists use mathematical functions to model the behavior of humans and human populations.

Finally, mathematical problem solving and modeling is essential to business planning and execution. For example, businesses rely on mathematical projections to plan business strategy. Additionally, stock market analysis and accounting rely on mathematical concepts.

The beginning teacher translates mathematical ideas between verbal and symbolic forms.

Mathematics is, in some ways, a formalization of language that concerns such concepts as quantity and organization. Mathematics often involves symbolic representations, which can help alleviate the ambiguities found in common language. Naturally, then, communication of mathematical ideas requires conversion back and forth from verbal and symbolic forms is a necessary skill. These two forms can often help to elucidate one another when an attempt is made to understand an idea that they represent.

Mathematical ideas and expressions may sometimes be simple to translate into language; for instance, basic arithmetic operations are usually fairly easy to express in everyday language (although complicated expressions may be less so). In some cases, common language more easily expresses certain ideas than does symbolic language (and sometimes vice versa). Much of the translation process is learned through practicing expression of mathematical ideas in verbal (or written) form and by translating verbal or written expressions into a symbolic form.

The following examples are just a few illustrations of the translation process. The material throughout this guide attempts to present mathematical ideas both in symbolic and written forms. Thus, practicing by carefully following the text and example problems and by attempting to articulate the various concepts both in English and in mathematical symbols should help the student (and teacher) of mathematics gain mastery of this skill.

Example: Find a symbolic expression for the area inside an elliptical region with a minor axis of length *a* and a major axis of length *b*.

Although this written expression of can simply be summed up in the expression given in the problem, the symbolic form is significantly more complicated. First, find the symbolic expression *f(x)* for a parabola in a coordinate plane (assume the major axis is coincident with the *x*-axis):

$$f(x) = \pm \frac{a}{2}\left(1 - \frac{4x^2}{b^2}\right)^{1/2}$$

The area *A* inside this region can be expressed using integral notation:

$$A = \int_{-b}^{b} \frac{a}{2}\left(1 - \frac{4x^2}{b^2}\right)^{1/2} dx$$

This integral form is one particular symbolic expression. Simpler (in some sense) forms could also be found—for instance, by evaluating the integral (although such results are not included here).

Example: Express the following in written language form:
$\{\ldots, -2, -1, 0, 1, 2, \ldots\}$.

This symbolic expression, in written language form, is simply "the set of integers."

The beginning teacher communicates mathematical ideas using a variety of representations (e.g., numeric, verbal, graphical, pictorial, symbolic, concrete).

Throughout this guide, mathematical operations and situations are represented through words, algebraic symbols, geometric diagrams and graphs. A few commonly used representations are discussed below.

The basic mathematical operations include addition, subtraction, multiplication and division. In word problems, these are represented by the following typical expressions.

Operation	Descriptive Words
Addition	"plus", "combine", "sum", "total", "put together"
Subtraction	"minus", "less", "take away", "difference"
Multiplication	"product", "times", "groups of"
Division	"quotient", "into", "split into equal groups",

Some verbal and symbolic representations of basic mathematical operations include the following:

7 added to a number	$n + 7$
a number decreased by 8	$n - 8$
12 times a number divided by 7	$12n \div 7$
28 less than a number	$n - 28$
the ratio of a number to 55	$\dfrac{n}{55}$
4 times the sum of a number and 21	$4(n + 21)$

Multiplication can be shown using arrays. For instance, 3×4 can be expressed as 3 rows of 4 each

In a similar manner, addition and subtraction can be demonstrated with symbols.

$\psi\ \psi\ \psi\ \xi\ \xi\ \xi\ \xi$
$3 + 4 = 7$
$7 - 3 = 4$

Fractions can be represented using pattern blocks, fraction bars, or paper folding.

Diagrams of arithmetic operations can present mathematical data in visual form. For example, a number line can be used to add and subtract, as illustrated below.

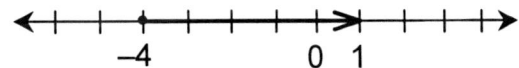

Five added to negative four on the number line or –4 + 5 = 1.

Pictorial representations can also be used to explain the arithmetic processes.

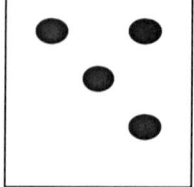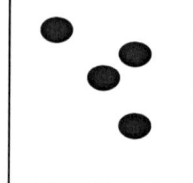

The diagram above shows two groups of four equal eight, or 2 x 4 = 8. The next diagram illustrates addition of two objects to three objects, resulting in five objects.

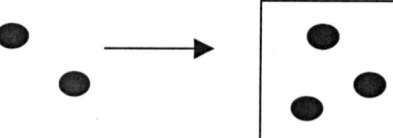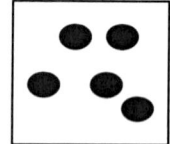

Concrete examples are real world applications of mathematical concepts. For example, measuring the shadow produced by a tree or building is a real-world application of trigonometric functions, acceleration or velocity of a car is an application of derivatives, and finding the volume or area of a swimming pool is a real-world application of geometric principles.

Pictorial illustrations of mathematic concepts help clarify difficult ideas and simplify problem solving. The following example illustrates the use of pictures.

Rectangle R represents the 300 students in School A. Circle P represents the 150 students that participated in band. Circle Q represents the 170 students that participated in a sport. 70 students participated in both band and a sport.

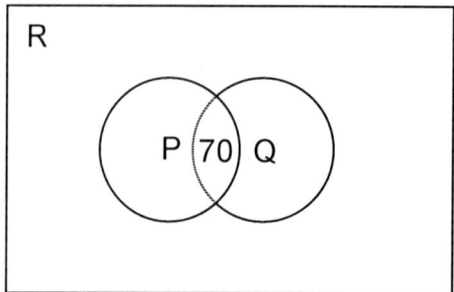

Symbolic representation is the basic language of mathematics. Converting data to symbols allows for easy manipulation and problem solving. Students should have the ability to recognize what the symbolic notation represents and convert information into symbolic form. For example, from the graph of a line, students should have the ability to determine the slope and intercepts and derive the line's equation from the observed data. Another possible application of symbolic representation is the formulation of algebraic expressions and relations from data presented in word-problem form.

The beginning teacher understands the use of visual media, such as graphs, tables, diagrams, and animations, to communicate mathematical information.

Although symbolic and verbal presentations of mathematical concepts and data can be both useful and informative, they are not always the most lucid representations of that information. The use of visual media can be helpful in numerous cases. For instance, graphs and tables of data can in many cases provide a clearer representation of that data than can, for instance, a symbolic expression (from some form of regression, for example). Diagrams can also be extremely helpful, especially in problem-solving contexts. One of the chief rules of solving a problem is to draw a diagram (where appropriate) illustrating the problem; this approach helps organize information and it provides a perspective that often makes the information more accessible than it would be from words and mathematical expressions alone. In a similar fashion, animations can also be helpful. Although animations generally require more technology (typically a computer) to construct than do diagrams, they provide an additional dimension to the visual presentation.

Throughout this guide, the concepts and examples that are presented often include graphs, tables, and diagrams to illustrate the information being presented.

The beginning teacher uses appropriate mathematical terminology to express mathematical ideas.

The above section dealing with various mathematical representations briefly covers mathematical terminology. Understanding jargon in any discipline, whether mathematics or another field, typically requires study of that particular discipline. Mathematics is a broad field covering a range of subareas, and each has its own particular terms (although there may be a large overlap of terminology with that of other areas of mathematics). The best way for a teacher to learn to use appropriate mathematical terminology is to have a solid understanding of both the fundamental and more advanced concepts of the field he is teaching. Throughout this guide, an attempt is made to present mathematical concepts both symbolically and in a written form that makes use of appropriate terminology. Thus, a review of the material should help reinforce knowledge and use of these terms.

TEACHER CERTIFICATION STUDY GUIDE

DOMAIN VI.	MATHEMATICAL LEARNING, INSTRUCTION, AND ASSESSMENT

Competency 020   **The teacher understands how children learn mathematics and plans, organizes, and implements instruction using knowledge of students, subject matter, and statewide curriculum (Texas Essential Knowledge and Skills [TEKS]).**

This competency reviews the skills that a teacher needs to prepare mathematics lessons appropriate to the skill level of the students, to illustrate mathematical principles using manipulative or other tools, to relate lessons to the real world, and to help students learn the skills required by TEKS.

The beginning teacher applies research-based theories of learning mathematics to plan appropriate instructional activities for all students.

The challenges involved in the teaching of math are well recognized in academia and have spawned many research projects aimed at improving the quality of mathematics education. Teachers have access to many resources that can help to keep them informed about current research and provide them with tools to implement new ideas in their teaching. These resources include the websites of the National Council of Teachers of Mathematics (NCTM) and other organizations. See the introduction to **Essential Tips for Every Math Teacher** at the end of this guide for links to professional development resources for math teachers.

Teachers can use theories of learning to plan curriculum and instructional activities. Research indicates that students learn math more easily in an applied, project-based setting. In addition, prior knowledge, learning, and self-taught understanding are important factors that dictate a student's ability to learn and his preferred method of learning.

Many educators believe that the best method of teaching math is **situated learning**. Proponents of situated learning argue that learning is largely a function of the activity, context, and environment in which it occurs. According to situated learning theory, students learn more easily from instruction involving relevant, real-world situations and applications rather than abstract thoughts and ideas.

Research or project-based learning is a product of situated learning theory. Open-ended research tasks and projects promote learning by engaging students on multiple levels. Such tasks require the use of multiple skills and reasoning strategies and help keep students focused and attentive. Additionally, projects promote active learning by encouraging the sharing of thoughts and ideas as well as teacher-student and student-student interaction.

The beginning teacher understands how students differ in their approaches to learning mathematics.

The cultural and ethnic background of a student greatly affects his or her approach to learning mathematics. In addition, factors such as gender, socioeconomic status, and learning disabilities can affect student learning styles. Teachers must have the ability to tailor their teaching style, methods, and curriculum to the varying learning styles present in a diverse classroom.

Many researches have studied diversity issues in teaching math. A few references are given below. High expectations and a high level of peer interaction through group study or peer tutoring has been found to be helpful. A relatively informal atmosphere that encourages questioning and guides students to find their own solutions to problems is conducive to math learning as well.

http://math.unipa.it/~grim/21_project/21_Charlotte_LongPaperEdit.pdf

http://www.unige.ch/math/EnsMath/Rome2008/WG3/Papers/BOALER.pdf

http://mathforum.org/~sarah/Discussion.Sessions/biblio.attitudes.html

The beginning teacher uses students' prior mathematical knowledge to build conceptual links to new knowledge and plan instruction that builds on students' strengths and addresses students' needs.

A popular theory of math learning is constructivism. Constructivists argue that prior knowledge greatly influences the learning of math and that learning is cumulative and vertically structured. Instruction must build on the innate knowledge of students and address any common misconceptions. Thus, it is important for teachers to ensure that students possess the prerequisite knowledge and ideas required to learn a particular topic.

Even without an appeal to constructivism, it is obvious that a student who does not understand the concept of percentage will not be able to do a problem involving interest rates.

Teachers can gain insight into the prior knowledge of students by beginning each lesson with open-ended questions that allow students to share their thoughts and ideas on the subject or topic. A short pre-test covering the prerequisite topics can also be useful as an assessment tool.

In order to identify the prerequisite skills needed to solve a particular kind of problem, the broad concepts that underlie the problem must first be identified. For each concept, one can then list the specific skills needed to perform the related mathematical operations. This kind of hierarchical analysis may be summarized in a tree diagram as shown below.

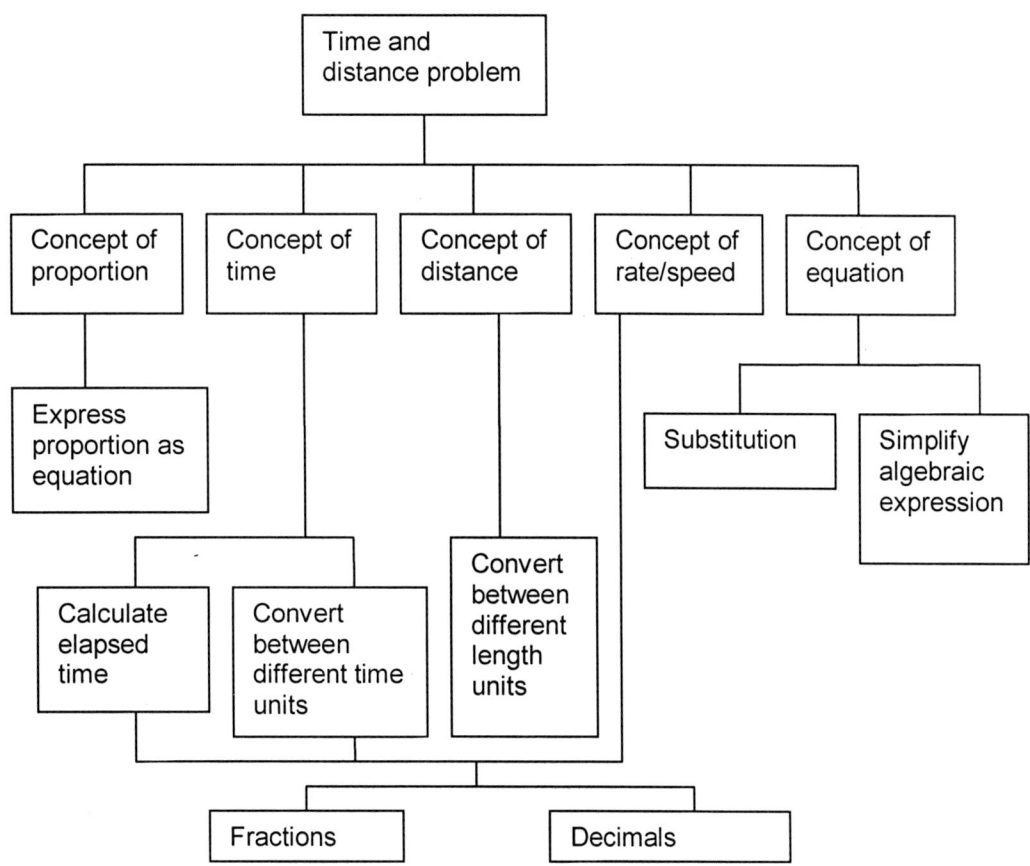

The above analysis can of course be done in different ways. The essential idea is to identify all the pieces that go into learning a topic.

Once the gaps and weaknesses in the student's prerequisite knowledge are identified, the teacher will need to review those topics before teaching the new content. This can be done through discussion, written exercises as well as through hands-on activities.

The beginning teacher understands how learning may be enhanced through the use of manipulatives, technology, and other tools (e.g., stop watches, scales, rulers).

The use of supplementary materials in the classroom can greatly enhance the learning experience by stimulating student interest and satisfying different learning styles. Manipulatives, models, and technology are examples of tools available to teachers.

Manipulatives are materials that students can physically handle and move. Manipulatives allow students to understand mathematic concepts by allowing them to see concrete examples of abstract processes. Manipulatives are attractive to students because they appeal to the students' visual and tactile senses. Available for all levels of math, manipulatives are useful tools for reinforcing operations and concepts. They are not, however, a substitute for the development of sound computational skills.

Models are another means of representing mathematical concepts by relating the concepts to real-world situations. Teachers must choose wisely when devising and selecting models because, to be effective, models must be applied properly. For example, a building with floors above and below ground is a good model for introducing the concept of negative numbers. It would be difficult, however, to use the building model in teaching subtraction of negative numbers.

Finally, there are many forms of **technology** available to math teachers. For example, students can test their understanding of math concepts by working on specific computer programs and websites. Graphing calculators can help students visualize the graphs of functions. Teachers can also enhance their lectures and classroom presentations by creating multimedia presentations.

See **Essential Tips for Every Math Teacher** at the end of this guide for ideas about using manipulatives, software, and other educational aids.

The beginning teacher understands how to provide instruction along a continuum from concrete to abstract.

According to Piaget, there are four primary cognitive structures or development stages: sensorimotor, preoperations, concrete operations, and formal operations. In the sensorimotor stage (0-2 years), intelligence takes the form of motor actions. In the preoperation stage (3-7 years), intelligence is intuitive in nature. Intelligence in the concrete operational stage (8-11 years) is logical but depends upon concrete referents. In the final stage of formal operations (12-15 years), thinking involves abstractions.

Even though middle school students are typically ready to approach mathematics in abstract ways, some of them still require concrete referents such as manipulatives. It is useful to keep in mind that the developmental stages of individuals vary. In addition, different people have different learning styles, some tending more towards the visual and others relatively verbal. Research has shown that learning is most effective when information is presented through multiple modalities or representations. Most mathematics textbooks now use this multi-modal approach.

When introducing a new mathematical concept to students, teachers should utilize the concrete-to-representational-to-abstract sequence of instruction. The first step of the instructional progression is the introduction of a concept modeled with concrete materials. The second step is the translation of concrete models into representational diagrams or pictures. The third and final step is the translation of representational models into abstract models using only numbers and symbols.

Teachers should first use concrete models to introduce a mathematical concept because they are easiest to understand. For example, teachers can allow students to use counting blocks to learn basic arithmetic. Teachers should give students ample time and many opportunities to experiment, practice, and demonstrate mastery with the concrete materials.

The second step in the learning process is the translation of concrete materials to representational models. For example, students may use tally marks or pictures to represent the counting blocks they used in the previous stage. Once again, teachers should give students ample time to master the concept on the representational level.

The final step in the learning process is the translation of representational models into abstract numbers and symbols. For example, students represent the processes carried out in the previous stages using only numbers and arithmetic symbols.

To ease the transition, teachers should associate numbers and symbols with the concrete and representational models throughout the learning progression.

The beginning teacher understands a variety of instructional strategies and tasks that promote students' abilities to do the mathematics described in the TEKS.

The Texas Essential Knowledge and Skills (TEKS) (http://ritter.tea.state.tx.us/teks/) are a comprehensive list of standards for subject matter learning. TEKS provide teachers with a framework for curriculum design and instructional method selection. The different skills described in TEKS require different teaching strategies and techniques.

The primary goals of middle school math instruction, as defined by TEKS, are the building of a strong foundation in mathematical concepts and the development of problem-solving and analytical skills. Direct teaching methods, including lecture and demonstration, are particularly effective in teaching basic mathematical concepts. To stimulate interest, accommodate different learning styles and enhance understanding, teachers should incorporate manipulatives and technology into their lectures and demonstrations. Indirect teaching methods, including cooperative learning, discussion and projects, promote the development of problem solving skills. Cooperative learning and discussion allow students to share ideas and strategies with their peers. In addition, projects require students to apply knowledge and develop and implement problem-solving strategies.

The TEKS provide teachers with a comprehensive list of skills that the state requires students to master. Utilizing the learning goals presented in TEKS, teachers can plan instruction to promote student understanding. In addition, teachers can deliver instruction in ways that are appropriate to the specific skill, classroom environment and student population. To assess the effectiveness of instruction, teachers can design tests that evaluate student mastery of specific skills. In grading such tests, teachers should look for patterns in student mistakes and errors that may indicate a deficiency in the instructional plan. In reevaluating instruction, teachers can attempt to use different instructional methods or shift areas of emphasis to meet the needs of the students.

Mathematics TEKS toolkit:
http://www.utdanacenter.org/mathtoolkit/

The beginning teacher understands how to create a learning environment that provides all students, including English Language Learners, with opportunities to develop and improve mathematical skills and procedures.

Teachers and school officials must understand the special needs of English Language Learners. Mathematic assessments may understate the abilities of English Language Learners because poor test scores may stem from difficulty in reading comprehension, not a lack of understanding of mathematic principles. Uncharacteristically poor performance on word problems by English Language Learners is a sign that reading comprehension, not mathematic understanding, is the underlying problem.

The TSU Math for English Language Learners Project provides helpful information on this topic: http://www.tsusmell.org/.

The beginning teacher understands a variety of questioning strategies to encourage mathematical discourse and to help students analyze and evaluate their mathematical thinking.

As the teacher's role in the classroom changes from lecturer to facilitator, the questions need to further stimulate students in various ways. Examples of potential questions in various contexts are offered below.

- Helping students work together:
  What do you think about what John said?
  Do you agree? Disagree?
  Can anyone explain that differently?

- Helping students determine for themselves if an answer is correct:
  Why do you think that is true?
  How did you get that answer?
  Do you think that is reasonable? Why?

- Helping students learn to reason mathematically:
  Will that method always work?
  Can you think of a case where it is not true?
  How can you prove that?
  Is that answer true in all cases?

- Helping student brainstorm and solve problems:
  Is there a pattern?
  What else can you do?
  Can you predict the answer?
  What if...?

- Helping students connect mathematical ideas:
  What did we learn before that is like this?
  Can you give an example?
  What math did you see on television last night? in the newspaper?

The beginning teacher understands how to relate mathematics to students' lives and a variety of careers and professions.

Teachers can increase student interest in math and promote learning and understanding by relating mathematical concepts to the lives of students. Instead of using only abstract presentations and examples, teachers should relate concepts to real-world situations to shift the emphasis from memorization and abstract application to understanding and applied problem solving. In addition, relating math to careers and professions helps illustrate the relevance of math and aids in the career exploration process.

Artists, musicians, scientists, social scientists, and business people use mathematical modeling to solve problems in their disciplines. These disciplines rely on the tools and symbols of mathematics to model natural events and manipulate data. Mathematics is a key aspect of visual art.

Artists use the geometric properties of shapes, ratios, and proportions in creating paintings and sculptures. For example, mathematics is essential to the concept of perspective. Artists must determine the appropriate lengths and heights of objects to portray three-dimensional distance in two dimensions.

Mathematics is also an important part of music. Many musical terms have mathematical connections. For example, the musical octave contains twelve notes and spans a factor of two in frequency. In other words, the frequency, the speed of vibration that determines tone and sound quality, doubles from the first note in an octave to the last. Thus, starting from any note we can determine the frequency of any other note with the following formula.

$$\text{Freq} = \text{note} \times 2^{N/12}$$

Where N is the number of notes from the starting point and note is the frequency of the starting note. Mathematical understanding of frequency plays an important role in the tuning of musical instruments.

In addition to the visual and auditory arts, mathematics is an integral part of most scientific disciplines. The uses of mathematics in science are almost endless. The following are but a few examples of how scientists use mathematics. Physical scientists use vectors, functions, derivatives, and integrals to describe and model the movement of objects. Biologists and ecologists use mathematics to model ecosystems and study DNA. Finally, chemists use mathematics to study the interaction of molecules and to determine proper amounts and proportions of reactants.

Many social science disciplines use mathematics to model and solve problems. Economists, for example, use functions, graphs, and matrices to model the activities of producers, consumers, and firms. Political scientists use mathematics to model the behavior and opinions of the electorate. Finally, sociologists use mathematical functions to model the behavior of humans and human populations.

Finally, mathematical problem solving and modeling is essential to business planning and execution. For example, businesses rely on mathematical projections to plan business strategy. Additionally, stock market analysis and accounting rely on mathematical concepts.

Competency 021 The teacher understands assessment and uses a variety of formal and informal assessment techniques to monitor and guide mathematics instruction and to evaluate student progress.

Assessment is an important part of mathematics instruction, both for the purposes of allowing the teacher to gauge student performance and of allowing the student to judge his own progress in the subject. This competency briefly reviews the development, purpose, and use of various assessment techniques as well as how to apply them in a manner that is both fair and accurate.

The beginning teacher understands the purpose, characteristics, and uses of various assessments in mathematics, including formative and summative assessments.

The primary purpose of student assessment is to evaluate the effectiveness of the curriculum and instruction by measuring student performance. Teachers and school officials use the results of student assessments to monitor student progress and modify and design curriculum to meet the needs of the students. Teachers and school officials carefully assess the results of tests to determine the parts of the curriculum that need altering. For example, the results of a test may indicate that the majority of the students in a class struggle with problems involving logarithmic functions. In response to such findings, the teacher would evaluate the method of logarithmic function instruction and make the necessary changes to increase student understanding.

Student assessment is an important part of the educational process. High quality assessment methods are necessary for the development and maintenance of a successful learning environment. Teachers must develop and implement assessment procedures that accurately evaluate student progress, test content areas of greatest importance, and enhance and improve learning. To enhance learning and accurately evaluate student progress, teachers should use a variety of assessment tasks to gain a better understanding a student's strengths and weaknesses. Finally, teachers should implement scoring patterns that fairly and accurately evaluate student performance.

A special type of student assessment, state **standardized testing**, is an important tool for curriculum design and modification. Most states have stated curriculum standards that mandate what students should know. Teachers can use the standards to focus their instruction and curriculum planning. State tests evaluate and report student performance on the specific curriculum standards. Thus, teachers can easily determine the areas that require greater attention.

Teachers should use a variety of assessment procedures to evaluate student knowledge and understanding. In addition to the traditional methods of performance assessment like multiple choice, true/false, and matching tests, there are many other methods of student assessment available to teachers. Alternative assessment is any type of assessment in which students create a response rather than choose an answer. It is sometimes know as **formative assessment**, due to the emphasis placed on feedback and the flow of communication between teacher and student. It is the opposite of **summative assessment**, which occurs periodically and consists of temporary interaction between teacher and student.

Short response and **essay** questions are alternative methods of performance assessment. In responding to such questions, students must utilize verbal, graphical, and mathematical skills to construct answers to problems. These multi-faceted responses allow the teacher to examine more closely a student's problem solving and reasoning skills.

Student **portfolios** are another method of alternative assessment. In creating a portfolio, students collect samples and drafts of their work, self-assessments, and teacher evaluations over a period of time. Such a collection allows students, parents, and teachers to evaluate student progress and achievements. In addition, portfolios provide insight into a student's thought process and learning style.

Projects, **demonstrations**, and **oral presentations** are means of alternative assessment that require students to use different skills than those used on traditional tests. Such assessments require higher order thinking, creativity, and the integration of reasoning and communication skills. The use of predetermined rubrics, with specific criteria for performance assessment, is the accepted method of evaluation for projects, demonstrations, and presentations.

One type of alternative assessment is **bundled testing**. Bundled testing is the grouping of different question formats for the same skill or competency. For example, a bundled test of exponential functions may include multiple choice questions, short response questions, word problems, and essay questions. The variety of questions tests different levels of reasoning and expression.

Scoring methods are an important, and often overlooked, part of effective assessment. Teachers can use a simple three-point scale for evaluating student responses. No answer or an inappropriate answer that shows no understanding scores zero points. A partial response showing a lack of understanding, a lack of explanation, or major computational errors scores one point. A somewhat satisfactory answer that answers most of the question correctly but contains simple computational errors or minor flaws in reasoning receives two points. Finally, a satisfactory response displaying full understanding, adequate explanation, and appropriate reasoning receives three points. When evaluating student responses, teachers should look for common error patterns and mistakes in computation. Teachers should also incorporate questions and scoring procedures that address common error patterns and misconceptions into their methods of assessment.

The beginning teacher understands how to select and develop assessments that are consistent with what is taught and how it is taught.

To assess whether a student has truly grasped the content of the curriculum being taught, teachers must ensure that tests not only evaluate isolated skills but also important mathematical concepts and thought processes. The assessment should enhance learning and serve as a tool that identifies areas of misunderstanding. The teacher must also take into account the learning styles of the students being assessed and how the material was approached in the classroom.

Before selecting a particular assessment method, a teacher should develop a list of criteria that reflect the goals of the assessment. The following website provides a variety of resources that may be helpful in planning, developing, and evaluating assessments:
http://mathforum.org/mathed/assessment.html

The beginning teacher understands how to develop a variety of assessments and scoring procedures consisting of worthwhile tasks that assess mathematical understanding, common misconceptions, and error patterns.

All too often, assessments test topics in isolation and lower-level procedural skills within each topic. Higher-level skills such as reasoning, problem solving, communicating, and connecting ideas are ignored as a result. To develop effective assessments that balance conceptual understanding, procedural knowledge, and problem solving and that focus on unearthing common misconceptions and error patterns, a predetermined assessment framework is needed as a guide for developing new assessments.

A good assessment framework must specify details such as goals of the assessment, background knowledge expected, type of guidance and instructions to be given as part of the assessment, and so on. A framework of this type is often built on an existing curriculum framework. Although each individual assessment may not do so, assessments over the duration of a class as a whole must reflect all facets of the framework.

Worthwhile assessment tasks intended to test higher-level skills must use connections between different areas of mathematics as well as real-world settings. Because the goal is to assess how a student thinks and not whether a student is familiar with a particular type of problem, non-routine problems must be used. One must ensure, however, that the student has been taught all the prerequisite skills, has experience solving novel problems, and that the task is explained clearly with all assumptions explicitly laid out. Tasks must also be structured so that students have the opportunity to show what they know even if they are unable to complete the whole problem.

The New Standards Project provides an assessment system that goes beyond standardized testing:
http://www.nctm.org/news/release.aspx?id=770.

The beginning teacher understands the relationship between assessment and instruction and knows how to evaluate assessment results to design, monitor, and modify instruction to improve mathematical learning for all students, including English Language Learners.

One way to use assessments to improve instruction is to integrate them into the teaching process. In iterative assessments, students receive feedback from the teacher at different stages while the task in progress. Thus, the assessment is used as a learning tool.

Even when the assessment is more traditional, the following criteria can help to ensure that it supports learning:
- Reports must provide feedback with regard to different aspects of learning, not just numerical scores
- Reports must be timely so the students have an opportunity to use the feedback
- Scoring rubrics must be specific and address different learning goals
- Students must be given the opportunity to explain their thinking

The following chapters from the book "Measuring What Counts: A Conceptual Guide for Mathematics Assessment" address the use of assessment in improving instruction:

http://www.nap.edu/openbook.php?record_id=2235&page=67
http://www.nap.edu/openbook.php?record_id=2235&page=91

TEACHER CERTIFICATION STUDY GUIDE

CURRICULUM AND INSTRUCTION

TEACHING METHODS – The art and science specific for high school mathematics

Some commonly used teaching techniques and tools are described below along with links to further information. These links provide a wealth of instructional ideas and materials. You should consider joining The National Council of Teachers of Mathematics, as this organization has many ideas about pedagogy and curriculum standards in its journals, and it publishes professional books that are also useful. You can write to the Council at 1906 Association Drive, Reston, VA 20191-1593. You can also order a starter kit for $9 that includes three recent journals by calling 800-235-7566 or emailing e-mailorders@nctm.org.

> A couple of resources for students to use at home:
> http://www.algebra.com/, http://www.mathsisfun.com/algebra/index.html and http://www.purplemath.com/. A helpful .pdf guide for parents is available at:
> http://my.nctm.org/ebusiness/ProductCatalog/product.aspx?ID=12931.
> A good website for understanding the causes of and how to prevent "math anxiety":
> http://www.mathgoodies.com/articles/math_anxiety.html

1. Classroom warm-up: Engage your students as soon as they walk in the door: provide an interesting short activity each day. You can make use of thought-provoking questions and puzzles. Also use relevant puzzles specific to topics you may be covering in your class. The following websites provide some ideas:
    http://www.math-drills.com/?gclid=CP-P0dzenJICFRSTGgodNjG0Zw
    http://www.mathgoodies.com/games/
    http://mathforum.org/k12/k12puzzles/
    http://mathforum.org/pow/other.html

2. Real-life examples: Connect math to other aspects of your students' lives by using examples and data from the real world whenever possible. It will not only keep them engaged, it will also help answer the perennial question "Why do we have to learn math?" The online resources below can get you started:

    > http://chance.dartmouth.edu/chancewiki/index.php/Main_Page has some interesting real-world probability problems (such as, "Can statistics determine if Robert Clemens used steroids?").
    > http://www.mathnotes.com/nos_index.html has all kinds of links between math and the real world that are suitable for high school students.
    > http://www.nssl.noaa.gov/edu/ideas/ uses weather to teach math.

MATHEMATICS 8-12

TEACHER CERTIFICATION STUDY GUIDE

http://standards.nctm.org/document/eexamples/index.htm#9-12 (Using Graphs, Equations, and Tables to Investigate the Elimination of Medicine from the Body: Modeling the Situation)
http://mathforum.org/t2t/faq/election.html presents election math in the context of the classroom.
http://www.education-world.com/a_curr/curr148.shtml offers examples of real-life problems such as calculating car payments, saving and investing, the world of credit cards, and other finance problems.
http://score.kings.k12.ca.us/real.world.html is a website connecting math to real jobs, elections, NASA projects and other situations.

3. Graphing and spreadsheets for enhancing math learning:
   http://www.cvgs.k12.va.us/digstats/
   http://score.kings.k12.ca.us/standards/probability.html for graphing and statistics.
   http://www.microsoft.com/education/solving.mspx for using spreadsheets and to solve polynomial problems.

4. Use technology, such as manipulatives, software and interactive online activities that can help all students learn, particularly those oriented more towards visual and kinesthetic learning. Here are some websites:
   http://illuminations.nctm.org/ActivitySearch.aspx has games for grades 9–12 that can be played against the computer or another student.
   http://nlvm.usu.edu/ The National Library of Virtual Manipulatives has resources for all grades on numbers and operations, algebra, geometry, probability and measurement.
   http://mathforum.org/pow/other.html has links to various math challenges, manipulatives and puzzles.
   http://www.etacuisenaire.com/algeblocks/algeblocks.jsp Algeblocks are blocks that use the relationship between algebra and geometry.

5. Word problem strategies – the hardest thing to do is take the English and turn it into math, but there are six key steps to teach students how to solve word problems:

   a. **Look for Key Words**: The problem will have key words to suggest the type of operation or operations that must be performed to solve the problem. For example, words such as "altogether" or "total" imply addition, and words such as "difference" or "How many more?" imply subtraction.
   b. **Use Pictures or Concrete Materials**: Math is very abstract; it is easier to solve a problem using pictures or concrete materials to illustrate the problem. Pictures and concrete materials allow students to manipulate the material to solve the problem by trial and error.

c. **Use Logic**: Ask your students if their answers make sense. Make them familiar with the process of deduction. Model the deduction process in deciding on the answer to a word problem. For example, solve a problem such as "Two consecutive numbers have a sum of 91. What are the numbers?" If the student arrives at an answer of 44 and 45, it is obvious that there was an error in the equation or in the calculation used, since 44 and 45 are consecutive but don't add up to 91. Let x be the first number and $(x + 1)$ be the second number, so that $x + (x + 1) = 91$ and $2x + 1 = 91$. Then, $2x = 90$, $x = 45$ and $x + 1 = 46$. The answer is 45 and 46.
d. ***Eliminate the possibilities and look for patterns, or work the problem backwards***
e. **Guess the Answer**: Students should guess an approximate answer that makes sense based on the problem. For example, if the student knows that the word problem implies addition, he should recognize that the answer must be greater than the numbers in the problem. Often, students are afraid of guessing because they don't want to get the wrong answer; nevertheless, encourage your students to guess and then double check the answer to see if it works. If the answer is incorrect, the student can try another strategy for finding the answer.
f. **Make a Table**: Selecting relevant information from a word problem and organizing the data is very helpful in solving word problems. Often, students become confused because there are too many numbers or variables.

http://www.purplemath.com/modules/translat.htm, http://math.about.com/library/weekly/aa071002a.htm and http://www.onlinemathlearning.com/algebra-word-problems.html are great resources for students to solve word problems.

6. Mental math practice
   Give students regular practice in doing mental math. The following websites offer many mental calculation tips and strategies:
   http://www.cramweb.com/math/index.htm
   http://mathforum.org/k12/mathtips/mathtips.html

Because frequent calculator use tends to deprive students of a sense of numbers and an ability to calculate on their own, they will often approach a sequence of multiplications and divisions the hard way. For instance, when asked to calculate 770 x 36 / 55, they may first multiply 770 and 36 and then do a long division with the 55. They may fail to recognize that both 770 and 55 can be divided by 11 and then by 5 to considerably simplify the problem. Give students plenty of practice in multiplying and dividing a sequence of integers and fractions so that they are comfortable with canceling top and bottom terms.

7. Math language
   Math vocabulary help is available for high school students on the web:

   http://www.amathsdictionaryforkids.com/ is a colorful website math dictionary.
   http://www.math.com/tables/index.html is a math dictionary in English and Spanish.

ERROR ANALYSIS

A simple method for analyzing student errors is to ask how the answer was obtained. The teacher can then determine if a common error pattern has resulted in the wrong answer. There is a value in having the students explain how they arrived at their answers, whether correct or incorrect.

Many errors are due to simple **carelessness**. Students need to be encouraged to work slowly and carefully; they should check their calculations by redoing the problem on another paper, not merely by looking at the work. Addition and subtraction problems need to be written neatly so the numbers line up. Students need to be careful regrouping in subtraction. Students must write clearly and legibly, including erasing fully. Use estimation to ensure that answers make sense.

Many students' computational skills exceed their **reading** level. Although they can understand basic operations, they fail to grasp the concept or completely understand the question. Students must read directions slowly.
Fractions are often a source of many errors. Students need to be reminded to use common denominators when adding and subtracting and to always express answers in simplest terms. Again, it is helpful to check by estimating.

The most common error that is made when working with **decimals** is failure to line up the decimal points when adding or subtracting or not moving the decimal point when multiplying or dividing. Students also need to be reminded to add zeroes when necessary. Reading aloud may also be beneficial. Estimation, as always, is especially important.

Students need to know that it is okay to make mistakes. The teacher must keep a positive attitude so they do not feel defeated or frustrated.

TEACHER CERTIFICATION STUDY GUIDE

THE ART OF TEACHING - PEDAGOGICAL PRINCIPLES
Maintain a supportive, non-threatening environment

The key to success in teaching goes beyond your mathematical knowledge and the desire to teach. Though important, knowledge and desire alone do not make you a good teacher. Being able to connect with your students is vital: learn their names immediately, have a seating chart the first day (even if you intend to change it) and learn about your students—their hobbies, phone number, parents' names and what they like and dislike about school, learning and math. Keep this information on each student and learn it: adapt your lessons, their required skill level and any other resources you may need to accommodate your students' individual strengths and weaknesses.

Learn to see math as your students see it. If you aren't able to connect with your students, no matter how good your lessons are and how well you know the material, you won't inspire them to learn math from you. As you demand respect, you must give respect, and as you demand their attention, they also need your attention and understanding. Talk to them with the same tone of voice as you would an adult, not in a tone that makes them feel like children. Look your students in the eye when you talk to them and encourage questions and comments. Take advantage of teachable moments and explain the rationale behind math rules.

Demonstrate respect, care and trust toward every student; assume the best. This does not mean becoming "friends" with your students; doing so can cause problems with discipline. You can be kind and firm at the same time. Have a fair and clear grading and discipline system that is posted, reviewed and made clear to your students. Consistency, structure and fairness are essential to earning their trust in you as a teacher. Always admit your mistakes and be available to your students certain days after school. Finally, demonstrate your care for them and your love of math and you will be a positive influence on their learning.

Below are websites to help make your teaching more effective and fun:

1. Teachers Helping Teachers has several resources for high school mathematics. (http://www.pacificnet.net/~mandel/)
2. Math Resources for Teachers is a resources for grades 7–10. (http://math.nie.edu.sg/bwjyeo/resources/index.htm#Part%20V)
3. Math is Marvelous Web Site is a fascinating website on the history of geometry. (http://www.people.memphis.edu/~brveteto/)
4. Math Archives K-12 provides resources for lesson plans and software. (http://archives.math.utk.edu/k12.html)
5. http://www.edhelper.com/algebra.htm covers Algebra I and II.
6. Math Goodies offers interactive lessons, worksheets and homework help. (http://www.mathgoodies.com/)

TEACHER CERTIFICATION STUDY GUIDE

7. Multicultural Lessons is an interesting site with lessons on Babylonian Square Roots, Chinese Fraction Reducing, Egyptian multiplication, etc. (http://www.deltacollege.edu/dept/basicmath/Multicultural_Math.htm)
8. http://www.goenc.com/ offers resources and professional development for teachers.
9. Math and Reading Help is a guide to math, reading, homework help, tutoring and earning a high school diploma. (http://math-and-reading-help-for-kids.org/index.html)
10. Purple Math.com is all about algebra and provides lessons, help for students and lots of other resources.
11. http://library.thinkquest.org/20991/home.html discusses algebra, geometry and pre-calc/calculus.
12. http://www.math.com/ discusses algebra, geometry, trigonometry, and calculus, plus it offers homework help.
13. http://www.math.armstrong.edu/MathTutorial/index.html is a tutorial in algebra.
14. http://www.wtamu.edu/academic/anns/mps/math/mathlab/beg_algebra/index.htm is a site that is helpful for those beginning algebra or as a refresher.
15. Math Complete offers radicals, quadratics and linear equation solvers. (http://www.mathcomplete.com/)
16. Math Tutor – PEMDAS & Integers discusses fractions, integers and information for parents and teachers. (http://www.squidoo.com/lensmasters/Rebecca_Newburn)
17. Matrix Algebra is all about matrix operations and applications. (http://www.sosmath.com/matrix/matrix.html)
18. Mr. Stroh's Algebra Site offers help for Algebra I and II. (http://www.homestead.com/stroh/MathPage.html)
19. Polynomials and Polynomial Functions covers everything from factoring to graphing, finding rational zeros and multiplying and adding and subtracting polynomials. (http://webpages.charter.net/thejacowskis/chapter6/)
20. Quadratic Formula is all about quadratics. (http://www.flashyapps.com/frameset/algebra/algebraframeset.htm)
21. Animated Pythagorean Theorem is a fun animated proof of the Pythagorean Theorem. (http://www.nadn.navy.mil/MathDept/mdm/pyth.html)
22. Brunnermath – Interactive Activities covers general math, algebra, geometry, trigonometry, statistics, calculus and calculators. (http://www.brunnermath.com/geometry.htm)
23. CoolMath4Kids Geometry teaches creation of art with math and geometry lessons. (http://coolmath4kids.com/geometrystuff.html)
24. The Curlicue Fractal The curlicue fractal is an exceedingly easy-to-make but richly complex pattern using trigonometry and calculus. (http://oolong.co.uk/curlicue.htm)

25. Euclid's Elements Interactive Euclid's *Elements* form one of the most beautiful and influential works of science in the history of humankind. (http://aleph0.clarku.edu/~djoyce/java/elements/elements.html)
26. Howe-Two Free Software provides software solutions for mathematics instruction. (http://www.howe-two.com/free/index.html)
27. http://regentsprep.org/regents/math/math-topic.cfm?TopicCode=syslin discusses systems of equations in lessons and in practice.
28. http://www.sparknotes.com/math/algebra1/systemsofequations/problems3.rhtml deals with word problems and systems of equations.
29. http://math.about.com/od/complexnumbers/Complex_Numbers.htm provides several complex number exercise pages.
30. http://regentsprep.org/Regents/math/ALGEBRA/AE3/PracWord.htm offers practice with word problems based on systems of inequalities.
31. http://regentsprep.org/regents/Math/solvin/PSolvIn.htm teaches how to solve inequalities.
32. http://www.wtamu.edu/academic/anns/mps/math/mathlab/beg_algebra/beg_alg_tut18_ineq.htm offers an inequality tutorial as well as examples and problems.
33. http://www.wtamu.edu/academic/anns/mps/math/mathlab/beg_algebra/beg_alg_tut24_ineq.htm provides a linear inequalities graphing tutorial.
34. http://www.wtamu.edu/academic/anns/mps/math/mathlab/col_algebra/col_alg_tut17_quad.htm offers a quadratic equations tutorial as well as examples and problems.
35. http://regentsprep.org/Regents/math/math-topic.cfm?TopicCode=factor lets students practice factoring.
36. http://www.wtamu.edu/academic/anns/mps/math/mathlab/col_algebra/col_alg_tut37_syndiv.htm provides a synthetic division tutorial.
37. http://www.tpub.com/math1/10h.htm offers synthetic division examples and problems.

DEVELOPMENTAL PSYCHOLOGY AND TEACHING MATHEMATICS
Things you may not know about your students

Studies show that health matters more than gender or social status when it comes to learning. Healthy girls and boys do equally well on most cognitive tasks. Boys perform better at manipulating shapes and analyzing problems, and girls perform better in processing speed and motor dexterity. No differences have been measured in calculation ability, meaning girls and boys have an equal aptitude for mathematics.

The following was written by Jay Giedd, M.D., who is a practicing Child and Adolescent Psychiatrist and Chief of Brain Imaging at the Child Psychiatry Branch of the National Institute of Mental Health:

http://nihrecord.od.nih.gov/newsletters/2005/08_12_2005/story04.htm

The most surprising thing has been how much the teen brain is changing. By age six, the brain is already 95 percent of its adult size. But the gray matter, or thinking part of the brain, continues to thicken throughout childhood as the brain cells get extra connections, much like a tree growing extra branches, twigs and roots...In the frontal part of the brain, the part of the brain involved in judgment, organization, planning, strategizing—those very skills that teens get better and better at—this process of thickening of the gray matter peaks at about age 11 in girls and age 12 in boys, roughly about the same time as puberty. After that peak, the gray matter thins as the excess connections are eliminated or pruned...

Contrary to what most parents have thought at least once, "teens really do have brains," quipped Dr. Jay Giedd, NIMH intramural scientist, in a lecture on the "Teen Brain Under Construction." His talk was the kick-off event for the recent NIH Parenting Festival. Giedd said scientists have only recently learned more about the trajectories of brain growth. One of the findings he discussed showed the frontal cortex area—which governs judgment, decision-making and impulse control—doesn't fully mature until around age 25. "That really threw us," he said. "We used to joke about having to be 25 to rent a car, but there's tons of data from insurance reports [showing] that 24-year-olds are costing them more than 44-year-olds.

So why is that? "It must be behavior and impulse control," Giedd said. "Whatever these changes are, the top 10 bad things that happen to teens involve emotion and behavior." Physically, according to Giedd, the teen years and early 20s represent an incredibly healthy time of life in terms of cancer, heart disease and other serious illnesses. Nevertheless, with accidents as the leading cause of death in adolescents, and suicide following close behind, "This isn't a great time emotionally and psychologically. This is the great paradox of adolescence: right at the time you should be on the top of your game, you're not."

The next step in Giedd's research is to learn more about what influences brain growth, for good or bad. "Ultimately, we want to use these findings to treat illness and enhance development."

One of the things scientists have come to understand, though, is that parents do have something to do with their children's brain development.

"From imaging studies, one of the things that seems intriguing is this notion of modeling...that the brain is pretty adept at learning by example," he said. "As parents, we teach a lot when we don't even know we're teaching, just by showing how we treat our spouses, how we treat other people, what we talk about in the car on the way home...things that a parent says in the car can stick with them for years.

They're listening even though it may appear they're not."

What can we do to change our kids? "Well, start with yourself in terms of what you show by example," Giedd concluded.

Maybe the parts of the brain doing geometry are different from the parts doing algebra; there is no definitive research to answer that question yet, but it is obviously what researchers are looking for.

<u>Time-Lapse Imaging Tracks Brain Maturation from ages 5 to 20</u>
Constructed from MRI scans of healthy children and teens, the time-lapse "movie", from which the images below were extracted, compresses 15 years of brain development (ages 5–20) into just a few seconds.

To view in color, go to the website below: Red indicates **more** gray matter, **blue less** gray matter. Gray matter wanes in a back-to-front wave as the brain matures and neural connections are pruned.
Source: Paul Thompson, Ph.D., UCLA Laboratory of Neuroimaging
http://www.loni.ucla.edu/%7Ethompson/DEVEL/PR.html

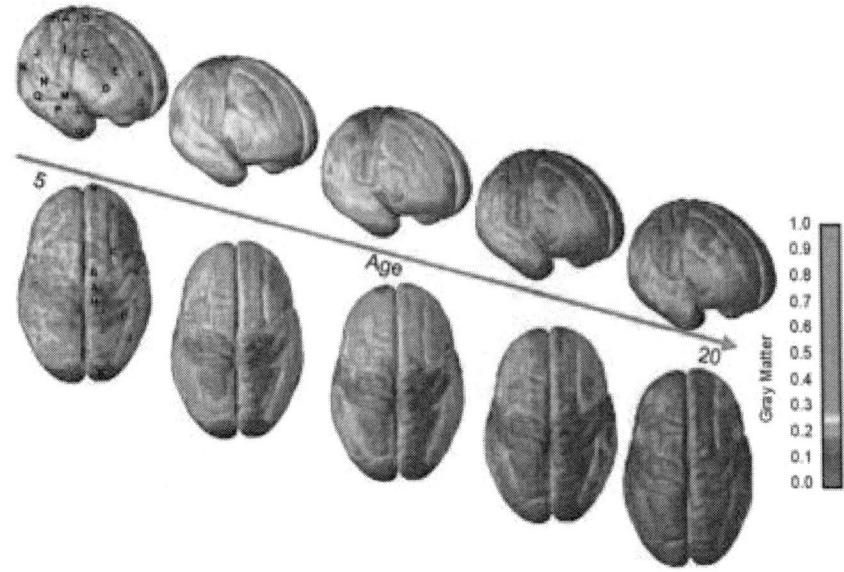

What are the implications of this fascinating study for teachers? It's unreasonable to expect teens to have adult levels of organizational or decision-making skills before their brains have completely developed. In teens, the frontal lobe, or the executive of the brain, is what handles organization, decision making, emotions, attention, shifting attention, planning and making strategies, and it is not fully developed until the early to mid twenties.

Perhaps since certain parts of the brain develop sooner than others, subjects should be taught in a different order. Until we know more, just understanding that the parts of teen brains related to decision making and emotions are still developing through the early twenties is important, and **that stressful situations lead to diminished ability to made good judgments**. For some children, just being called on in class is stressful. At this age, social relationships become very important and **teachers need to be sensitive to this aspect of teen development as it relates to stress and decision-making.** The immaturity of this part of the teen brain might explain why the teen crash rate is **four** times that of adults…

Sample Test

Directions: Read each item and select the best response.

1. Convert $.\overline{63}$ into a fraction in simplest form.
   (Average Rigor)
   (Competency 001)

   A) $\dfrac{63}{100}$
   B) $\dfrac{7}{11}$
   C) $6\dfrac{3}{10}$
   D) $\dfrac{2}{3}$

2. Which of the following illustrates an inverse property?
   (Easy) (Competency 001)

   A) $a + b = a - b$
   B) $a + b = b + a$
   C) $a + 0 = a$
   D) $a + (-a) = 0$

3. What is the total cost of a suit for $295.99 and a pair of shoes for $69.95 including 6.5% sales tax?
   (Average Rigor)
   (Competency 001)

   A) $389.73
   B) $398.37
   C) $237.86
   D) $315.23

4. Which of the following sets is closed under division?
   (Average Rigor)
   (Competency 001)

   I) {1/2, 1, 2, 4}
   II) {–1, 1}
   III) {–1, 0, 1}

   A) I only
   B) II only
   C) III only
   D) I and II

5. The conjugate of $4 + 5i$ is
   (Easy) (Competency 002)

   A) $-4 + 5i$
   B) $4 - 5i$
   C) $4i + 5$
   D) $4i - 5$

6. Simplify: $(6 + 3i) - (4 - 2i)$
   (Easy) (Competency 002)

   A) $2 + 5i$
   B) $2 + i$
   C) $10 + 5i$
   D) $2 - 2i$

7. Simplify: $\dfrac{10}{1+3i}$
   (Average Rigor)
   (Competency 002)

   A) $-1.25(1 - 3i)$
   B) $1.25(1 - 3i)$
   C) $1 + 3i$
   D) $1 - 3i$

8. Find the LCM of 27, 90 and 84.
   (Easy) (Competency 003)

   A) 90
   B) 3,780
   C) 204,120
   D) 1,260

9. Which of the following is always composite if x is odd, y is even, and both x and y are greater than or equal to 2?
   (Average Rigor) (Competency 003)

   A) $x + y$
   B) $3x + 2y$
   C) $5xy$
   D) $5x + 3y$

10. What is the smallest number that is divisible by 3 and 5 and leaves a remainder of 3 when divided by 7?
    (Average Rigor) (Competency 003)

    A) 15
    B) 18
    C) 25
    D) 45

11. The volume of water flowing through a pipe varies directly with the square of the radius of the pipe. If the water flows at a rate of 80 liters per minute through a pipe with a radius of 4 cm, at what rate would water flow through a pipe with a radius of 3 cm?
    (Rigorous) (Competency 003)

    A) 45 liters per minute
    B) 6.67 liters per minute
    C) 60 liters per minute
    D) 4.5 liters per minute

12. Given the series of examples below, what is $5 \not\subset 4$?
    (Average Rigor) (Competency 004)

    $4 \not\subset 3 = 13 \qquad 7 \not\subset 2 = 47$
    $3 \not\subset 1 = 8 \qquad 1 \not\subset 5 = -4$

    A) 20
    B) 29
    C) 1
    D) 21

13. What is the sum of the first 20 terms of the geometric sequence (2, 4, 8, 16, 32,…)?
    (Average Rigor)
    (Competency 004)

    A) 2,097,150
    B) 1,048,575
    C) 524,288
    D) 1,048,576

14. Find the sum of the first one hundred terms in the progression.
    (−6, −2, 2 . . .)
    (Rigorous) (Competency 004)

    A) 19,200
    B) 19,400
    C) −604
    D) 604

15. Evaluate $x^2 - 3x + 7$ when $x = 2$.
    (Easy) (Competency 005)

    A) 7
    B) 5
    C) 3
    D) 9

16. Given $f(x) = 3x - 2$, $f^{-1}(x) =$
    (Average Rigor)
    (Competency 005)

    A) $3x + 2$
    B) $\dfrac{x}{6}$
    C) $2x - 3$
    D) $\dfrac{x+2}{3}$

17. State the domain of the function $f(x) = \dfrac{3x-6}{x^2 - 25}$.
    (Average Rigor)
    (Competency 005)

    A) $x \neq 2$
    B) $x \neq 5, -5$
    C) $x \neq 2, -2$
    D) $x \neq 5$

18. Given $f(x) = 3x - 2$ and $g(x) = x^2$, determine $g(f(x))$.
    (Average Rigor)
    (Competency 005)

    A) $3x^2 - 2$
    B) $9x^2 + 4$
    C) $9x^2 - 12x + 4$
    D) $3x^3 - 2$

19. Solve for v_0: $d = at(v_t - v_0)$
    (Average Rigor)
    (Competency 005)

    A) $v_0 = atd - v_t$
    B) $v_0 = d - atv_t$
    C) $v_0 = atv_t - d$
    D) $v_0 = (atv_t - d)/at$

20. The formula for solving a quadratic equation is
(Easy) (Competency 006)

A) $x = \dfrac{-b \pm \sqrt{b^2 - 4ac}}{2a}$

B) $x = \dfrac{-b \pm \sqrt{b^2 - 4a}}{2a}$

C) $x = \dfrac{b \pm \sqrt{b^2 - 4ac}}{2a}$

D) $x = \dfrac{b \pm \sqrt{b^3 - 4ac}}{2a}$

21. Solve for x by factoring
$2x^2 - 3x - 2 = 0$.
(Average Rigor)
(Competency 006)

A) $x = (-1, 2)$
B) $x = (0.5, -2)$
C) $x = (-0.5, 2)$
D) $x = (1, -2)$

22. Which graph represents the equation $y = x^2 + 3x$?
(Average Rigor)
(Competency 006)

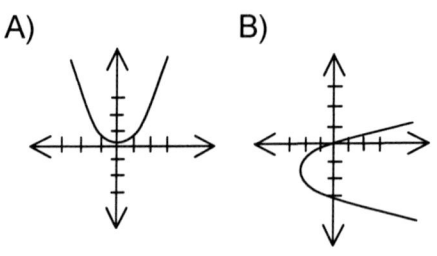

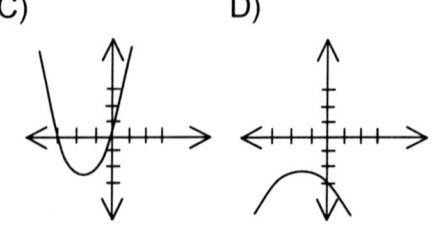

23. Which of the following is a factor of the expression
$9x^2 + 6x - 35$?
(Rigorous) (Competency 006)

A) $3x - 5$
B) $3x - 7$
C) $x + 3$
D) $x - 2$

24. Solve the system of equations for x, y and z.
(Rigorous) (Competency 006)

$3x + 2y - z = 0$
$2x + 5y = 8z$
$x + 3y + 2z = 7$

A) $(-1, 2, 1)$
B) $(1, 2, -1)$
C) $(-3, 4, -1)$
D) $(0, 1, 2)$

25. How does the function $y = x^3 + x^2 + 4$ behave from x = 1 to x = 3?
(Average Rigor)
(Competency 007)

A) increasing, then decreasing
B) increasing
C) decreasing
D) neither increasing nor decreasing

26. Solve for x: $18 = 4 + |2x|$
    (Rigorous) (Competency 007)

    A) $\{-11, 7\}$
    B) $\{-7, 0, 7\}$
    C) $\{-7, 7\}$
    D) $\{-11, 11\}$

27. Find the zeros of $f(x) = x^3 + x^2 - 14x - 24$
    (Rigorous) (Competency 007)

    A) 4, 3, 2
    B) 3, −8
    C) 7, −2, −1
    D) 4, −3, −2

28. Evaluate $3^{1/2}(9^{1/3})$
    (Rigorous) (Competency 008)

    A) $27^{5/6}$
    B) $9^{7/12}$
    C) $3^{5/6}$
    D) $3^{6/7}$

29. Which of the following is incorrect?
    (Rigorous) (Competency 008)

    A) $(x^2 y^3)^2 = x^4 y^6$
    B) $m^2(2n)^3 = 8m^2 n^3$
    C) $(m^3 n^4)/(m^2 n^2) = mn^2$
    D) $(x + y^2)^2 = x^2 + y^4$

30. The exponential equation $2^5 = 32$ can be written as
    (Average Rigor) (Competency 008)

    A) $\log_2(5) = 32$
    B) $\log_{10}(32) = 5$
    C) $\log_5(32) = 2$
    D) $\log_2(32) = 5$

31. Which equation corresponds to the logarithmic statement: $\log_x k = m$?
    (Rigorous) (Competency 008)

    A) $x^m = k$
    B) $k^m = x$
    C) $x^k = m$
    D) $m^x = k$

32. Solve for x: $10^{x-3} + 5 = 105$.
    (Rigorous) (Competency 008)

    A) 3
    B) 10
    C) 2
    D) 5

33. Determine the measures of angles A and B.
    (Average Rigor)
    (Competency 009)

    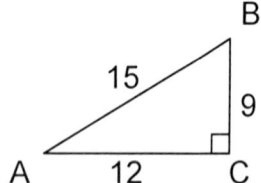

    A) A = 30°, B = 60°
    B) A = 60°, B = 30°
    C) A = 53°, B = 37°
    D) A = 37°, B = 53°

34. Which expression is not identical to sin x?
    (Average Rigor)
    (Competency 009)

    A) $\sqrt{1-\cos^2 x}$
    B) $\tan x \cos x$
    C) $1/\csc x$
    D) $1/\sec x$

35. Which expression is equivalent to $1-\sin^2 x$?
    (Rigorous) (Competency 009)

    A) $1-\cos^2 x$
    B) $1+\cos^2 x$
    C) $1/\sec x$
    D) $1/\sec^2 x$

36. For an acute angle x, sin x = 3/5. What is cot x?
    (Rigorous) (Competency 009)

    A) 5/3
    B) 3/4
    C) 1.33
    D) 1

37. Find the area under the function $y = x^2 + 4$ from x = 3 to x = 6.
    (Average Rigor)
    (Competency 010)

    A) 75
    B) 21
    C) 96
    D) 57

38. If the velocity of a body is given by v = 16 – t^2, find the distance traveled from t = 0 until the body comes to a complete stop.
    (Average Rigor)
    (Competency 010)

    A) 16
    B) 43
    C) 48
    D) 64

39. Find the following limit:
    $\lim_{x \to 0} \dfrac{\sin 2x}{5x}$
    (Rigorous) (Competency 010)

    A) Infinity
    B) 0
    C) 1.4
    D) 1

40. Find the first derivative of the function:
    $f(x) = x^3 - 6x^2 + 5x + 4$
    (Rigorous) (Competency 010)

    A) $3x^2 - 12x^2 + 5x$
    B) $3x^2 - 12x - 5$
    C) $3x^2 - 12x + 9$
    D) $3x^2 - 12x + 5$

41. Evaluate $\int_0^2 (x^2 + x - 1)dx$
    (Rigorous) (Competency 010)

    A) 11/3
    B) 8/3
    C) –8/3
    D) –11/3

42. Find the antiderivative for the function $y = e^{3x}$.
    (Rigorous) (Competency 010)

    A) $3x(e^{3x}) + C$
    B) $3(e^{3x}) + C$
    C) $1/3(e^x) + C$
    D) $1/3(e^{3x}) + C$

43. How does the function $y = x^3 + x^2 + 4$ behave from $x = 1$ to $x = 3$?
    (Average Rigor) (Competency 010)

    A) increasing, then decreasing
    B) increasing
    C) decreasing
    D) neither increasing nor decreasing

44. Find the surface area of a box that is 3 feet wide, 5 feet tall, and 4 feet deep.
    (Easy) (Competency 011)

    A) 47 sq. ft.
    B) 60 sq. ft.
    C) 94 sq. ft
    D) 188 sq. ft.

45. If a ship sails due south 6 miles, then due west 8 miles, how far was it from the starting point?
    (Average Rigor) (Competency 011)

    A) 100 miles
    B) 10 miles
    C) 14 miles
    D) 48 miles

46. Find the area of the figure pictured below.
    (Rigorous) (Competency 011)

    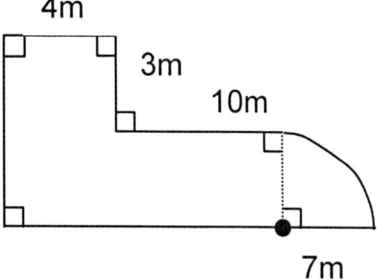

    A) 136.47 m²
    B) 148.48 m²
    C) 293.86 m²
    D) 178.47 m²

47. If the area of the base of a cone is tripled, the volume will be
    (Rigorous) (Competency 011)

    A) the same as the original
    B) 9 times the original
    C) 3 times the original
    D) 3 π times the original

48. The length of a picture frame is 2 inches greater than its width. If the area of the frame is 143 square inches, what is its width?
(Rigorous) (Competency 011)

A) 11 inches
B) 13 inches
C) 12 inches
D) 10 inches

49. When you begin by assuming the conclusion of a theorem is false, then show that through a sequence of logically correct steps you contradict an accepted fact, this is known as
(Easy) (Competency 012)

A) inductive reasoning
B) direct proof
C) indirect proof
D) exhaustive proof

50. Which term most accurately describes two coplanar lines without any common points?
(Easy) (Competency 012)

A) perpendicular
B) parallel
C) intersecting
D) skew

51. Which theorem can be used to prove $\triangle BAK \cong \triangle MKA$?
(Average Rigor) (Competency 012)

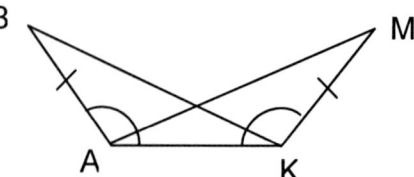

A) SSS
B) ASA
C) SAS
D) AAS

52. Choose the diagram which illustrates the construction of a perpendicular to the line at a given point on the line.
(Rigorous) (Competency 012)

A)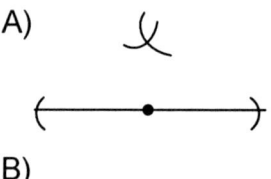

B)

C)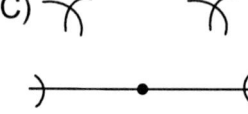

D)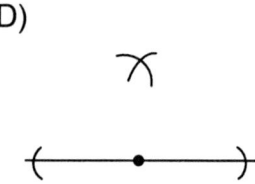

MATHEMATICS 8-12

53. Compute the area of the shaded region, given a radius of 5 meters. Point O is the center.
    (Rigorous) (Competency 013)

    A) 7.13 cm²
    B) 7.13 m²
    C) 78.5 m²
    D) 19.63 m²

    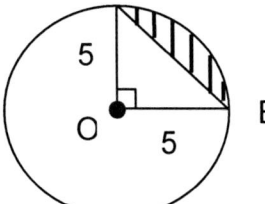

54. Given that QO⊥NP and QO=NP, quadrilateral NOPQ can most accurately be described as a
    (Easy) (Competency 013)

    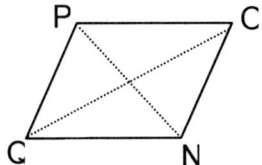

    A) parallelogram
    B) rectangle
    C) square
    D) rhombus

55. Which of the following statements about a trapezoid is incorrect?
    (Average Rigor) (Competency 013)

    A) It has one pair of parallel sides
    B) The parallel sides are called bases
    C) If the two bases are the same length, the trapezoid is called isosceles
    D) The median is parallel to the bases

56. What is the degree measure of an interior angle of a regular 10-sided polygon?
    (Rigorous) (Competency 013)

    A) 18°
    B) 36°
    C) 144°
    D) 54°

57. What is the measure of minor arc AD, given measure of arc PS is 40° and $m < K = 10°$?
    (Rigorous) (Competency 013)

    A) 50°
    B) 20°
    C) 30°
    D) 25°

    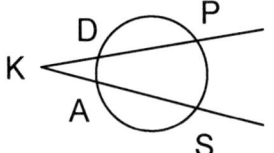

58. Determine the area of the shaded region of the trapezoid in terms of x and y.
    (Rigorous) (Competency 013)

    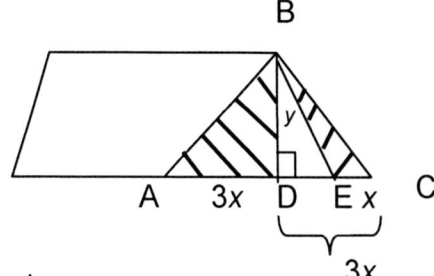

    A) $4xy$
    B) $2xy$
    C) $3x^2y$
    D) There is not enough information given.

59. Given a vector with horizontal component 5 and vertical component 6, determine the length of the vector.
(Average Rigor) (Competency 014)

A) 61
B) $\sqrt{61}$
C) 30
D) $\sqrt{30}$

60. Compute the distance from (–2, 7) to the line $x = 5$.
(Average Rigor) (Competency 014)

A) –9
B) –7
C) 5
D) 7

61. Given K(–4, y) and M(2, –3) with midpoint M(x, 1), determine the values of x and y.
(Rigorous) (Competency 014)

A) $x = -1, y = 5$
B) $x = 3, y = 2$
C) $x = 5, y = -1$
D) $x = -1, y = -1$

62. Find the length of the major axis of $x^2 + 9y^2 = 36$.
(Rigorous) (Competency 014)

A) 4
B) 6
C) 12
D) 8

63. Which equation represents a circle with a diameter whose endpoints are (0, 7) and (0, 3)?
(Rigorous) (Competency 014)

A) $x^2 + y^2 + 21 = 0$
B) $x^2 + y^2 - 10y + 21 = 0$
C) $x^2 + y^2 - 10y + 9 = 0$
D) $x^2 - y^2 - 10y + 9 = 0$

64. Compute the standard deviation for the following set of temperatures.
(37, 38, 35, 37, 38, 40, 36, 39)
(Easy) (Competency 015)

A) 37.5
B) 1.5
C) 0.5
D) 2.5

65. Which type of graph uses symbols to represent quantities?
(Average rigor) (Competency 015)

A) Bar graph
B) Line graph
C) Pictograph
D) Circle graph

66. Half the students in a class scored 80% on an exam, most of the rest scored 85% except for one student who scored 10%. Which would be the best measure of central tendency for the test scores?
(Rigorous) (Competency 015)

A) mean
B) median
C) mode
D) either the median or the mode because they are equal

67. A jar contains 3 red marbles, 5 white marbles, 1 green marble and 15 blue marbles. If one marble is picked at random from the jar, what is the probability that it will be red?
(Easy) (Competency 016)

A) 1/3
B) 1/8
C) 3/8
D) 1/24

68. A die is rolled several times. What is the probability that a 3 will not appear before the third roll of the die?
(Rigorous) (Competency 016)

A) 1/3
B) 25/216
C) 25/36
D) 1/216

69. If there are three people in a room, what is the probability that at least two of them will share a birthday? (Assume a year has 365 days)
(Rigorous) (Competency 016)

A) 0.67
B) 0.05
C) 0.008
D) 0.33

70. Which of the following is not a valid method of collecting statistical data?
(Average Rigor) (Competency 017)

A) Random sampling
B) Systematic sampling
C) Cluster sampling
D) Cylindrical sampling

71. To determine the odds for or against a given deviation from expected statistical distribution statisticians use
(Average Rigor) (Competency 017)

A) The t-test
B) Linear regression
C) The chi-square test
D) Exponential regression

72. If the correlation between two variables is given as zero, the association between the two variables is
(Rigorous) (Competency 017)

A) negative linear
B) positive linear
C) quadratic
D) random

73. About two weeks after introducing formal proofs, several students in your geometry class are having a difficult time remembering the names of the postulates. They cannot complete the reason column of the proof and as a result are not even attempting the proofs. What would be the best approach to help students understand the nature of geometric proofs?
(Average Rigor) (Competency 018)

A) Give them more time; proofs require time and experience.
B) Allow students to write an explanation of the theorem in the reason column instead of the name.
C) Have the student copy each theorem in a notebook.
D) Allow the students to have open book tests.

74. Identify the correct sequence of subskills required for solving and graphing inequalities involving absolute value in one variable, such as $|x+1| \leq 6$.
(Average Rigor) (Competency 018)

A) understanding absolute value, graphing inequalities, solving systems of equations
B) graphing inequalities on a Cartesian plane, solving systems of equations, simplifying expressions with absolute value
C) plotting points, graphing equations, graphing inequalities
D) solving equations with absolute value, solving inequalities, graphing conjunctions and disjunctions

75. Mr. Lacey is using problem solving to help students develop their math skills. He gives the class a box of pencils. He says that the pencils have to be divided so that each student has the same number of pencils. What step should come first in problem solving?
(Rigorous) (Competency 018)

    A) Find a strategy to solve the problem
    B) Identify the problem
    C) Count the number of pencils
    D) Make basic calculations

76. Kindergarten students are doing a butterfly art project. They fold paper in half. On one half, they paint a design. Then they fold the paper closed and reopen. The resulting picture is a butterfly with matching sides. What math principle does this demonstrate?
(Rigorous) (Competency 019)

    A) Slide
    B) Rotate
    C) Symmetry
    D) Transformation

77. Students are working with a set of rulers and various small objects from the classroom. Which concept are these students exploring?
(Average rigor) (Competency 020)

    A) Volume
    B) Weight
    C) Length
    D) Temperature

78. Third grade students are looking at a circle graph. Most of the graph is yellow. A small wedge of the graph is blue. Each colored section also has a number followed by a symbol. What are the students most likely learning about?
(Rigorous) (Competency 020)

    A) Addition
    B) Venn diagrams
    C) Percent
    D) Pictographs

79. Which of the following is the best example of the value of personal computers in advanced high school mathematics?
    (Easy) (Competency 020)

    A) Students can independently drill and practice test questions.
    B) Students can keep an organized list of theorems and postulates on a word processing program.
    C) Students can graph and calculate complex functions to explore their nature and make conjectures.
    D) Students are better prepared for business because of mathematics computer programs in high school.

80. What would be the least appropriate use for handheld calculators in the classroom?
    (Average Rigor) (Competency 020)

    A) practice for standardized tests
    B) integrating algebra and geometry with applications
    C) justifying statements in geometric proofs
    D) applying the law of sines to find dimensions

81. A group of students working with trigonometric identities have concluded that cos 2x = 2 cos x. How could you best lead them to discover their error?
    (Average Rigor) (Competency 020)

    A) Have the students plug in values on their calculators.
    B) Direct the student to the appropriate chapter in the text.
    C) Derive the correct identity on the board.
    D) Provide each student with a table of trig identities.

82. $-3 + 7 = -4$ $6(-10) = -60$
    $-5(-15) = 75$ $-3+-8 = 11$
    $8-12 = -4$ $7-(-8) = 15$

    Which best describes the type of error observed above?
    (Easy) (Competency 020)

    A) The student is incorrectly multiplying integers.
    B) The student has incorrectly applied rules for adding integers to subtracting integers.
    C) The student has incorrectly applied rules for multiplying integers to adding integers.
    D) The student is incorrectly subtracting integers.

83. **Which of the following statements is untrue?**
    (Easy) (Competency 021)

    A) A teacher may use a variety of formal and informal assessment methods to evaluate a student's progress
    B) A multiple-choice test is a type of formative assessment
    C) Alternative assessment is any type of assessment in which students create a response rather than choose an answer
    D) Summative assessment consists of temporary interaction between teacher and student

84. **A student portfolio is**
    (Easy) (Competency 021)

    A) a collection of a student's work over a period of time to help the teacher and student assess progress
    B) a collection of test papers to help the teacher and student assess progress
    C) a collection of student art
    D) a collection of assignments given by the teacher

85. **Higher order thinking, creativity, and the integration of reasoning and communication skills are most demonstrated by**
    (Easy) (Competency 021)

    A) Multiple-choice and true/false tests
    B) Projects, demonstrations and oral presentations
    C) Essay questions
    D) Portfolios

ANSWER KEY

1)	B	18)	C	35)	D	52)	D	69)	C
2)	D	19)	D	36)	B	53)	B	70)	D
3)	A	20)	A	37)	A	54)	C	71)	C
4)	B	21)	C	38)	B	55)	C	72)	D
5)	B	22)	C	39)	C	56)	C	73)	B
6)	A	23)	A	40)	D	57)	B	74)	D
7)	D	24)	A	41)	B	58)	B	75)	B
8)	B	25)	B	42)	D	59)	B	76)	C
9)	C	26)	C	43)	B	60)	D	77)	C
10)	D	27)	D	44)	C	61)	A	78)	C
11)	A	28)	B	45)	B	62)	C	79)	C
12)	D	29)	D	46)	B	63)	B	80)	C
13)	A	30)	D	47)	C	64)	B	81)	C
14)	A	31)	A	48)	A	65)	C	82)	C
15)	B	32)	D	49)	C	66)	B	83)	B
16)	D	33)	D	50)	B	67)	B	84)	A
17)	B	34)	D	51)	C	68)	B	85)	B

TEACHER CERTIFICATION STUDY GUIDE

Rigor Table

	Easy 20%	Average Rigor 40%	Rigorous 40%
Question #	2, 5, 6, 8, 15, 20, 44, 49, 50, 54, 64, 67, 79, 82, 83, 84, 85	1, 3, 4, 7, 9, 10, 12, 13, 16, 17, 18, 19, 21, 22, 25, 30, 33, 34, 37, 38, 43, 45, 51, 55, 59, 60, 65, 70, 71, 73, 74, 77, 80, 81	11, 14, 23, 24, 26, 27, 28, 29, 31, 32, 35, 36, 39, 40, 41, 42, 46, 47, 48, 52, 53, 56, 57, 58, 61, 62, 63, 66, 68, 69, 72, 75, 76, 78

TEACHER CERTIFICATION STUDY GUIDE

Rationales with Sample Questions

1. Convert $.\overline{63}$ into a fraction in simplest form.
   (Average Rigor)(Competency 001)

   A) $\dfrac{63}{100}$

   B) $\dfrac{7}{11}$

   C) $6\dfrac{3}{10}$

   D) $\dfrac{2}{3}$

Answer: B

Let N = .636363…. Then, multiplying both sides of the equation by 100 or 10^2 (because there are two repeated numbers) yields 100N = 63.636363… Next, subtract the two equations to get 99N = 63 or N = $\dfrac{63}{99} = \dfrac{7}{11}$.

2. Which of the following illustrates an inverse property?
   (Easy) (Competency 001)

   A) $a + b = a - b$
   B) $a + b = b + a$
   C) $a + 0 = a$
   D) $a + (-a) = 0$

Answer: D

The correct answer is D because $a + (-a) = 0$ is a statement of the additive inverse property.

TEACHER CERTIFICATION STUDY GUIDE

3. **What is the total cost of a suit for $295.99 and a pair of shoes for $69.95 including 6.5% sales tax?**
   (Average Rigor) (Competency 001)

A) $389.73
B) $398.37
C) $237.86
D) $315.23

Answer: A

Before the tax, the total comes to $365.94. Then, the tax is .065($365.94) = $23.79. With the tax added on, the total bill is $365.94 + $23.79 = $389.73. (The same answer can be found in a quicker way: 1.065($365.94) = $389.73.)

4. **Which of the following sets is closed under division?**
   (Average Rigor) (Competency 003)

   I) {1/2, 1, 2, 4}
   II) {−1, 1}
   III) {−1, 0, 1}

A) I only
B) II only
C) III only
D) I and II

Answer: B

Set I is not closed because $\frac{4}{.5} = 8$ and 8 is not in the set. Set III is not closed because $\frac{1}{0}$ is undefined. Set II is closed because $\frac{-1}{1} = -1, \frac{1}{-1} = -1, \frac{1}{1} = 1, \frac{-1}{-1} = 1$, and all the quotients are in the set.

TEACHER CERTIFICATION STUDY GUIDE

5. **The conjugate of** $4 + 5i$ **is**
   (Easy) (Competency 002)

A) $-4 + 5i$
B) $4 - 5i$
C) $4i + 5$
D) $4i - 5$

Answer: B

By definition, the conjugate of a complex number is obtained by changing the sign of its imaginary part.

6. **Simplify:** $(6 + 3i) - (4 - 2i)$
   (Easy) (Competency 002)

A) $2 + 5i$
B) $2 + i$
C) $10 + 5i$
D) $2 - 2i$

Answer: A

Use the rules of addition and subtraction for complex numbers.

$(6 + 3i) - (4 - 2i) = 6 + 3i - 4 + 2i = 2 + 5i$

7. **Simplify:** $\dfrac{10}{1+3i}$
   (Average Rigor) (Competency 002)

A) $-1.25(1 - 3i)$
B) $1.25(1 - 3i)$
C) $1 + 3i$
D) $1 - 3i$

Answer: D

Multiplying numerator and denominator by the conjugate yields

$$\frac{10}{1+3i} \times \frac{1-3i}{1-3i} = \frac{10(1-3i)}{1-9i^2} = \frac{10(1-3i)}{1-9(-1)} = \frac{10(1-3i)}{10} = 1-3i$$

8. **Find the LCM of 27, 90 and 84.**
   (Easy) (Competency 003)

A) 90
B) 3,780
C) 204,120
D) 1,260

Answer: B

To find the LCM of the above numbers, factor each into its prime factors and multiply each common factor the maximum number of times it occurs. Thus,

$$27 = 3 \times 3 \times 3$$
$$90 = 2 \times 3 \times 3 \times 5$$
$$84 = 2 \times 2 \times 3 \times 7$$
$$LCM = 2 \times 2 \times 3 \times 3 \times 3 \times 5 \times 7 = 3,780.$$

9. **Which of the following is always composite if x is odd, y is even, and both x and y are greater than or equal to 2?**
   (Average Rigor) (Competency 003)

A) $x + y$
B) $3x + 2y$
C) $5xy$
D) $5x + 3y$

Answer: C

A composite number is a number that is not prime. The prime number sequence begins 2,3,5,7,11,13,17,... To determine which of the expressions is <u>always</u> composite, experiment with different values of x and y, such as $x = 3$ and $y = 2$, or $x = 5$ and $y = 2$. It turns out that $5xy$ will always be an even number and is therefore composite if $y = 2$.

10. What is the smallest number that is divisible by 3 and 5 and leaves a remainder of 3 when divided by 7?
    (Average Rigor) (Competency 003)

A) 15
B) 18
C) 25
D) 45

Answer: D

To be divisible by both 3 and 5, the number must be divisible by 15. Inspecting the first few multiples of 15, you will find that 45 is the first of the sequence that is 4 greater than a multiple of 7.

11. The volume of water flowing through a pipe varies directly with the square of the radius of the pipe. If the water flows at a rate of 80 liters per minute through a pipe with a radius of 4 cm, at what rate would water flow through a pipe with a radius of 3 cm?
    (Rigorous) (Competency 007)

A) 45 liters per minute
B) 6.67 liters per minute
C) 60 liters per minute
D) 4.5 liters per minute

Answer: A

Set up the direct variation: $\frac{V}{r^2} = \frac{V}{r^2}$. Substitution yields $\frac{80}{16} = \frac{V}{9}$. Solve for V to get 45 liters per minute.

12. Given the series of examples below, what is $5 \not\subset 4$?
    (Average Rigor) (Competency 004)

    $4 \not\subset 3 = 13 \quad 7 \not\subset 2 = 47$
    $3 \not\subset 1 = 8 \quad 1 \not\subset 5 = -4$

A) 20
B) 29
C) 1
D) 21

Answer: D

By inspection of the examples given, $a \not\subset b = a^2 - b$. Therefore, $5 \not\subset 4 = 25 - 4 = 21$.

13. What is the sum of the first 20 terms of the geometric sequence (2, 4, 8, 16, 32,…)?
    (Average Rigor) (Competency 004)

A) 2,097,150
B) 1,048,575
C) 524,288
D) 1,048,576

Answer: A

For a geometric sequence $a, ar, ar^2, \ldots, ar^n$, the sum of the first *n* terms is given by $\frac{a(r^n - 1)}{r - 1}$. In this case $a = 2$ and $r = 2$. Thus, the sum of the first 20 terms of the sequence is $\frac{2(2^{20} - 1)}{2 - 1} = 2,097,150$.

14. Find the sum of the first one hundred terms in the progression.
(−6, −2, 2,…)
(Rigorous) (Competency 004)

A) 19,200
B) 19,400
C) −604
D) 604

Answer: A

To find the 100th term: t_{100} = −6 + 99(4) = 390. To find the sum of the first 100 terms, use S = $\frac{100}{2}(-6+390) = 19,200$.

15. Evaluate $x^2 - 3x + 7$ when x = 2.
(Easy) (Competency 005)

A) 7
B) 5
C) 3
D) 9

Answer: B

Substitute x = 2 in the expression to get $2^2 - 3 \times 2 + 7 = 4 - 6 + 7 = 5$.

TEACHER CERTIFICATION STUDY GUIDE

16. Given $f(x) = 3x - 2$, $f^{-1}(x) =$

    (Average Rigor) (Competency 005)

A) $3x + 2$
B) $\dfrac{x}{6}$
C) $2x - 3$
D) $\dfrac{x + 2}{3}$

Answer: D

To find the inverse, $f^{-1}(x)$, of the given function, reverse the variables in the given equation, $y = 3x - 2$, to get $x = 3y - 2$. Then solve for *y* as follows:

$x + 2 = 3y$

$y = \dfrac{x + 2}{3}$

17. State the domain of the function $f(x) = \dfrac{3x - 6}{x^2 - 25}$.

    (Average Rigor) (Competency 005)

A) $x \neq 2$
B) $x \neq 5, -5$
C) $x \neq 2, -2$
D) $x \neq 5$

Answer: B

The values of 5 and –5 must be omitted from the domain of all real numbers because if x took on either of those values, the denominator of the fraction would have a value of 0, and therefore the fraction would be undefined.

18. Given $f(x) = 3x - 2$ and $g(x) = x^2$, determine $g(f(x))$.

    (Average Rigor) (Competency 005)

A) $3x^2 - 2$
B) $9x^2 + 4$
C) $9x^2 - 12x + 4$
D) $3x^3 - 2$

Answer: C

The composite function $g(f(x))$ is

$$g(f(x)) = (3x - 2)^2 = 9x^2 - 12x + 4$$

19. Solve for v_0: $d = at(v_t - v_0)$

    (Average Rigor) (Competency 005)

A) $v_0 = atd - v_t$
B) $v_0 = d - atv_t$
C) $v_0 = atv_t - d$
D) $v_0 = (atv_t - d)/at$

Answer: D

Using the distributive property and other properties of equality to isolate v_0 yields

$$d = atv_t - atv_0$$
$$atv_0 = atv_t - d$$
$$v_0 = \frac{atv_t - d}{at}$$

TEACHER CERTIFICATION STUDY GUIDE

20. The formula for solving a quadratic equation is
    (Easy) (Competency 006)

A) $x = \dfrac{-b \pm \sqrt{b^2 - 4ac}}{2a}$

B) $x = \dfrac{-b \pm \sqrt{b^2 - 4a}}{2a}$

C) $x = \dfrac{b \pm \sqrt{b^2 - 4ac}}{2a}$

D) $x = \dfrac{b \pm \sqrt{b^3 - 4ac}}{2a}$

Answer: A

Option B is missing the factor *c* from the term 4*ac* within the square root. Option C does not have the minus sign with the term *b* in the numerator. Option D has *b* cubed instead of squared within the square root symbol. A is thus the correct choice.

21. Solve for *x* by factoring $2x^2 - 3x - 2 = 0$.
    (Average Rigor) (Competency 006)

A) $x = (-1, 2)$
B) $x = (0.5, -2)$
C) $x = (-0.5, 2)$
D) $x = (1, -2)$

Answer: C

Rewrite the expression as follows.

$$0 = 2x^2 - 3x - 2 = 2x^2 - 4x + x - 2$$
$$0 = 2x(x-2) + (x-2) = (2x+1)(x-2)$$

Thus, $x = -0.5$ or 2.

MATHEMATICS 8-12

22. Which graph represents the equation $y = x^2 + 3x$?
    (Average Rigor) (Competency 006)

A)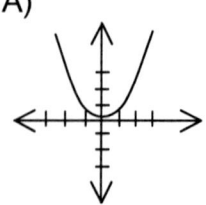
B)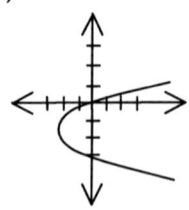
C)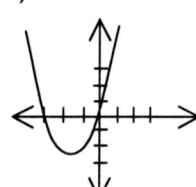
D)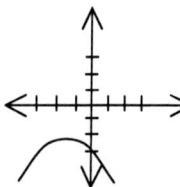

Answer: C

Answer B is not the graph of a function. Answer D is the graph of a parabola where the coefficient of x^2 is negative. Answer A appears to be the graph of $y = x^2$. To find the x-intercepts of $y = x^2 + 3x$, set $y = 0$ and solve for x: $0 = x^2 + 3x = x(x + 3)$. This expression yields $x = 0$ or $x = -3$. Therefore, the graph of the function intersects the x-axis at $x = 0$ and $x = -3$. The correct answer is C.

23. Which of the following is a factor of the expression $9x^2 + 6x - 35$?
    (Rigorous) (Competency 006)

A) $3x - 5$
B) $3x - 7$
C) $x + 3$
D) $x - 2$

Answer: A

Recognize that the given expression can be written as the sum of two squares and use the formula $a^2 - b^2 = (a+b)(a-b)$.

$$9x^2 + 6x - 35 = (9x^2 + 6x + 1) - 36 = (3x+1)^2 - 36 = (3x+1+6)(3x+1-6)$$
$$= (3x+7)(3x-5)$$

TEACHER CERTIFICATION STUDY GUIDE

24. **Solve the system of equations for *x*, *y* and *z*.**
    (Rigorous) (Competency 006)

    $$3x + 2y - z = 0$$
    $$2x + 5y = 8z$$
    $$x + 3y + 2z = 7$$

    A) (−1, 2, 1)
    B) (1, 2, −1)
    C) (−3, 4, −1)
    D) (0, 1, 2)

    Answer: A

    Multiplying the first equation by 2 and the second equation by −3, and then adding the results together yields −11*y* + 22*z* = 0. Solving for *y* yields *y* = 2*z*. In the meantime, multiplying the third equation by −2 and adding it to the second equation yields −*y* − 12*z* = −14. Then, substituting 2*z* for *y*, yields the result *z* = 1. Subsequently, one can easily find that *y* = 2, and *x* = −1.

25. **How does the function $y = x^3 + x^2 + 4$ behave from *x* = 1 to *x* = 3?**
    (Average Rigor) (Competency 007)

    A) increasing, then decreasing
    B) increasing
    C) decreasing
    D) neither increasing nor decreasing

    Answer: B

    To find critical points, take the derivative, set it equal to 0, and solve for x. $y' = 3x^2 + 2x = x(3x + 2) = 0$. The critical points are at *x* = 0 and *x* = −2/3. Neither of these critical points is on the interval from *x* = 1 to *x* = 3. Test the endpoints: *y* = 6 at *x* = 1, and *y* = 38 at *x* = 3. Since the derivative is positive for all values of *x* from *x* = 1 to *x* = 3, the curve is increasing on the entire interval.

26. **Solve for *x*:** $18 = 4 + |2x|$

    (Rigorous) (Competency 007)

    A) $\{-11, 7\}$
    B) $\{-7, 0, 7\}$
    C) $\{-7, 7\}$
    D) $\{-11, 11\}$

    Answer: C

    Using the definition of absolute value, two equations are possible: $18 = 4 + 2x$ and $18 = 4 - 2x$. Solving for *x* in both cases yields $x = 7$ and $x = -7$.

27. **Find the zeros of** $f(x) = x^3 + x^2 - 14x - 24$

    (Rigorous) (Competency 007)

    A) 4, 3, 2
    B) 3, –8
    C) 7, –2, –1
    D) 4, –3, –2

    Answer: D

    Possible rational roots of the equation $0 = x^3 + x^2 - 14x - 24$ are all the positive and negative factors of 24. By substituting into the equation, we find that –2 is a root, and therefore $x + 2$ is a factor. By performing the long division $(x^3 + x^2 - 14x - 24)/(x + 2)$, we can find that another factor of the original equation is $x^2 - x - 12$ or $(x - 4)(x + 3)$. Therefore, the zeros of the original function are –2, –3, and 4.

TEACHER CERTIFICATION STUDY GUIDE

28. Evaluate $3^{1/2}(9^{1/3})$

    (Rigorous) (Competency 007)

A) $27^{5/6}$
B) $9^{7/12}$
C) $3^{5/6}$
D) $3^{6/7}$

Answer: B

Getting the bases the same yields $3^{\frac{1}{2}}3^{\frac{2}{3}}$. Adding exponents yields $3^{\frac{7}{6}}$. Additional manipulation of exponents produces $3^{\frac{7}{6}} = 3^{\frac{14}{12}} = (3^2)^{\frac{7}{12}} = 9^{\frac{7}{12}}$.

29. Which of the following is incorrect?
    (Rigorous) (Competency 007)

A) $(x^2y^3)^2 = x^4y^6$
B) $m^2(2n)^3 = 8m^2n^3$
C) $(m^3n^4)/(m^2n^2) = mn^2$
D) $(x+y^2)^2 = x^2 + y^4$

Answer: D

Using FOIL to do the expansion, we get the following:

$(x + y^2)^2 = (x + y^2)(x + y^2)$
$= x^2 + 2xy^2 + y^4$

Thus, answer D is the correct choice.

30. The exponential equation $2^5 = 32$ can be written as
    (Average Rigor) (Competency 008)

A) $\log_2(5) = 32$
B) $\log_{10}(32) = 5$
C) $\log_5(32) = 2$
D) $\log_2(32) = 5$

Answer: D

By definition, $\log_2(32) = 5$ corresponds to $2^5 = 32$. Answer D is thus correct.

31. Which equation corresponds to the logarithmic statement: $\log_x k = m$?
    (Rigorous) (Competency 008)

A) $x^m = k$
B) $k^m = x$
C) $x^k = m$
D) $m^x = k$

Answer: A

By definition of log form and exponential form, $\log_x k = m$ corresponds to $x^m = k$.

32. **Solve for x:** $10^{x-3} + 5 = 105$
    (Rigorous) (Competency 008)

A) 3
B) 10
C) 2
D) 5

Answer: D

Note that $10^{x-3} = 100$. Taking the logarithm to base 10 of both sides yields

$$(x-3)\log_{10} 10 = \log_{10} 100$$
$$x - 3 = 2$$
$$x = 5$$

Answer D is correct.

33. **Determine the measures of angles A and B.**
    (Average Rigor) (Competency 009)

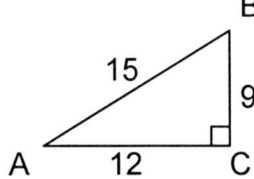

A) A = 30°, B = 60°
B) A = 60°, B = 30°
C) A = 53°, B = 37°
D) A = 37°, B = 53°

Answer: D

Using the diagram, note that

$$\tan A = \frac{9}{12} = 0.75$$
$$\arctan 0.75 = 37°$$

Since angle B is complementary to angle A, the measure of angle B is therefore 53 degrees.

34. Which expression is not identical to sin x?
    (Average Rigor) (Competency 009)

A) $\sqrt{1-\cos^2 x}$
B) $\tan x \cos x$
C) $1/\csc x$
D) $1/\sec x$

Answer: D

Using the basic definitions of the trigonometric functions and the Pythagorean identity, we see that the first three options are all identical to sin x. But, $\sec x = \dfrac{1}{\cos x}$ is not the same as sin x.

35. Which expression is equivalent to $1-\sin^2 x$?
    (Rigorous) (Competency 009)

A) $1-\cos^2 x$
B) $1+\cos^2 x$
C) $1/\sec x$
D) $1/\sec^2 x$

Answer: D

Using the Pythagorean Identity, we know $\sin^2 x + \cos^2 x = 1$. Thus,

$$1-\sin^2 x = \cos^2 x = \dfrac{1}{\sec^2 x}$$

The correct answer is D.

36. For an acute angle x, $\sin x = 3/5$. What is $\cot x$?
    (Rigorous) (Competency 009)

A) 5/3
B) 3/4
C) 1.33
D) 1

Answer: B

Using the Pythagorean Identity, we know $\sin^2 x + \cos^2 x = 1$. Thus,

$$\cos x = \sqrt{1 - \frac{9}{25}} = \frac{4}{5}$$

$$\cot x = \frac{\cos x}{\sin x} = \left(\frac{4}{5}\right)\left(\frac{5}{3}\right) = \frac{4}{3}$$

37. Find the area under the function $y = x^2 + 4$ from $x = 3$ to $x = 6$.
    (Average Rigor) (Competency 010)

A) 75
B) 21
C) 96
D) 57

Answer: A

To find the area set up the definite integral:

$$\int_3^6 (x^2 + 4)\,dx = \left(\frac{x^3}{3} + 4x\right)\Big|_3^6$$

Evaluate the resulting expression at $x = 6$ and at $x = 3$ to get

$$\left(\frac{x^3}{3} + 4x\right)\Big|_3^6 = \frac{6^3}{3} + 4(6) - \left[\frac{3^3}{3} + 4(3)\right] = 72 + 24 - 9 - 12 = 75$$

The correct answer is A.

38. If the velocity of a body is given by $v = 16 - t^2$, find the distance traveled from $t = 0$ until the body comes to a complete stop.
(Average Rigor) (Competency 010)

A) 16
B) 43
C) 48
D) 64

Answer: B

Recall that the derivative of the distance function is the velocity function. Conversely, the integral of the velocity function is the distance function. To find the time needed for the body to come to a stop (when $v = 0$), solve for t:

$$v(t) = 16 - t^2 = 0$$
$$t = 4$$

Thus, the body travels from time $t = 0$ to time $t = 4$. The distance function is (excluding the constant of integration, which is unneeded here)

$$s(t) = \int v(t)\,dt = 16t - \frac{t^3}{3}$$

At $t = 4$,

$$s(4) = 16(4) - \frac{4^3}{3} = \frac{128}{3} \approx 42.7$$

Thus, the body travels approximately 42.7 units.

39. **Find the following limit:** $\lim_{x \to 0} \dfrac{\sin 2x}{5x}$

   (Rigorous) (Competency 010)

   A) Infinity
   B) 0
   C) 1.4
   D) 1

   ### Answer: C

   Since substitution of $x = 0$ into the expression yields an undefined answer, we can use L'Hospital's rule and take derivatives of both the numerator and denominator to find the limit.

   $$\lim_{x \to 0} \dfrac{\sin 2x}{5x} = \lim_{x \to 0} \dfrac{2\cos 2x}{5}$$

   Now substitute $x = 0$:

   $$\lim_{x \to 0} \dfrac{2\cos 2x}{5} = \dfrac{2}{5} = 1.4$$

40. **Find the first derivative of the function:** $f(x) = x^3 - 6x^2 + 5x + 4$

   (Rigorous) (Competency 010)

   A) $3x^2 - 12x^2 + 5x$
   B) $3x^2 - 12x - 5$
   C) $3x^2 - 12x + 9$
   D) $3x^2 - 12x + 5$

   ### Answer: D

   Use the power rule for polynomial differentiation: if $y = ax^n$, then $y' = nax^{n-1}$. Then,

   $$f'(x) = 3x^2 - 12x + 5$$

41. Evaluate $\int_0^2 (x^2 + x - 1)dx$

    (Rigorous) (Competency 010)

A) 11/3
B) 8/3
C) –8/3
D) –11/3

Answer: B

Use the fundamental theorem of calculus to find the definite integral: given a continuous function *f* on an interval [a,b], then $\int_a^b f(x)dx = F(b) - F(a)$, where *F* is an antiderivative of *f*.

$$\int_0^2 (x^2 + x - 1)dx = (\frac{x^3}{3} + \frac{x^2}{2} - x)$$

Evaluate the expression at x = 2 and at x = 0, then subtract to get 8/3 + 4/2 – 2 – 0 = 8/3.

42. Find the antiderivative for the function $y = e^{3x}$.

    (Rigorous) (Competency 010)

A) $3x(e^{3x}) + C$
B) $3(e^{3x}) + C$
C) $1/3(e^x) + C$
D) $1/3(e^{3x}) + C$

Answer: D

Use the rule for integration of functions of e:

$$\int e^x dx = e^x + C$$

$$\int e^{3x} dx = \frac{1}{3} \int e^u du = \frac{1}{3} e^u + C = \frac{1}{3} e^{3x} + C$$

The correct answer is thus D.

43. How does the function $y = x^3 + x^2 + 4$ behave from $x = 1$ to $x = 3$?
   (Average Rigor) (Competency 010)

A) increasing, then decreasing
B) increasing
C) decreasing
D) neither increasing nor decreasing

Answer: B

To find critical points, take the derivative of the function, set it equal to 0, and solve for *x*.

$f'(x) = 3x^2 + 2x = x(3x+2) = 0$

The critical points are at *x* = 0 and *x* = –2/3. Neither of these CP is on the interval from *x* = 1 to *x* = 3. Testing the endpoints, *y* = 6 at *x* = 1 and *y* = 38 at *x* = 3. Since the derivative is positive for all values of *x* from *x* = 1 to *x* = 3, the curve is increasing on the entire interval.

44. **Find the surface area of a box that is 3 feet wide, 5 feet tall, and 4 feet deep.**
   (Easy) (Competency 011)

A) 47 sq. ft.
B) 60 sq. ft.
C) 94 sq. ft.
D) 188 sq. ft.

Answer: C

Let's assume the base of the rectangular solid (box) is 3 by 4 and the height is 5. Then, the surface area of the top and bottom together is 2(12) = 24. The sum of the areas of the front and back are 2(15) = 30, and the sum of the areas of the sides are 2(20) = 40. The total surface area is therefore 94 square feet.

45. If a ship sails due south 6 miles, then due west 8 miles, how far was it from the starting point?
   (Average Rigor) (Competency 011)

   A) 100 miles
   B) 10 miles
   C) 14 miles
   D) 48 miles

Answer: B

Draw a right triangle with legs of 6 and 8. Find the hypotenuse using the Pythagorean Theorem:

$$6^2 + 8^2 = c^2 = 36 + 64 = 100$$

Therefore, $c = 10$ miles.

46. Find the area of the figure pictured below.
   (Rigorous) (Competency 011)

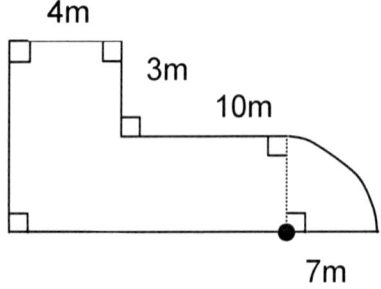

A) 136.47 m²
B) 148.48 m²
C) 293.86 m²
D) 178.47 m²

Answer: B

Divide the figure into two rectangles and one quarter circle. The tall rectangle on the left has dimensions 10m by 4m and thus an area of 40m². The rectangle in the center has dimensions 7m by 10m and thus an area of 70m². The quarter circle has an area of $.25(\pi)(7m)^2 = 38.48m^2$. The total area is therefore approximately 148.48m².

47. **If the area of the base of a cone is tripled, the volume will be**
    (Rigorous) (Competency 011)

    A) the same as the original
    B) 9 times the original
    C) 3 times the original
    D) 3π times the original

Answer: C

The formula for the volume of a cone is $V = \frac{1}{3}Bh$, where B is the area of the circular base and h is the height. If the area of the base is tripled, the volume becomes $V = \frac{1}{3}(3B)h = Bh$, or three times the original area.

48. **The length of a picture frame is 2 inches greater than its width. If the area of the frame is 143 square inches, what is its width?**
    (Rigorous) (Competency 011)

    A) 11 inches
    B) 13 inches
    C) 12 inches
    D) 10 inches

Answer: A

First set up the equation for the problem. If the width of the picture frame is w, then $w(w+2) = 143$. Next, solve the equation to obtain w. Using the method of completing squares we have

$$w^2 + 2w + 1 = 144$$
$$(w+1)^2 = 144$$
$$w + 1 = \pm 12$$

Thus w is 11 or -13. Since the width cannot be negative, the correct answer is 11 inches.

49. When you begin by assuming the conclusion of a theorem is false, then show that through a sequence of logically correct steps you contradict an accepted fact, this is known as
(Easy) (Competency 012)

A) inductive reasoning
B) direct proof
C) indirect proof
D) exhaustive proof

Answer: C

By definition this describes the procedure of an indirect proof.

50. Which term most accurately describes two coplanar lines without any common points?
(Easy) (Competency 012)

A) perpendicular
B) parallel
C) intersecting
D) skew

Answer: B

By definition, parallel lines are coplanar lines without any common points.

51. Which theorem can be used to prove $\triangle BAK \cong \triangle MKA$?
    (Average Rigor) (Competency 012)

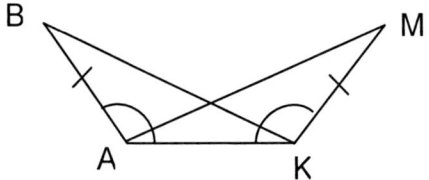

A) SSS
B) ASA
C) SAS
D) AAS

Answer: C

Since side AK is common to both triangles, the triangles can be proved congruent by using the Side-Angle-Side Postulate.

52. Choose the diagram which illustrates the construction of a perpendicular to the line at a given point on the line.
(Rigorous) (Competency 012)

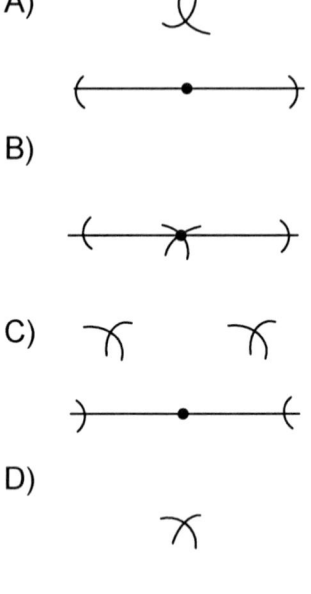

A)

B)

C)

D)

Answer: D

Given a point on a line, place the compass point there and draw two arcs intersecting the line in two points, one on either side of the given point. Then using any radius larger than half the new segment produced, and with the pointer at each end of the new segment, draw arcs which intersect above the line. Connect this new point with the given point.

53. Compute the area of the shaded region, given a radius of 5 meters. Point O is the center.
   (Rigorous) (Competency 011)

   A) 7.13 cm²
   B) 7.13 m²
   C) 78.5 m²
   D) 19.63 m²

   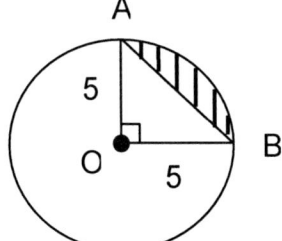

   Answer: B

   The area of triangle AOB is .5(5)(5) = 12.5 square meters. Since $\frac{90}{360} = .25$, the area of sector AOB (the pie-shaped piece) is approximately $.25(\pi)5^2 = 19.63$. Subtracting the triangle area from the sector area to get the area of segment AB, we get approximately 19.63 − 12.5 = 7.13 square meters.

54. Given that QO⊥NP and QO=NP, quadrilateral NOPQ can most accurately be described as a
   (Easy) (Competency 013)

   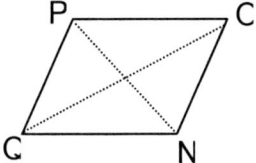

   A) parallelogram
   B) rectangle
   C) square
   D) rhombus

   Answer: C

   In an ordinary parallelogram, the diagonals are not perpendicular or equal in length. In a rectangle, the diagonals are not necessarily perpendicular. In a rhombus, the diagonals are not equal in length. In a square, the diagonals are both perpendicular and congruent.

55. **Which of the following statements about a trapezoid is incorrect?**
    (Average Rigor) (Competency 013)

A) It has one pair of parallel sides
B) The parallel sides are called bases
C) If the two bases are the same length, the trapezoid is called isosceles
D) The median is parallel to the bases

Answer: C

A trapezoid is isosceles if the two legs (not bases) are the same length.

56. **What is the degree measure of an interior angle of a regular 10-sided polygon?**
    (Rigorous) (Competency 013)

A) 18°
B) 36°
C) 144°
D) 54°

Answer: C

Formula for finding the measure of each interior angle of a regular polygon with n sides is $\frac{(n-2)180}{n}$. For $n = 10$, we get $\frac{8(180)}{10} = 144$ degrees.

57. **What is the measure of minor arc AD, given measure of arc PS is 40° and $m < K = 10°$?**
    (Rigorous) (Competency 013)

A) 50°
B) 20°
C) 30°
D) 25°

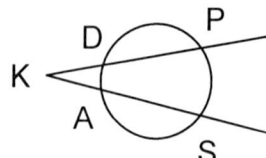

Answer: B

The formula relating the measure of angle K and the two arcs it intercepts is $m\angle K = \frac{1}{2}(mPS - mAD)$. Substituting the known values, we get $10 = \frac{1}{2}(40 - mAD)$. Solving for mAD yields an answer of 20 degrees.

TEACHER CERTIFICATION STUDY GUIDE

58. Determine the area of the shaded region of the trapezoid in terms of *x* and *y*.
    (Rigorous) (Competency 013)

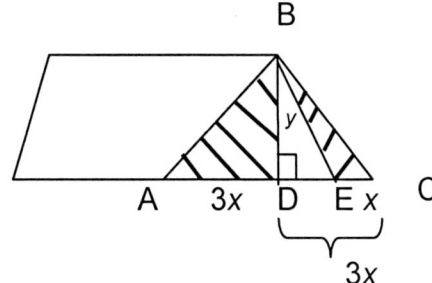

A) $4xy$
B) $2xy$
C) $3x^2 y$
D) There is not enough information given.

Answer: B

To find the area of the shaded region, find the area of triangle ABC and then subtract the area of triangle DBE. The area of triangle ABC is $.5(6x)(y) = 3xy$. The area of triangle DBE is $.5(2x)(y) = xy$. The difference is $2xy$.

59. Given a vector with horizontal component 5 and vertical component 6, determine the length of the vector.
    (Average Rigor) (Competency 014)

A) 61
B) $\sqrt{61}$
C) 30
D) $\sqrt{30}$

Answer: B

Using the Pythagorean Theorem, we get a length of $\sqrt{36+25} = \sqrt{61}$.

60. Compute the distance from (–2, 7) to the line x = 5.
    (Average Rigor) (Competency 014)

A) –9
B) –7
C) 5
D) 7

Answer: D

The line x = 5 is a vertical line passing through (5,0) on the Cartesian plane. By observation, the distance along the horizontal line from the point (–2,7) to the line x = 5 is 7 units.

61. Given K(–4, y) and M(2, –3) with midpoint M(x, 1), determine the values of x and y.
    (Rigorous) (Competency 014)

A) $x = -1, y = 5$
B) $x = 3, y = 2$
C) $x = 5, y = -1$
D) $x = -1, y = -1$

Answer: A

The formula for finding the midpoint (a, b) of a segment passing through the points (x_1, y_1) and (x_2, y_2) is $\left(\dfrac{x_1 + x_2}{2}, \dfrac{y_1 + y_2}{2}\right)$. Setting up the corresponding equations from this information yields $x = \dfrac{-4 + 2}{2}$ and $1 = \dfrac{y - 3}{2}$. Solving for x and y yields x = –1 and y = 5.

62. **Find the length of the major axis of $x^2 + 9y^2 = 36$.**
    (Rigorous) (Competency 014)

A) 4
B) 6
C) 12
D) 8

Answer: C

Dividing by 36, we get $\frac{x^2}{36} + \frac{y^2}{4} = 1$, which tells us that the ellipse intersects the x-axis at 6 and –6, and therefore the length of the major axis is 12. (The ellipse intersects the y-axis at 2 and –2).

63. **Which equation represents a circle with a diameter whose endpoints are (0, 7) and (0, 3)?**
    (Rigorous) (Competency 014)

A) $x^2 + y^2 + 21 = 0$
B) $x^2 + y^2 - 10y + 21 = 0$
C) $x^2 + y^2 - 10y + 9 = 0$
D) $x^2 - y^2 - 10y + 9 = 0$

Answer: B

With a diameter going from (0,7) to (0,3), the diameter of the circle must be 4, the radius must be 2, and the center of the circle must be at (0,5). Using the standard form for the equation of a circle, we get $(x - 0)^2 + (y - 5)^2 = 2^2$. Expanding, we get $x^2 + y^2 - 10y + 21 = 0$.

64. Compute the standard deviation for the following set of temperatures. (37, 38, 35, 37, 38, 40, 36, 39)
    (Easy) (Competency 015)

A) 37.5
B) 1.5
C) 0.5
D) 2.5

Answer: B

Find the mean: 300/8 = 37.5. Then, using the formula for standard deviation,

$$\sqrt{\frac{2(37.5-37)^2 + 2(37.5-38)^2 + (37.5-35)^2 + (37.5-40)^2 + (37.5-36)^2 + (37.5-39)^2}{8}}$$

This expression has a value of 1.5.

65. Which type of graph uses symbols to represent quantities?
    (Average rigor) (Competency 015)

A) Bar graph
B) Line graph
C) Pictograph
D) Circle graph

Answer: C

A pictograph shows comparison of quantities using symbols. Each symbol represents a number of items.

66. Half the students in a class scored 80% on an exam, most of the rest scored 85% except for one student who scored 10%. Which would be the best measure of central tendency for the test scores?
(Rigorous) (Competency 015)

A) mean
B) median
C) mode
D) either the median or the mode because they are equal

Answer: B

In this set of data, the median would be the most representative measure of central tendency because the median is independent of extreme values. Because of the 10% outlier, the mean (average) would be disproportionately skewed. In this data set, it is true that the median and the mode (number which occurs most often) are the same, but the median remains the best choice because of its special properties.

67. A jar contains 3 red marbles, 5 white marbles, 1 green marble and 15 blue marbles. If one marble is picked at random from the jar, what is the probability that it will be red?
(Easy) (Competency 016)

A) 1/3
B) 1/8
C) 3/8
D) 1/24

Answer: B

The total number of marbles is 24 and the number of red marbles is 3. Thus, the probability of picking a red marble from the jar is 3/24 = 1/8.

68. A die is rolled several times. What is the probability that a 3 will not appear before the third roll of the die?
(Rigorous) (Competency 016)

A) 1/3
B) 25/216
C) 25/36
D) 1/216

Answer: B

The probability that a 3 will not appear before the third roll is the same as the probability that the first two rolls will consist of numbers other than 3. Since the probability of any one roll resulting in a number other than 3 is 5/6, the probability of the first two rolls resulting in a number other than 3 is (5/6) x (5/6) = 25/36.

69. If there are three people in a room, what is the probability that at least two of them will share a birthday? (Assume a year has 365 days)
(Rigorous) (Competency 016)

A) 0.67
B) 0.05
C) 0.008
D) 0.33

Answer: C

The best way to approach this problem is to use the fact that the probability of an event plus the probability of the event not happening is unity. First, find the probability that no two people will share a birthday and then subtract the result from 1. The probability that two of the people will not share a birthday is 364/365 (because the second person's birthday can be one of the 364 days other than the birthday of the first person). The probability that the third person will also not share either of the first two birthdays is (364/365) * (363/365) = 0.992. Therefore, the probability that at least two people will share a birthday is 1 – 0.992 = 0.008.

TEACHER CERTIFICATION STUDY GUIDE

70. **Which of the following is not a valid method of collecting statistical data?**
    (Average Rigor) (Competency 017)

    A) Random sampling
    B) Systematic sampling
    C) Cluster sampling
    D) Cylindrical sampling

    Answer: D

    There is no such method as cylindrical sampling.

71. **To determine the odds for or against a given deviation from expected statistical distribution statisticians use**
    (Average Rigor) (Competency 017)

    A) The t-test
    B) Linear regression
    C) The chi-square test
    D) Exponential regression

    Answer: C

    The chi-square test is a method of determining the odds for or against a given deviation from expected statistical distribution.

72. **If the correlation between two variables is given as zero, the association between the two variables is**
    (Rigorous) (Competency 017)

    A) negative linear
    B) positive linear
    C) quadratic
    D) random

    Answer: D

    A correlation of 1 indicates a perfect positive linear association, a correlation of –1 indicates a perfect negative linear association, and a correlation of zero indicates a random relationship between the variables.

TEACHER CERTIFICATION STUDY GUIDE

73. About two weeks after introducing formal proofs, several students in your geometry class are having a difficult time remembering the names of the postulates. They cannot complete the reason column of the proof and as a result are not even attempting the proofs. What would be the best approach to help students understand the nature of geometric proofs?
(Average Rigor) (Competency 018)

A) Give them more time; proofs require time and experience.
B) Allow students to write an explanation of the theorem in the reason column instead of the name.
C) Have the student copy each theorem in a notebook.
D) Allow the students to have open book tests.

Answer: B

Since the purpose of the reason column is to provide an explanation for the corresponding step in the proof, it is fully acceptable for students to write out an explanation instead of using the shorthand method of naming the postulate.

74. Identify the correct sequence of subskills required for solving and graphing inequalities involving absolute value in one variable, such as $|x+1| \leq 6$.
(Average Rigor) (Competency 018)

A) understanding absolute value, graphing inequalities, solving systems of equations
B) graphing inequalities on a Cartesian plane, solving systems of equations, simplifying expressions with absolute value
C) plotting points, graphing equations, graphing inequalities
D) solving equations with absolute value, solving inequalities, graphing conjunctions and disjunctions

Answer: D

The steps listed in answer D would look like this for the given example:
If $|x+1| \leq 6$, then $-6 \leq x+1 \leq 6$, which means $-7 \leq x \leq 5$. Then the inequality would be graphed on a number line and would show that the solution set is all real numbers between -7 and 5, including -7 and 5.

75. **Mr. Lacey is using problem solving to help students develop their math skills. He gives the class a box of pencils. He says that the pencils have to be divided so that each student has the same number of pencils. What step should come first in problem solving?** *(Rigorous) (Competency 018)*

A) Find a strategy to solve the problem
B) Identify the problem
C) Count the number of pencils
D) Make basic calculations

Answer: B

The first step in problem solving is always to identify the problem.

76. **Kindergarten students are doing a butterfly art project. They fold paper in half. On one half, they paint a design. Then they fold the paper closed and reopen. The resulting picture is a butterfly with matching sides. What math principle does this demonstrate?** *(Rigorous) (Competency 019)*

A) Slide
B) Rotate
C) Symmetry
D) Transformation

Answer: C

By folding the painted paper in half, the design is mirrored on the other side, creating symmetry and reflection. The butterfly design is symmetrical about the center.

77. Students are working with a set of rulers and various small objects from the classroom. Which concept are these students exploring?
(Average Rigor) (Competency 019)

A) Volume
B) Weight
C) Length
D) Temperature

Answer: C

The use of a ruler indicates that the activity is based on exploring length.

78. Third grade students are looking at a circle graph. Most of the graph is yellow. A small wedge of the graph is blue. Each colored section also has a number followed by a symbol. What are the students most likely learning about?
(Rigorous) (Competency 019)

A) Addition
B) Venn diagrams
C) Percent
D) Pictographs

Answer: C

The symbol after the numbers of the sections indicates that students are learning about percents instead of an exact number.

79. **Which of the following is the best example of the value of personal computers in advanced high school mathematics?**
    (Easy) (Competency 020)

    A) Students can independently drill and practice test questions.
    B) Students can keep an organized list of theorems and postulates on a word processing program.
    C) Students can graph and calculate complex functions to explore their nature and make conjectures.
    D) Students are better prepared for business because of mathematics computer programs in high school.

    Answer: C

    The activities mentioned in options A, B and D can be carried out in other ways without using a personal computer. It would be extremely difficult to graph and calculate complex functions without computers. Thus, C is the correct answer.

80. **What would be the least appropriate use for handheld calculators in the classroom?**
    (Average Rigor) (Competency 020)

    A) practice for standardized tests
    B) integrating algebra and geometry with applications
    C) justifying statements in geometric proofs
    D) applying the law of sines to find dimensions

    Answer: C

    There is no need for calculators when justifying statements in a geometric proof.

81. A group of students working with trigonometric identities have concluded that cos 2x = 2 cos x. How could you best lead them to discover their error?
(Average Rigor) (Competency 020)

A) Have the students plug in values on their calculators.
B) Direct the student to the appropriate chapter in the text.
C) Derive the correct identity on the board.
D) Provide each student with a table of trig identities.

Answer: C

Option C is the right choice because it will show the students how to correctly manipulate trigonometric functions and help them identify the point at which they made a mistake.

82. $-3 + 7 = -4 \quad 6(-10) = -60$
$-5(-15) = 75 \quad -3 + -8 = 11$
$8 - 12 = -4 \quad 7 - (-8) = 15$

Which best describes the type of error observed above?
(Easy) (Competency 001)

A) The student is incorrectly multiplying integers.
B) The student has incorrectly applied rules for adding integers to subtracting integers.
C) The student has incorrectly applied rules for multiplying integers to adding integers.
D) The student is incorrectly subtracting integers.

Answer: C

The errors are in the following: $-3+7=-4$ and $-3 + -8 = 11$, where the student seems to be using the rules for signs when multiplying, instead of the rules for signs when adding.

83. **Which of the following statements is untrue?**
    (Easy) (Competency 021)

A) A teacher may use a variety of formal and informal assessment methods to evaluate a student's progress
B) A multiple-choice test is a type of formative assessment
C) Alternative assessment is any type of assessment in which students create a response rather than choose an answer
D) Summative assessment consists of temporary interaction between teacher and student

Answer: B

In formative assessment, emphasis is placed on feedback and the flow of communication between teacher and student—something that does not happen in a multiple-choice test.

84. **A student portfolio is**
    (Easy) (Competency 021)

A) a collection of a student's work over a period of time to help the teacher and student assess progress
B) a collection of test papers to help the teacher and student assess progress
C) a collection of student art
D) a collection of assignments given by the teacher

Answer: A

A portfolio is a collection of samples of a student's work over a period of time to help the teacher and student assess progress. The work to be included is selected by the student or teacher or by both together.

85. Higher order thinking, creativity, and the integration of reasoning and communication skills are most demonstrated by
(Easy) (Competency 021)

A) Multiple-choice and true/false tests
B) Projects, demonstrations and oral presentations
C) Essay questions
D) Portfolios

Answer: B

Projects, demonstrations and oral presentations call upon a variety of skills and allow for the greatest use of creativity and higher order thinking.

TEACHER CERTIFICATION STUDY GUIDE

XAMonline, INC. 25 First St. Suite 106 Cambridge MA 02141

Toll Free number 800-509-4128

TO ORDER Fax 781-662-9268 OR www.XAMonline.com

<u>TEXAS EXAMINATION OF EDUCATOR STANDARD-EXAMINATION FOR THE CERTIFICATION OF EDUCATORS   - TEXES/EXCET - 2009</u>

PO#   Store/School:

Address 1:

Address 2 (Ship to other):

City, State Zip

Credit card number_____-_____-_____-_____   expiration_____

EMAIL _____

PHONE   FAX

ISBN	TITLE	Qty	Retail	Total
978-1-58197-925-1	ExCET ART SAMPLE TEST (ALL-LEVEL-SECONDARY) 005 006		$15.00	
978-1-58197-926-8	ExCET FRENCH SAMPLE TEST (SECONDARY) 048		$15.00	
978-1-58197-723-3	ExCET SPANISH (SECONDARY) 047		$59.95	
978-1-58197-580-2	TExES PRINCIPAL 068		$59.95	
978-1-58197-929-9	TExES PEDAGOGY AND PROFESSIONAL RESPONSIBILITIES 4-8 110		$24.95	
978-1-58197-899-5	TExES PEDAGOGY AND PROFESSIONAL RESPONSIBILITIES EC-4 100		$24.95	
978-1-58197-271-9	TExES GENERALIST 4-8 111		$59.95	
978-1-58197-945-9	TExES GENERALIST EC-4 101		$59.95	
978-1-58197-948-0	TExES MATHEMATICS-SCIENCE 4-8 114		$73.50	
978-1-58197-295-5	TExES MATHEMATICS 4-8 114-115		$59.95	
978-1-58197-297-9	TExES SCIENCE 4-8 116		$59.95	
978-1-58197-931-2	TExES SCIENCE 8-12 136		$59.95	
978-1-58197-772-1	TExES ENGLISH LANG-ARTS AND READING 4-8 117		$59.95	
978-1-58197-771-4	TExES ENGLISH LANG-ARTS AND READING 8-12 131		$59.95	
978-1-58197-661-8	TExES SOCIAL STUDIES 4-8 118		$59.95	
978-1-58197-621-2	TExES SOCIAL STUDIES 8-12 132		$59.95	
978-1-58197-339-6	TExES MATHEMATICS 8-12 135		$59.95	
978-1-58197-618-2	TExES LIFE SCIENCE 8-12 138		$59.95	
978-1-58197-949-7	TExES CHEMISTRY 8-12 140		$59.95	
978-1-58197-939-8	TExES MATHEMATICS-PHYSICS 8-12 143		$73.50	
978-1-58197-940-4	TExES SCHOOL LIBRARIAN 150		$59.95	
978-1-58197-941-1	TExES READING SPECIALIST 151		$59.95	
978-1-58197-719-6	TExES SCHOOL COUNSELOR 152		$59.95	
978-1-58197-620-5	TExES PHYSICAL EDUCATION EC-12 158		$59.95	
978-1-58197-262-7	TExES SPECIAL EDUCATION EC-12 161		$73.50	
978-1-58197-606-9	THEA TEXAS HIGHER EDUCATOR ASSESSMENT		$21.95	
			SUBTOTAL	
	$8.25 for one book, $11.00 for two books, $15.00 for three plus		**Ship**	
			TOTAL	

Printed in the United States
151069LV00003BA/1/P